Sustainable Irrigation Management, Technologies and Policies II

WIT*PRESS*

WIT Press publishes leading books in Science and Technology.
Visit our website for the current list of titles.
www.witpress.com

WIT*eLibrary*

Home of the Transactions of the Wessex Institute.
Papers presented at Sustainable Irrigation II are archived in the WIT eLibrary in volume 112 of WIT Transactions on Ecology and the Environment (ISSN 1743-3541).
The WIT eLibrary provides the international scientific community with immediate and permanent access to individual papers presented at WIT conferences.
Visit the WIT eLibrary at www.witpress.com.

SECOND INTERNATIONAL CONFERENCE ON SUSTAINABLE IRRIGATION MANAGEMENT, TECHNOLOGIES AND POLICIES

Sustainable Irrigation II

CONFERENCE CHAIRMEN

Y. Villacampa Esteve
University of Alicante, Spain

C.A. Brebbia
Wessex Institute of Technology, UK

D. Prats Rico
IWES University of Alicante, Spain

INTERNATIONAL SCIENTIFIC ADVISORY COMMITTEE

Organised by

Wessex Institute of Technology, UK
University of Alicante, Spain

Sponsored by

WIT Transactions on Ecology and the Environment
Institute for Water and Environmental Sciences of the University of Alicante
Ministerio de Educación y Ciencia, España (CTM2007-29892-E/TECNO)
Instituto del Agua y de las Ciencias Ambientales
Patronato Municipal de Turismo de Alicante
Grupo de Investigación Modelización Matemática de Sistemas, University of Alicante
Consolider-Tragua

WIT Transactions

Transactions Editor

Sustainable Irrigation Management, Technologies and Policies II

Editors

Y. Villacampa Esteve
University of Alicante, Spain

C.A. Brebbia
Wessex Institute of Technology, UK

D. Prats Rico
IWES University of Alicante, Spain

WITPRESS Southampton, Boston

Y. Villacampa Esteve
University of Alicante, Spain

C.A. Brebbia
Wessex Institute of Technology, UK

D. Prats Rico
IWES Unviversity of Alicante, Spain

Published by

WIT Press
Ashurst Lodge, Ashurst, Southampton, SO40 7AA, UK
Tel: 44 (0) 238 029 3223; Fax: 44 (0) 238 029 2853
E-Mail: witpress@witpress.com
http://www.witpress.com

For USA, Canada and Mexico

Computational Mechanics Inc
25 Bridge Street, Billerica, MA 01821, USA
Tel: 978 667 5841; Fax: 978 667 7582
E-Mail: infousa@witpress.com
http://www.witpress.com

British Library Cataloguing-in-Publication Data

A Catalogue record for this book is available
from the British Library

ISBN: 978-1-84564-116-0
ISSN: 1746-448X (print)
ISSN: 1743-3541 (online)

The texts of the papers in this volume were set individually by the authors or under their supervision. Only minor corrections to the text may have been carried out by the publisher.

Printed in Great Britain by Athenaeum Press Ltd.

Preface

This book contains most of the papers presented at the Second International Conference on Sustainable Irrigation, Management, Technologies and Policies, held at the University of Alicante in 2008. The meeting follows the success of the first Conference which was organised in Bologna in 2006.

The objective of the Conference is to discuss ways to achieve the most efficient and equitable use of water resources used in irrigation, at the same time as ensuring their sustainability.

Over-exploitation of fresh water resources is leading to damaging long lasting environmental effects with considerable depletion of groundwater and surface water sources. The problem of contamination adds to these effects, effectively reducing the availability of clean water. The Conference also discussed how irrigation ought to be used to avoid deterioration of crops and soils.

The papers presented in this volume have been classified in the following sections: Sustainable Irrigation and Economic Instruments; Irrigation Management; Irrigation Modelling; Irrigation Systems and Planning; Re-use of Water.

The Conference was organised by the University of Alicante and the Wessex Institute of Technology and sponsored by the Institute for Water and Environmental Sciences of the University of Alicante; the Ministerio de Educación y Ciencia, España; the Instituto del Agua y de las Ciencias Ambientales; the Patronato Municipal de Turismo de Alicante; Grupo de Investigación Modelización Matemática de Sistemas of the University of Alicante; and the company, Consolider-Tragua. The organisers are grateful for their support.

The Editors are indebted to all contributors for their papers and to the members of the International Scientific Advisory Committee and other colleagues who helped in the selection procedure.

The Editors
Alicante, 2008

Contents

Section 2: Irrigation management

Section 3: Irrigation modelling

Section 1
Sustainable irrigation and economic instruments
Special session organised by Henning Bjornlund

Economic instruments and irrigation water management – a comparative study of private and district irrigators in Alberta, Canada

H. Bjornlund[1,2], L. Nicol[1] & K. Klein[1]
[1]*Department of Economics, University of Lethbridge, Lethbridge, Alberta, Canada*
[2]*University of South Australia, Adelaide, Australia*

Abstract

Irrigation activity in Alberta accounts for 71% of consumptive use of surface water in the province. Pressures on water resources are acute and are expected to intensify. Alberta's answer to its water problems is contained in the Water for Life strategy which aims for a 30% increase in water use efficiency and productivity and the implementation of economic instruments if necessary. Irrigators' contribution towards this endeavour will be imperative. But the foundation of irrigation activity in Alberta is grounded in a private and irrigation district water management system that has resulted in the development of two very distinct irrigation groups. The differences in the production activity and water management practices between private and district irrigators are striking. This study attempts to identify these distinguishing characteristics relating specifically to the adoption of irrigation technology and management practices and ascertain the effect of economic instruments which Alberta, until recently, has largely avoided using.

Keywords: economic instruments, water efficiency, water productivity, irrigation, private irrigation, irrigation districts.

1 Introduction

The majority of irrigation activity in Canada is concentrated in Alberta (64%) (Statistics Canada [8]). Within the province itself, irrigation is by far the major water consumer, accounting for 71% of consumptive use of surface water

www.witpress.com, ISSN 1743-3541 (on-line)
doi:10.2495/SI080011

(AENV [3]). For purposes of irrigation, the South Saskatchewan River Basin (SSRB), located in the southernmost area of the province is the most important of the province's seven river basins. 82% of the irrigated area lies within the SSRB.

In Alberta, water is managed under a licensing system. Licenses are legally tied to specific parcels of land and historically have remained with the property when the land is sold. The date the license was issued establishes its seniority and the first-in-time, first-in-right principle ensures that during times of water shortages, licence holders obtain access to their water in accordance with their seniority.

For irrigation purposes, licenses are held by two distinct types of irrigators: private irrigators and irrigation districts. Private irrigators are issued individual licenses just like other private or public landowners needing water for purposes such as golf courses or parks. Licensing is regulated by Alberta Environment. Almost 2,900 private irrigators exist in Alberta, accounting for 18% of the irrigated area (AAFRD [1]). Water for private irrigation is mainly extracted from ten different rivers within the SSRB, along which most of the private irrigators are located (Figure 1). Private irrigators are responsible for the installation and maintenance of the infrastructure needed to pump the water from the river and convey it from the river to the field as well as the irrigation equipment used on the field itself. Private irrigators do not pay for the water. When a license is approved, a one-time payment of the license is levied, based on the volume of water involved. These irrigators are governed by the Water Act (1999).

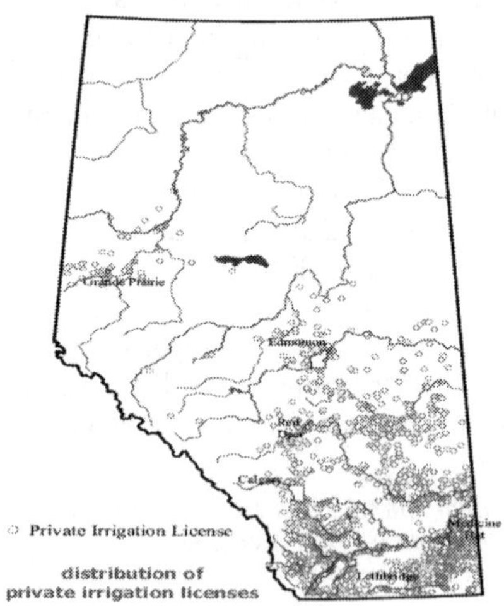

Figure 1: Private licences. Source: AAFRD, 2000.

The provinces' 13 irrigation districts, which account for 82% of the irrigated area, operate very differently from private irrigators (Figure 2). The districts hold water licenses and the irrigators within them have their irrigable area on the district's assessment role (acres approved for irrigation for which an annual water rate is paid to the district). These irrigators constitute the district's ratepayers. Irrigators pay a flat fee per hectare for administration costs and some rehabilitation of infrastructure, varying from as high as almost $45.00 per hectare in the St. Mary River Irrigation District to as low as $18.50 per hectare in the Eastern Irrigation District (AAFRD [1]). The variation in rates is reflective of whether or not the irrigators have piped and pressurized water supply and whether the districts have access to other sources of income. Irrigators do not pay for the water itself and they also do not pay for the cost of head works and the supply infrastructure delivering the water to the districts off-take from the river. Some districts also supply water to municipalities, golf courses, feedlots, as well as oil and gas and other industries, resulting in a complex fee structure among districts. The districts are governed by the Irrigation Districts Act (2000).

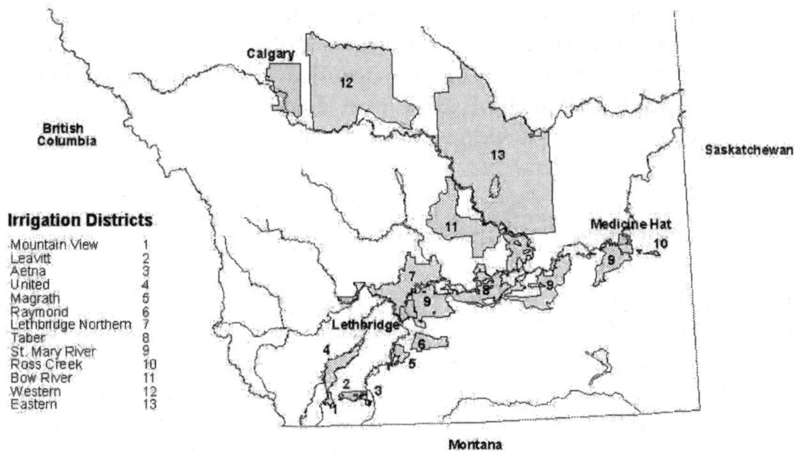

Figure 2: Irrigation districts in Alberta. Source: http://www.agric.gov.ab.ca/irrigate/irrbase.htm.

Water use efficiency in Alberta, measured as the fraction of water delivered to the farm that actually reaches the root zone of crops, has improved over time through the gradual movement away from surface irrigation (30% water use efficiency) to wheel-move and ultimately to the most efficient, low pressure centre pivots (80% water use efficiency). Under existing irrigation techniques efficiency overall was estimated in 1999 at 71% across the irrigation districts (AIPA [5]).

Economic instruments found their way tentatively into water management in Alberta when the new Water Act in 1999 and Irrigation Districts Act in 2000 provided for the introduction of trading in water rights and allocations. During the drought of 2001 the value of allocation trading proved itself when it assisted irrigators in the St. Mary River Irrigation District to manage the drought conditions (Nicol and Klein [6]). Discussions with irrigation district managers reveal, however, that little or no allocation trading has taken place since then. Water right trading has also not been widely used, mainly due to the complexity of administrative procedures and lack of effective communication systems between buyers and sellers (Nicol et al. [7]).

Shortly after the introduction of this legislation the government embarked on a public review process with the view of establishing a long-term provincial water management strategy. This review process occurred between November, 2001 and June, 2002. The result was the Water for Life strategy that was released in November 2003 (AENV [2]). The strategy confirms that water resources in the SSRB are fully or over committed and that demand for water is likely to continue to increase due to Alberta's population and economic growth as well as increased demand from in-stream uses. The strategy projects that water use efficiency and productivity can be increased by 30% by 2015 and that economic instruments will be used to achieve this if necessary. Although water markets and water pricing are two main economic instruments often discussed in the literature (Johannson et al. [12]; Briscoe [9]; Haddad [11]; Dinar et al. [10]), many other options exists such as various types of subsidies and taxes that can be used to provide incentives for the introduction of best management practices or more efficient technologies. Except for trading of water allocations and water rights, use of other economic instruments applied to irrigation water has been largely avoided in Alberta.

2 Survey design and methods

In 2005, the managers and board members of the 13 irrigation districts were surveyed to determine the irrigation industry's perception of the Water for Life strategy and its main objectives and policy instruments (Bjornlund et al. [4]). That study found that there was very little support for the use of economic instrument and the expectation of further efficiency gains was far below the 30% target outlined in the Water for Life strategy. It also was found that further adoption of improved technologies or management practices likely would vary across the 13 districts depending on physical production conditions and other factors.

That survey was followed up by two additional surveys which obtained the views and practices of private irrigators and irrigators within two irrigation districts. The surveys, the results of which are presented here, delved more deeply into issues relating to water use efficiency measures and reaction to the use of economic instruments. The survey asked irrigators: 1) what they have done in the past and what they intend to do in the future to improve water use efficiency on their farms; 2) what were the drivers and impediments to make such improvements; 3) what has influenced their decision making about improved technologies and management practices; and 4) their likely responses to economic instruments including subsidies, greater opportunities to obtain specialty crop contracts, and trading water allocation and water rights. Many of the questions sought information on irrigator's historical, recent past and future intended technological and management practices in three distinct time periods: historical (prior to 2001), recent past (2001–2006), and future (2007–2012).

The private irrigators' survey was conducted by telephone between March 1 and March 31, 2006. A list of private irrigator names and locations was obtained from Alberta Environment, Lethbridge office. Names were randomly selected from the list. One hundred and fifty surveys were conducted. The irrigation district survey was conducted by mail with questionnaires sent on December 14, 2006 to 810 irrigators: 320 in the Raymond Irrigation District and 490 in the Taber Irrigation District. A reminder postcard was mailed 10 days later. One hundred and fifty questionnaires were returned, representing a 19% response rate. The Raymond and Taber Irrigation Districts were chosen because they have quite distinct production characteristics and technology adoption patterns due to differences in soil type, number of frost-free days, heat units, and geography.

Data were entered into SPSS for analyses. Frequencies and descriptive statistics were produced. To identify statistically significant differences in survey outcomes cross-tabulation between private and district irrigators were performed using Pearson's Chi-Square tests or Fishers Exact tests. Outcomes are reported in the tables.

3 Private and district irrigator features

Study results reveal striking differences in production and personal characteristics between private and district irrigators. A factor of particular importance, one liable to affect virtually all decision making, is the prominence of dryland farming among private irrigators (Table 1). A significantly higher percentage of private irrigators have large amounts of land under dryland farming compared to district irrigators – 39% of private compared to 16% of district irrigators have dryland area greater than 260 hectares (Pearsons Chi-Square $p<0.01$). Alternatively, only 13% of private irrigators have irrigated areas greater than 260 hectares compared to 24% of district irrigators (Pearsons Chi-Square $p<0.05$). This suggests that private irrigators are probably less reliant on irrigation to generate farm revenue.

Table 1: Dryland and Irrigation Farming: Private and Irrigation District Irrigators (%).

	Dryland[1]		Irrigation[2]	
Size (hectares)	Private	District	Private	District
< 65	38	60	55	42
65 and <130	9	16	20	16
130 and <260	14	9	13	18
260 and >	39[1]	16	13	24

[1] Sign. different at the 0.01 level; Sign. different at the 0.05 level

There is also a noticeable difference in production. A significantly higher proportion of private diverters grow forage (Pearsons Chi-Square *$p<0.01$*). The vast majority of private irrigators use their relatively small irrigated area to grow forages in support of secondary production – primarily cow-calf and feedlot operations. Fully 93% of irrigators indicated this is the case. 55% of private irrigators indicate all their irrigated area is dedicated to forage production. So prominent is forage production that 64% and 87% of private irrigators indicate they have no irrigated area dedicated to cereal or specialty crop production, respectively. While 31% of district irrigators dedicate all their area to forage, a significantly higher proportion have more land in specialty crops and cereal (for both, Pearsons Chi-Square $p<0.01$). About 40% indicate they have some irrigated area under both cereals and specialty crop production. Reflecting these differences it is not surprising that 43% of private irrigators report that the weather is the most important factor determining when to irrigate, while 57% of district irrigators report that the crop growth stage is the most important factor – a significantly higher percentage than for private irrigators.

Comparing additional production and personal characteristics between the two groups' results indicate that, compared to district irrigators, private irrigators seem to:

- have significantly more land under less advanced surface ($p<0.01$) and wheel move irrigation systems ($p<0.05$) – 21% have all irrigated land under surface irrigation compared to 8% of district irrigators; 28% have all irrigated land under wheel move irrigation compared to 21% of district irrigators; 5% of have all irrigated land under high pressure pivot compared to 8% of district irrigators ($p<0.01$);
- have less off-farm work (52% private, 61% district, $p<0.1$) and depend less on off-farm income (21% of private irrigators derive over 75% of household income from off-farm work compared to 40% of district irrigators, $p<0.1$));
- have less formal education (29% have college education compared to 32% irrigation district; 6% have a university graduate degree compared to 14% irrigation district, $p<0.1$));
- are significantly older (67% are 55 years or older compared to 45% of district irrigators, $p<0.01$);
- have a significantly higher level of family background in farming (91% private, 84% district, $p<0.05$);
- have significantly greater expectation of family continuity of the farm (33% v. 48% do not expect family continuity, $p<0.01$);
- participate almost equally with irrigation district irrigators in government programs (44% of private irrigators on average compared to 42% irrigation district irrigators).

4 Adoption of irrigation technology and management practices

Relative to district irrigators, private irrigators have invested less in irrigation technology in the past and have less intention to invest in the future. As table 2 shows, across most initiatives, a significantly lower percentage of private irrigators have undertaken these measures compared to irrigation district irrigators. For the time periods of before 2001 and 2001–2006, the percentage of district irrigators implementing these measures was significantly higher, in many cases more than double the percentage of private irrigators. The rate of adoption for three measures slowed considerable for both groups from the first to the second time period – converting from surface to wheel move, wheel move to pivot and surface to pivot. The percentages for the remaining two measures – converting from high to low pressure and purchasing a computer panel – increased for both groups, probably prompted by increasing energy costs and a desire to adopt recent computer innovations. In the future, the rate of adoption will slow considerably for both groups, with the highest percentage continuing to change to low pressure pivots and purchase a computer panel. These percentages are low but district irrigators will continue to adopt most of these measures at a significantly higher pace.

Table 2: Implementing irrigation technologies, percentage of irrigators.

Type	Before 2001		2001-2006		2007-2012	
	Private	District	Private	District	Private	District
Surface to wheel move	20[1]	37	1[1]	6	2	3
Wheel move to pivot	13[1]	33	7[2]	20	2[1]	10
Surface to pivot	7	10	1	3	1	2
High to low pressure	5[1]	16	10[1]	23	3[1]	12
Purchase computer panel	5	7	6[1]	16	1[1]	12

[1] Sign. different at the 0.01 level; [2] Sign. different at the 0.05 level

The disparity in adoption of improved management practices between the two groups (Table 3) is even more apparent than the adoption of improved technologies. In virtually all categories across all time periods, the differences are statistically significant. It appears that private irrigators started to adopt improved management practices quite aggressively prior to 2001 then the number dropped off substantially from 2001 to 2006 followed by very little activity planned for the future. Compared to district irrigators, private irrigators therefore had a much greater rate of adoption before 2001 across all measures except hand auger and feel method. During the 2001–2006 period, like private irrigators, the rate of initiation for district irrigators also dropped but only for two measures -initiating visual monitoring and using the hand auger and feel method. For the additional four initiatives – starting to use monitoring instruments, computers or phones to change the position of pivots, website sources like AIMM (Alberta Irrigation Management Model) or IMCIN (Irrigation Management Climate Information Network) and private consultants – the percentage increased. Thus the percentage of district irrigators who started these practices in the 2001 to 2006 period surpassed the percentage of private irrigators across all measures. Noticeably similar, however, is the relatively small percentage of both groups who had undertaken three relatively new measures during the 2001–2006 period – the use of monitoring instruments, computer or phone to change the position of pivots, and internet sources such as AIMM (Alberta Irrigation Management Model) or IMCIN (Irrigation Management Climate Information Network), suggesting a reluctance to try new methods. In the near future, a larger percentage of district irrigators plan to initiate all measures compared to private irrigators, with the highest percentage planning to begin using monitoring instruments (9%) and computers and phones to change the position of pivots (10%). For private irrigators, the percentages for the future are very small, are in all but one instance significantly different from district irrigators, and involve 3% or less of those irrigators.

Table 3: Start to implement improved management practices, percentage of irrigators.

Type	Before 2001		2001-2006		2007-2012	
	Private	District	Private	District	Private	District
Visual Monitoring	74[1]	47	3[2]	9	1	1
Hand auger and feel method	27[3]	36	3[1]	19	1	1
Monitoring instruments	11[2]	3	2	5	1[1]	9
Computer/Phone	3	1	2[3]	6	1[1]	10
AIMM/IMCIN	6[3]	2	1[1]	7	1[3]	5
Private consultants	19[2]	10	2[1]	9	1[3]	4

[1]Sign. different at the 0.01 level; [2] Sign. different at the 0.05 level; [3] Sign different at the 0.10 level.

When asked to identify from a list of reasons, the most important reason for improving water management, the differences in ratings were all statistically significant. The vast majority of irrigation district irrigators, almost 60%, identified “to improve crop yield and quality”. For private irrigators, the reasons not only related to improving crop yield and quality, where 30% identified this reason as being the most important, an additional 27% identified “to reduce labour costs” and a further 24% identified “to reduce energy costs”. This response suggests that while irrigation district irrigators are undoubtedly cognizant of costs, the extra energy and labour costs involved in water management for private irrigators would result in those factors being particularly relevant in their decision making.

When asked to identify from a list of reasons, the main reason inhibiting them from improving water management, the differences between district and private irrigators are again notable. The largest percentage of district irrigators, 25% identify “I already use all the water saving practices that are practical”. This factor is rated as most important by a significantly lower number of private irrigators, just 15%. This suggests that private irrigators believe they have more room to improve water saving practices which is not unexpected since a significantly lower proportion of these irrigators has or plans to invest in improved irrigation efficiency. The main reason inhibiting private irrigators is that "physical field conditions limit system improvements" which was rated by 28%. The second reason noted by both district and private irrigators relates to finances – for 21% of district irrigators it was that "my financial situation does not permit the investment" and for 16% of private irrigators it was that "improvements will reduce costs but not enough to cover installation costs".

5 Economic instruments

5.1 Financial incentives

Irrigators were first asked to identify which option they would choose if financial assistance were available to assist in investing in more efficient irrigation equipment. Of the three options – cash subsidy, subsidizing borrowing rates and accelerated depreciation – both private and district irrigators by far favour cash subsidy – 74% and 71% respectively. In terms of the amount of assistance needed to do so, private irrigators appear to require more economic incentive. Although the differences are not statistically different, between about 60% and 70% of private irrigators would require the highest level of subsidization, compared to about one-third of district irrigators (table 4). Close to a majority of district irrigators would require the mid-range level of subsidization.

Table 4: Level of subsidization based on 65 hectares (percent of irrigators).

Improve existing equipment			Invest in new low pressure pivot		
Level	Private	District	Level	Private	District
<$5,000	23	18	<$10,000	12	20
$5,000-$10,000	18	47	$10,000-$30,000	16	49
>$10,000	59	34	>$10,000	68	31

5.2 Processing opportunities

The data suggests financial constraints exist for both groups of irrigators but especially private irrigators. Irrigators when first asked if new processors were to locate nearby and contracts were available, would they increase production of high value speciality crops. Irrigators were then queried as to whether such opportunities would lead to investment in improved water use efficiency measures. Of seven specialty crops listed, an average of 12% of private irrigators would begin or increase production of these types of crops compared to 19% of district irrigators, perhaps reflecting restrictions to irrigation of such crops due to physical field constraints and the need to continue to dedicate land to growing forages in support of their cow-calf and feedlot operations. Not surprisingly, less private irrigators, about half, would therefore invest in improved water use efficiency measures – only 28% of private irrigators compared to 53% of district irrigators. Although not statistically significant, this result underlines differences in the characteristics of private and irrigation district irrigators.

5.3 Water allocation and rights trading

As noted, activity in the water allocation and water rights market among irrigators in Alberta is very limited. However, if supply pressures intensify, such activity may grow. Irrigators were asked whether if someone was available to purchase their water allocation for a year and the offered price made economic sense, they would consider selling it. An almost equal number, approximately 37% of private and district irrigators indicated they would. But when asked if such circumstances existed for the selling of a permanent water license, a much higher percentage of private irrigators, 22%, compared to district irrigators, 7%, would do so and convert to dryland. This is perhaps not an unexpected result given private irrigators' experience and custom of dryland farming and reduced reliance on irrigation. Finally, irrigators were asked if they were in a situation where they wanted to expand or maintain their irrigated area, whether they would consider buying additional water licenses to do so. Not surprisingly, less private irrigators, 42% indicated they would do so, compared to 61% of district irrigators.

6 Conclusions

The differences between private and district irrigators extend to numerous facets of personal characteristics, production methods, water management, motives for decisions, and ultimately, responses to the use of economic instruments. Compared to district irrigators, private irrigators seem to be much more conservative and grounded in traditional methods of farming. They are older, have less formal education, depend much less on off-farm income, more often have parents involved in farming and have greater expectation of family taking over the farm. They have large dryland farming areas and much smaller areas of irrigated land which is dedicated primarily to forage production in support of cow-calf or feedlot operations. District irrigators have a significantly higher proportion of their land in specialty production reflecting their better growing conditions, more reliable water supply and the presence of processing facilities Given the prominence of dryland farming relative to irrigation farming, it is not surprising that private irrigators are much less inclined to adopt more advanced irrigation technology relative to irrigation district irrigators. Private irrigators adoption of improved management practices seems to have been undertaken prior to 2001 with very little intent to adopt in the future. These irrigators, and perhaps to a lesser but still substantial extent, district irrigators, do not plan on initiating improved water management practices. The highest percentage is 10% who plan to begin using computer or phone to change the position of pivots. This suggests there may be potential for extension work to promote greater water efficiency through use of management practices which often involve minimal cost to the user. In the case of private irrigators, where a large number believe they do not use all the water saving practices that are practical, this approach may be particularly fruitful.

Left on their own recognisance, adoption of more efficient technologies in the future will be modest. This is especially true for private irrigators. Physical field conditions for private irrigators seem to be a major factor inhibiting adoption. In addition lack of financial capability is a factor. Generally, private irrigators will be a difficult group to motivate unless monetary incentives are very generous. Processing opportunities that allow for more speciality crop production and enhanced finances to purchase improved technologies, appear to be a non-starter. Few would venture into, or increase production of, specialty crops, resulting in a limited number investing in improved water use efficiency measures. Given an opportunity to sell their water license, a higher number of private irrigators than district irrigators would do so and relatively less would buy additional water licenses to expand or maintain their irrigated area, reinforcing what appears to be private irrigator's proclivity to dryland farming.

References

[1] AAFRD. Alberta Agriculture, Food and Rural Development, *Alberta Irrigation Districts – Alberta Irrigation Information for the Year 2006,* Lethbridge, Alberta: 2007.

[2] AENV, Alberta Environment, W*ater for life – Alberta's strategy for sustainability,* Alberta Environment, Edmonton, 2003.

[3] AENV, Alberta Environment, *Water for Life – Facts and Information on Water in Alberta,,* publication #I/930. Edmonton, Alberta: 2002.

[4] Bjornlund, H., Nicol, L. & Klein, K., Challenges in implementing economic instruments to manage irrigation water on farms in southern Alberta, *Journal of Agricultural Water Management*, 92, 131–141, 2007.

[5] AIPA, Alberta Irrigation Projects Association, *Irrigation in the 21st Century, Volume 1: Summary Report.* Lethbridge, Alberta: 2002.

[6] Nicol, L., & K. Klein, Water market characteristics: Results from a survey of Southern Alberta irrigators. *Canadian Water Resources Journal,* 31(2): 91–104, 2006.

[7] Nicol, L., Klein, K., & Bjornlund, H. Permanent transfers of water rights: A study of the southern Alberta market. *Prairie Forum*, in press.

[8] Statistics Canada, 2001 Census of Agriculture, www40.statcan.ca/l01/cst01/agrc05f.htm, 2006.

[9] Briscoe, J., *Managing water as an economic good: Rules for reformers,* paper presented to The International Committee on Irrigation and Drainage Conference on Water as an Economic Good, Oxford, September, 1997.

[10] Dinar, A., Rosegrant, M., & Meinzen-Dick, R., Water Allocation Mechanisms – Principles and Examples, (World Bank Working Paper, No.1779), Washington D.C.: World Bank, 1997.

[11] Haddad, B. *Rivers of gold – Designing markets to allocate water in California,* Washington D.C: Island Press, 2000.

[12] Johansson, R., Tsur, Y., Roe, T., Doukkali, R., & Dinar, A., Pricing Irrigation Water: A Review of Theory and Practice. *Water Policy*, 4 (2) 173-199, 2002.

Application of economic instruments, tradable licenses and good governance for sustainable water conservation

W. L. Nieuwoudt[1] & G. R. Backeberg[2]
[1]University of KwaZulu Natal, South Africa
[2]Water Research Commission, South Africa

Abstract

According to the National Water Policy in South Africa the objectives are to achieve equitable access to water, and to promote efficient and sustainable use for optimum social and economic benefit. The National Water Act provides the legal framework and the National Water Resource Strategy explains the ways in which water resources will be managed. Several inter-related actions are taken to implement water conservation and demand management. The results of completed and ongoing research projects in South Africa demonstrate the application of technologies for direct and indirect water measurement in rivers, canals and pipelines. Decision support systems are available to determine the cost of irrigation and assess the risks of agricultural water management on farms. Empirical analyses have been done of temporary or permanent transfers and lease or trade of water use entitlements on irrigation schemes with stable and variable water supply. The establishment and functioning of catchment management agencies (CMA's) for local water management by different water use sectors including irrigation have been evaluated. Based on these studies it is concluded that volumetric charging and cost recovery of water supply services through a two-part charging system is practically feasible. Complete description of water use entitlements and provision of information must improve to increase security and reduce the risks associated with market trades. Training and capacity building is essential to support participation by all water users and maintain standards of accountability for irrigation management.

Keywords: water user charges, water marketing, water governance.

WIT Transactions on Ecology and the Environment, Vol 112, © 2008 WIT Press
www.witpress.com, ISSN 1743-3541 (on-line)
doi:10.2495/SI080021

1 Introduction

Several studies have been undertaken to assist with the implementation of the new National Water Act (NWA) of 1998 and the National Water Resource Strategy (NWRS) of 2004 [1] by the Department of Water Affairs and Forestry (DWAF). The NWA brought about a new framework for water resource management in South Africa. Water resources should accordingly be allocated in a way that will ensure its “best possible use”. The “best possible use” entails more than the productive use of water since in addition social, economic and environmental factors must be included to achieve the objectives of equity, efficiency and sustainability of water use. These issues will be analysed and discussed in the paper. The paper is mainly based on completed and ongoing research projects managed and funded by the Water Research Commission.

2 Some theoretical and practical considerations

Water markets are based on a system of water law that displays three attributes: security, stability, and flexibility in protecting transferability of property rights. Security is the ability to identify and gain protection for the right of use. Stability assumes that the right of use will continue to be recognised. Flexibility allows the right of use to be transferred to another use. The flexibility of being able to transfer a water right adds value to it because the market value of the right reflects not only the value of current use but also that of future opportunities. Security and stability in water markets are important, as it will affect investment.

An economic explanation of “best use” will be given as it was attempted to study this empirically in projects reported in this paper. The economic interpretation of efficient use of water is that the return per cubic meter of water must be maximized. The full economic cost of water consists of financial, opportunity and external costs. In a water market the rent return to water is maximized as water moves to a better use. In the rent return to water, risk is reflected as a cost (opportunity cost). The economic meaning of efficiency of water use and the rent return to water are thus synonymous concepts. Supply risk is often high where not sufficient storage dams exist. This is important in South Africa as in the irrigation areas of the Crocodile River, water use rights moved from a high risk environment to an environment where lower risk crops such as sugar-cane can be produced.

It is also often stated that water markets are not effective because of few trades. This statement may miss probably the biggest contribution of a water market: It discovers the opportunity cost of water and all water users, including the non buyers/sellers, are faced by this opportunity cost and are provided incentives to conserve water resources. However, participation in the water market is only possible by water users who have existing rights and have access to funds. Due to the history of South Africa, unequal allocation of use rights was caused by racially discriminatory legislation. This unequal distribution cannot be corrected by the market process alone. Political and legal negotiations are necessary to achieve equity through water allocation reform (Backeberg [2]).

3 Economic instruments for water conservation

3.1 Charges for water resource development and use

A user charge is a non-market economic instrument, similar in nature to regulation or administrative control of a resource and is used to encourage the conservation of water resources. The charge is levied by Government or a Catchment Management Agency (CMA) or a Water User Association (WUA) to recover the costs for providing a service to supply the allocated water. There is an exchange relationship between e.g. Government as 'supplier' of the service and the user as the 'buyer'. In the case of a quasi-collective service such as the supply of irrigation water, the farming public pays a 'price' for such a service known as a user charge (Gildenhuys [3]). Such instruments may ration or change resource use in the intended direction but is classified as ineffective if the market for allocation of water resources is not functioning.

The NWA provides for three types of water user charges to achieve water conservation and demand management (DWAF [1]): (a) water resource management charges to fund the controlling, monitoring and protection of water resources in a catchment; (b) water resource development charges which recover the cost of planning, designing, constructing, operating and maintaining water supply schemes; and (c) charges for achieving equitable and efficient water allocation which relate to the value of water. The latter charge has not been applied administratively (Genesis Analytics [4]). The purpose of the first two financial charges is cost recovery while the objective of the third economic charge is purportedly to provide incentives to shift water from lower to higher value uses. The imposition of user charges such as under (a) and (b) is justified since the subsidisation of irrigation schemes in the long run creates distortions by not relating actual costs to water supply. Charges to cover costs therefore make business and economic sense.

For commercial farmers the subsidy on operation and maintenance (O&M) costs has therefore been phased out. In contrast, for emerging and subsistence farmers the O&M charges for water supply will be subsidised at a reducing scale over five years. Currently user charges are levied on an area basis but with measurement of water use, volumetric charges will be instituted. The theoretical soundness of proposing a charge as envisaged under (c) in order to achieve equitable and efficient water allocation is questionable. Such a charge is in fact a tax which reduces the expected returns for productive use of water and is therefore rather a disincentive to transfer water. It is a fallacy to argue that economic charges will promote further conservation of water as resources are allocated based on opportunity cost and not financial cost. In fact the user charges do not change the opportunity cost faced by the irrigator as the sum of the tax and the water rent (which is lower because of tax) will stay unchanged. Where the characteristics of a resource is conducive to the formation of a market, such as in the case of explicit, exclusive, enforceable and transferable water use entitlements, the most effective and efficient mechanism to promote resource conservation is to promote markets.

3.2 Costing of water use

In addition to user charges for water services, provision has to be made for on-farm cost of water abstraction, storage, distribution and application. These irrigation cost form a significant (15-27%) of the variable cost of commercial crop production. Before an investment decision is made, the capital and operating cost of irrigation equipment must therefore be evaluated. The program IrriCost has been developed to estimate the annual capital and operating cost of irrigation (Meiring *et al.* [5]). This tool can be used to do cost comparisons of alternative designs, analyse annual cost of water use and compile budgets of irrigation costs. Apart from escalating costs, farmers are confronted with changing yields and prices. The model RiskMan was developed to provide information for risk management in irrigation farming. With the aid of this model, information can be processed to be useful for decision-making at enterprise and whole-farming level. These costing procedures taking risk into account have been applied and tested for small- and large-scale commercial farmers in the Nkomazi region of the Komati and Crocodile Rivers (Oosthuizen *et al.* [6, 7]). For small-scale farmers a subsidy on capital is crucial for financial survival while for large-scale farmers the simultaneous replacement of orchards and irrigation equipment has a severe effect on financial feasibility and riskiness of farming. The full financial cost of irrigation, i.e. capital, operating and maintenance cost were similarly evaluated in a separate case study of smallholder subsistence farming (Perret and Geyser [8]). The results show high cost of irrigation services in comparison with income from irrigation. This finding supports the approach adopted by DWAF to gradually increase charges for water development and use by subsistence farmers.

4 Hydrological issues

4.1 Metering and complete description of water use entitlements

The security of water use rights implies that it can be monitored and enforced which further implies that it can be measured. A process is underway to measure or meter water in South Africa. In this regard it has been shown that technologies are available for direct and indirect measurement of water conveyance in rivers, canals and pipelines. The challenge in practice is managed implementation of the water measuring system, both by individual farmers and by WUA's (Van der Stoep *et al.* [9]). Various conditions for licensing of water use, including installation of water measuring devices, are furthermore being implemented by DWAF [10]. As a whole these actions to measure and monitor should contribute to a clearer specification and enforcement of water use entitlements. Illegal use of water is an impediment to a water market especially in two water stressed areas namely the Olifants River (East) and Crocodile River (East) recently visited as part of a Water Research Commission project (2006/7). In the Olifants River water is metered and monitored in the Loskop and Blyde River irrigation areas and there is no room for illegal use.

4.2 Water quality impacts

Not only the volume of water must be conserved but its quality must be protected. With pollution of water, external costs are generated. The Berg River in the Cape was recently (2007) visited. Stakeholders concerned with the quality have formed an action group in this river. Water quality is also a concern in the Olifants River (East), one of the main rivers in South Africa. Coal mines on this river are allegedly blamed for discharges in the river. The policy recommendations to improve the water quality in this river were highlighted after meeting stakeholders of the Olifants Forum during 2006. Strong support from these stakeholders was received for policy options such as pollution permit trading and environmental offsets. The catchment surface of the Olifants River is fractured by mining activities, runoff decreases and water is drained into underground aquifers which then seeps into streams. A waste discharge charge system is proposed by DWAF [11] but at present, discharges in the catchment are not levied. It is recommended that polluters should pay a discharge rate, in the same way as water abstraction users pay water charges.

As in the case of a water market it is proposed that a market be established for the discharge of pollutants and that this market is used to discover the optimum price for pollutant disposal. This proposal is supported by representatives of some mines (Lodewijks [12]). All markets operate within certain rules. In a pollution permit trading market, rules that may be considered are that discharges in the river are only allowed when flow is sufficiently high and that trades may only occur within certain parameters. A permit trading program could complement desalination plants as some costs of these plants may be variable (reservoirs where the pollutants solidify fill up). Apart from a pollution trading program it is suggested that bio-diversity offsets be created to provide incentives for cooperation amongst stakeholders which may be mines, developers, environmental groups, farmers and public land agencies. Expert opinion is that the main source of pollution in the Loskop Dam is the leakage from abandoned old mines (pre-1956). The problem with the defunct mines is that they leak pollutants all the time including during the period when river flow is low. DWAF has apparently accepted responsibility for these mines but they may not have the appropriate technology, which is also expensive, to desalinate the effluent. In an offsetting arrangement, incentives can be provided to existing mines to desalinate water from these defunct mines by allowing them to discharge a given amount in the Olifants when the water flow is sufficiently high. The above arrangement will cost the taxpayer nothing while discharge during low flow periods is reduced.

5 Empirical results of water marketing studies

5.1 Efficiency of water use

Studies undertaken in several rivers in South Africa showed that water market trading promotes the more efficient use of water (Gillitt [13], Armitage [14]).

Water market trading will promote some of the objectives as stated in the Water Allocation Reform document [15] by supporting growth and development and by maximising the return, adjusted for risk per unit of water.

A discriminant analysis of water market transfers in the Lower Orange River showed that water user rights were transferred to farmers with the highest return per unit of water applied, those producing table grapes, and with a high potential arable 'outer land' without water entitlements (Armitage [14]). In this analysis the return per unit of water applied was the most significant of the variables studied and also had the highest standardised regression coefficient. It is concluded that the market promoted the more efficient use of water. Buyers of water entitlements only modestly used more water conservation technology as both buyers and sellers face almost the same opportunity cost. The opportunity costs faced by buyers are slightly more because of transaction cost. Only unused water was transferred, while water saved (through adoption of conservation practices) was retained possibly for security purposes.

A second study by Armitage [14] in the Nkwaleni Valley in northern KwaZulu-Natal found that no water market had emerged despite the scarcity of water in the area. No willing sellers of water rights existed. Transaction costs appear larger than benefits from trading. Farmers generally retain surplus rights as security against drought because of unreliable river flow while crop profitability in this area is similar for buyers and sellers (they grow the same crops). If potential buyers are compared with potential sellers then the most important variable that discriminated between them was that buyers produced sugar-cane. The reason being that sugar-cane can better withstand drought than competing crops. This may be attributed to the irregular river flow, a finding that was also observed in the Crocodile River (Gillitt [13]).

A follow-up study by Gillitt [13] was undertaken among irrigation farmers in the Boegoeberg and Kakamas Irrigation Schemes along the Orange River of South Africa who had transferred water entitlements between January 1998 and August 2003. A total of 37 farmers were interviewed. A Principal Component Analysis of factors associated with buyers in the Orange River was conducted. This indicated that buyers of water entitlements have a higher income per cubic meter of water applied, a larger percentage of cropped area planted to lucrative export table grapes and horticultural crops, larger advanced irrigation technology while it has a negative loadings with the percentage of cropland planted to other grapes and percentage of planted to field crops (lower return). This indicates that a water market promotes efficiency in water use and that water is transferred to high income crops (table grapes and horticultural crops). These relationships were confirmed in regression models (Ridge Regression, Logit Regression).

5.2 Risk in water marketing

Policy risk and risk aversion appear to be important in explaining future investment in irrigation farming in the Lower Orange River. These farmers are also highly risk averse especially regarding downside risk. Important policy implications are that farmers should be better informed about the practical implications of the National Water Act and specifically water licenses. The characteristics of buyers and sellers of water differ in the Crocodile and Orange rivers. In the Orange River where water supply appears more stable (due to large irrigation dams) water is transferred from farmers where the return per cubic meter of water is low to farmers where the return is high. In the Lower Crocodile River where water supply is highly irregular, water is transferred from farmers where risk is high to farmers where lower risk crops such as sugar-cane can be produced.

The risk aversion of irrigation farmers was measured by the Arrow/Pratt Absolute Risk Aversion Coefficient (APAR) (standardised for scale and range of data) (Nieuwoudt *et al.* [16]). The empirical investment model shows that farmers who are more risk averse, expect to invest less in the future. Farmers are more risk averse (down-side risk) than anticipated in the questionnaire as almost all the farmers picked the most risk averse category.

6 Equity with water allocation

One of the main objectives of the National Water Policy in South Africa is to achieve equitable access to water. A target has been set that at least 30% of water must move towards Previous Disadvantaged Individuals (PDI's) due to racially discriminatory legislation. If irrigation water is provided to PDI's then they will still have to be provided with suitable irrigable land which is a problem as most of the suitable land is already under cultivation. It does not make sense if water is moved from a developed farm which has little production potential without water and channelling it to a new farm that must still be developed. The development cost of irrigation farming is high and providing them with water only does not make sense. It is proposed that the most effective way to redistribute water to PDI's is through the Government programs of restitution and redistribution of land. The value of water is capitalized in the value of a farm and empowering a PDI to own a farm also provides him access to the water rights of the farm. PDI's not only need water and land but other support services to build their capacity which includes training. This implies coordination of actions from the government departments of Land Affairs, Water Affairs and Forestry, Agriculture (extension service) while other stakeholders need to be included such as the Land Bank of South Africa (financing) and commercial farmers.

7 Governance in water management

Water governance refers to the range of political, social, economic and administrative systems that are in place to develop and manage water resources, and the delivery of water services (Rogers and Hall [17]). "Good governance" depends upon the principles of predictability, inclusivity, representivity, accountability, efficiency, effectiveness, social equity and justice. Other principles such as transparency are necessary to ensure safeguards in the system, while cooperation is necessary in a highly complex system. Causes of "ineffective governance" include corruption, inadequate financial resources, inadequate labour and managerial skills, low prioritization and poor communication (Moss *et al.* [18]).

According to Pegram *et al.* [19], "coherent governance of the water environment has been simplified by the definition of water resources under the NWA to include a "watercourse, surface water, estuary, or aquifer", … while resource quality refers to the quantity, quality, habitat and biota of a water resource. These broad definitions together with the broad definition of water use to include abstraction, storage, streamflow reduction, waste discharge (including sea outfalls), waste disposal (with impact on water), in stream activities and recreation, provide a relatively integrated basis for water resources management.

South Africa's water resources policy and legislation is firmly grounded in the principles of the Constitution, considers international best practice around integrated water resources management (IWRM) (including decentralization and participation) within the historical context of South Africa requiring redress. The NWA develops a coherent and integrated governance framework around these concepts, addressing catchment level strategic planning, allocation, protection, development and utilization of the water resources and charging for water. It further provides for decentralized organizational framework for water resource management, based on the establishment of Catchment Management Agencies (CMAs).

However, the development and implementation of the necessary regulatory enabling framework for legislation and regulation has been slow, particularly in the delayed establishment of CMAs; transformation of WUAs; establishment of a water resources classification system; reallocation of water use entitlements, including compulsory licensing; development of catchment management strategies; authorisation of water use within a catchment paradigm; development of economic instruments under the charging strategy, such as the waste discharge charge system."

The Catchment Management Agencies (CMAs) are statutory bodies established in terms of chapter 7 of the NWA for the management for water resources. CMAs are responsible for the planning, implementation and management of water resources. Secondly, they are established to coordinate the water related activities of other organizations and water users. CMAs also play a role as organizations to which certain functions currently performed at national level may be delegated at regional or catchment level. The initial functions of the CMAs are mainly centered on managing the regional water resources and ensuring stakeholder participation within a Water Management Area (WMA).

WUAs are cooperative associations of water users established under the NWA to undertake water related activities for the mutual benefit of all its members within a WMA. Within the associations, members cooperate and pool resources to address local water related needs and priorities. WUAs are therefore mainly established to manage local water infrastructure, e.g. irrigation water supply schemes and to implement management decisions agreed upon between the members.

The Inkomati CMA establishment process began in 2000 and it was finally launched (the first in South Africa) during 2006. The Inkomati WMA is a combination of the Komati, Crocodile and Sabie-Sand Catchments. Within its jurisdiction falls the Crocodile Main Irrigation Board which still has to be transformed to a WUA. At present commercial agriculture, with crops such as sugar-cane, citrus and other sub-tropical fruit, is the main water user in the catchment.

Pegram *et al.* [19] further state that "delays in the development of key regulatory instruments have meant that the institutional and practical implementation of the policy and legal framework for water governance is not well developed. While this is not ideal, it has allowed improved understanding and implicit change within the sector and implies that the fundamental regulatory change may be introduced in a coherent manner over the next few years."

8 Conclusion

Considerable progress has been made with implementation of new approaches to water management according to the legal framework in South Africa. It is anticipated that acceptance of regulations to mandate water measurement will enable volumetric charging. This is an important economic instrument to relate financial water costs to water use and to provide incentives for efficient irrigation. At the same time a change from a unitary to a two-part charging system should be introduced. This will facilitate both demand management of water and balancing the budgets of CMA's and WUA's. Correct incentives for conservation and allocation of water will be instituted by a combination of accelerated compulsory licensing to achieve equity and promotion of water markets to achieve efficiency. Key requirements are restitution or redistribution of land together with water use rights and clear specification of the volume and reliability of available water attached to the use entitlement. Decentralisation of water management functions from DWAF central offices to CMA's finally requires cooperation and participation by all water users to ensure effective performance and governance.

References

[1] Department of Water Affairs and Forestry, *National Water Resource Strategy* (First Edition), Pretoria, 2004.

[2] Backeberg, G.R., *Allocation of water use rights in irrigated agriculture: Experience with designing institutions and facilitating market processes in South Africa*, USCID 4th International Conference on Irrigation and Drainage, 3–5 October 2007, Sacramento, California, 2007.

[3] Gildenhuys, J.S.H., *Public Financial Management*, JL Van Schaik Publishers, Pretoria, 1997.

[4] Genesis Analytics, Part 1 The Contribution, Costs and Development Opportunities of Forestry, Timber, Pulp and Paper Industries in South Africa, Private Bag X1, Melrose Arch, Johannesburg, 2005

[5] Meiring, J.A., Oosthuizen, L.K., Botha, P.W. & Crow, C.I., *The implementation of the FARMS system for management decision-making in*

irrigation farming: Summary reports and user guides, WRC Report No 894/1-4/02, Pretoria, 2002.

[6] Oosthuizen, L.K., Botha, P.W., Grové, B., Meiring, J.A., Monkhei M.M. & Pretorius, I.G., *Cost Estimating Procedures for micro-, drip- and furrow-irrigation systems for large- and small-scale farmers in the Onderberg/Nkomazi region*, WRC Report No 974/1/05, Pretoria, 2005.

[7] Oosthuizen, L.K., Botha, P.W., Grové, B. & Meiring, J.A., *Cost-estimating procedures for drip-, micro, and furrow-irrigation systems*, Water SA Vol 1 No 3 July 2005.

[8] Perret, S. & Geyser, M. *The full financial costs of irrigation services: A discussion on existing guidelines and implications for smallholder irrigation in South Africa*, Water SA Vol 33 No 1, January 2007.

[9] Van der Stoep, I., Benadé, N., Smal, H.S. & Reinders, F.B., *Guidelines for irrigation water measurement in practice*, WRC Report No TT 248/05, Pretoria, 2005.

[10] Department of Water Affairs and Forestry, *Water Conservation and Water Demand management Conditions: Use of water for domestic, agriculture and industrial sectors,* (Final report version 1.0), Directorate: Water Use Efficiency, Pretoria, 2006.

[11] Havenga, B., Personal Communication. 9 October 2006. DWAF. Pretoria.

[12] Lodewijks, H., Personal Comment, Anglo Coal, 2006.

[13] Gillitt, C.G., Water markets in irrigation areas of the lower Orange and Crocodile rivers of South Africa, Report No KV 160/04, Pretoria, 2005.

[14] Armitage, R.M., *An economic analysis of surface irrigation water rights transfers in selected areas of South Africa*, WRC Report No 870/1/99, Pretoria, 1999.

[15] Department of Water Affairs and Forestry, *Guidelines for water allocation in South Africa: A framework for water allocation reform,* (Draft 2.2), Directorate: Water Allocation, Pretoria, 2006.

[16] Nieuwoudt, W.L., Gillitt, C.G. & Backeberg, G.R., Implication of risk in irrigation water transfers in South Africa, Proceedings of International Conference on Water Economics, Statistics and Finance, Ed. KP, 2005.

[17] Rogers, P. & Hall, A., *Effective water governance: Global Water Partnership*, TEC Background Paper No 7, November 2002, in Pegram *et al.* 2006.

[18] Moss, J., *et al. Valuing water for better governance* CEO Panel: Business and Industry, 2003, in Pegram *et al.* 2006.

[19] Pegram, G., Mazibuko, G. Hollingworth, B. & Anderson, E. *Strategic review of current and emerging governance systems related to water in the environment in South Africa,* WRC Report No 1514/1/06, Pretoria, 2006.

Sustainable irrigation and the role of economic instruments and their supporting institutions

K. W. Easter
Department of Applied Economics, University of Minnesota, USA

Abstract

This paper identifies a number of institutions that are critical for the financial sustainability of irrigation projects. The framework of analysis is based on Williamson's four-levels of institutions (Williamson, O.E., The new institutional economics: taking stock, looking ahead, *J. of Economic Literature*, **38 (September)**, pp. 595–613, 2000) which are used to highlight the importance of institutions and the problems that arise when implementing institutional change. Examples are provided of institutional reforms and changes that have helped different countries raise both cost recovery and collection rates. A key objective in designing water instruments is to provide farmers and managers assurance regarding the actions of others in the system. Without the appropriate institutions, it is difficult to effectively use economic instruments such as water prices, taxes, or markets to improve the financial sustainability of irrigation projects.

Keywords: cost recovery, economic instruments, water pricing, institutional arrangements, financial sustainability.

1 Introduction

Strong finances to support and maintain both irrigation and its associated drainage system are essential for sustainable irrigation. Historically, the lack of adequate finances has resulted in inadequate system operation and maintenance (O&M) and caused many irrigation systems to be built with inadequate control structures and, in many cases, no facilities for drainage. The end result has been projects that decline rapidly in their ability to provide adequate and timely water delivery. In a few years these same projects also face declining irrigated acreage as water logging and salinity problems force land out of production. Thus, once it has been determined that it is appropriate to build an irrigation system of a

WIT Transactions on Ecology and the Environment, Vol 112, © 2008 WIT Press
www.witpress.com, ISSN 1743-3541 (on-line)
doi:10.2495/SI080031

given size and design, we need to determine how to appropriately fund the project over time. The key questions we will try to address in this paper are 1) how much of the financing can reasonably come from water users and 2) how this share can be effectively collected from water users on a sustainable basis.

To set up a system that will provide sustainable funds for irrigation and the necessary drainage will involve establishing effective institutional arrangements to support efforts to collect water charges from users. Institutions can be thought of as "rules of the game" while the "players or groups of players" are the organizations, firms, and individuals [2]. Institutions are important in structuring incentives as well as providing order and predictability, particularly regarding the actions of others. Livingston and Garrido [3] argue that institutions are important for effective water management. "Institutional arrangements are critical in creating incentives because they: 1) define who has access to water resources, 2) establish the range of (legal) options open to legitimate water users, and 3) determine who can claim income from water use and who will bear the costs of water use." Effective institutional arrangements will need to be in place to have sustainable finances for irrigation and to sustain the irrigation system.

The remainder of the paper will start with a brief description of the institutional framework used in the analysis. This is followed by a section concerning fee collection and the determination of what share users will pay. Next, is a discussion of water pricing mechanisms, followed by examples of projects where new institutional arrangements have helped improve cost recovery and project sustainability. This leads to a section that suggests how institutions can be combined to provide a stable source of funding for O&M. The final section provides a brief summary and conclusion.

2 Institutional setting

A good way to think about institutions and how they influence outcomes is to use Williamson's [1] four levels of nested institutions. They include, first, the informal institutions such as social norms, customs, religion, etc., which change very slowly. These norms and customs act as constraints to what you can do at the other three levels. For example, strongly held customs or mores regarding free access to water may have a big impact on who gets water and how much they pay. It also may prevent, or make it difficult, to introduce private property rights for water use and to introduce water markets.

The second level of institutions is the formal rules of the game or the policies that guide water use and allocation. To make changes at this level will usually take several years to over a decade. For example, if you want to establish a water market, one of the key changes needed is to establish and allocate water rights, or water use rights, to individual water users. Such changes in property rights, laws, or policy can be difficult to make. For existing systems they will only be changed after a number of years of hard negotiating and bargaining or a significant change in a country's economic policy that favors markets, as happened in Chile [4]. The content of water policies and laws are addressed at this level including whether or not water can be sold separately from land and for what uses. In terms of financing irrigation systems, it is at this level where water policies are crafted that specify who pays for water projects and their operation.

Level three focuses on governance structures for transactions. At this level decisions are made about the mechanism for allocating water, e.g., hierarchy vs. markets or contracts. There will also be concerns about mechanisms for enforcing water allocations and for resolving conflicts that are likely to arise. The complexity of the governance structures will increase as water scarcity, its value, and conflicts increase. Birner and Wittmer [5] argue that the most efficient form of governance and its structure and complexity will depend on the characteristics of the resource, e.g., water scarcity and the social and political characteristics of a country.

The fourth level of institutions falls in the domain of neoclassical economics where governance is ignored and the emphasis is on the firm as a production unit. Institutions are generally assumed to be fixed and treated as exogenous constraints. At this stage questions occur regarding the firm's ability to pay for water and how water charges or fees may change water use or the adaption of new water-saving technology. Level-four institutions are very important in determining the actual level of water charges paid by individual water users and the services they are provided by the irrigation system.

3 Financial failures in public irrigation

Traditionally, both developed and developing countries have found it difficult to establish a sustainable source of funding for operating and maintaining their irrigation projects. In his 1995 study, Jones [6] illustrates how cost recovery and charges for irrigation water have been a problem for decades. In many countries less than twenty percent of the cost of irrigation projects has been recovered from water users [7]. This is the result of poor rates of collection combined with relatively low water fees. The end result has been a large public subsidy for water users, particularly irrigated farms.

There are many reasons for this poor record of cost recovery in public irrigation projects. Although the reasons vary among countries and individual projects, Easter and Liu [7] list some of the most important ranging from: "1) no link between fees collected and funds allocated to a given irrigation project, and 2) lack of farmer participation in planning and management of projects, …. to 4) poor delivery of water services (timing, duration, and quantity are inadequate), …. and 9) corruption among irrigation officials and those collecting water charges." They go on to make it clear that the basic underlying causes for the poor cost recovery stem from "the collective good nature of water projects, combined with open access to water resources, the principal-agent problems and rent seeking activities of irrigation officials. It also can be thought of as an assurance problem: assurance for managers concerning what water users will do and assurance for water users concerning what water managers and their staff will actually do as opposed to what they say they will or can do given the existing project design and technology [8]."

Another part of the problem is that we do not think about sustainable finances early enough in a project's development. During the planning stage, we need to decide on the source of finances to effectively operate and maintain the project once it is built. As part of this financial planning we need to determine to what extent water users should be the major source of funding and how the cost will be allocated among the various users, e.g., farmers, hydropower users, domestic water users, commercial and industrial water users, and those protected from floods. The allocation of costs is an important issue because many of the water projects are multipurpose, particularly in Asia where 90% of the dams for irrigation are multipurpose. There is also a good argument to be made that some of the costs should be allocated to consumers who benefit from lower food costs, particularly in developing countries. In fact, if commodity markets are poorly developed, the increased production may mean farmers receive only modest increases in net returns because of the drop in commodity prices caused by increased production in the irrigated area. Yet many times decisions have been made to allocate most of the cost to farmers, since they receive much of the water. A better cost allocation might be to allocate costs based on direct project benefits or on direct and indirect project benefits. Easter and Liu [7] show it makes a significant difference. For example, if the costs to be recovered are allocated based on water delivered in the Sriram Sagar project in India, farmers must pay 95% of the costs. If the costs are allocated based on direct benefits, farmers have to pay 88% of the costs [9]. The percentage of cost allocated to farmers would drop even more if the allocation was based on direct and indirect benefits, probably to something less than 75% of the costs.

Once a reasonable allocation of costs has been decided on for farmers, the next question is: what economic instruments can be used to effectively collect the water fees necessary to cover the allocated costs? The other important question is: have the critical institutional arrangements been put in place, at the planning stage, that will make the fee collection effective?

4 Economic instruments

The approach one selects for charging the water users will depend a lot on what institutions already exist and the size of the project, both in terms of hectares irrigated and the numbers of farmers actually served. For large projects in developing countries with limited farmer participation, <u>area-irrigated</u> based fees have generally been used. The problem is that this means there is no relationship between what the farmer pays and the amount of water he or she receives. This also means that the water charges will have no effect on water used. A better alternative, maybe, is to vary the charge by type of crop grown with higher per hectare charges for crops such as rice and sugarcane that use more water. In some cases the water charge might vary based on irrigation technology used. If farmers adopted improved irrigation technology such as sprinkler irrigation, which distributes water more uniformly across the field than flood irrigation and uses less per hectare, they would be charged a lower water fee per hectare.

For smaller projects, particularly in more developed countries that face a high level of water scarcity, we need to be moving toward <u>volumetric-based</u> water charges. With the improvements in technology the argument that water use or delivery is hard to measure is no longer very convincing. The big issue may be cost, but even this may not be a true constraint given the new technology now available [10]. Volumetric-based water charges have two clear advantages. First, farmers know how much water they receive and that the charge will be based on this quantity. Second, it gives farmers an incentive to not over irrigate and to conserve water. This will require not only water measurement but also infrastructure and staff to effectively control water deliveries.

Block water charges can also be used with volumetric-based fees as a means to provide a minimum amount of water at low rates while charging much higher rates for use exceeding a set volume of water per hectare. This type of charge has several benefits. First, it gives irrigation managers at least three instruments to change cost recovery: the levels of the first and second block charges and the quantity at which the second block charge starts (e.g., 3,000 m^3/ha vs. 4,000 m^3/ha.). Second, the second block charge or even a third block charge can be set at the marginal cost of developing new water supplies, which will encourage farmers to conserve water. In addition, this higher charge will have less of an adverse impact on farmers' income since it is only charged for the units of water used beyond the first block.

A two-part charge can be used, combining volumetric charges with a fixed charge. Such a system of charges may be necessary if water availability varies a lot from season to season and/or year to year. The fixed charge allows water managers to obtain a basic amount of funding even in years of low water supplies when many farmers do not receive much water. If the water charges were based only on volumetric charges, then in the years of low supplies the volumetric charges might have to be very high to cover costs. This method of charging for water also recognizes that there is a large fixed component in operation and maintenance costs (O&M), which does not depend on the amount of water delivered. These costs need to be paid even when no water is delivered.

A final option would be to introduce a water market. This could be in addition to the water charges for O&M. Water trading would improve allocation efficiency by increasing water's value to farmers and shifting water to its highest valued uses. In turn, this would increase the user's ability to pay since water would be used to produce higher valued crops. As discussed above, to introduce markets a number of key institutions need to be in place, including a water law that allows the sale of water independent of land and a system to resolve disputes over water rights and third-party effects.

5 Institutions to improve cost recovery

One good measure of the success of an irrigation system in sustaining its financial base is to look at the collection rate from its water users. In many systems only 10 to 60% of the users pay their fees [11]. Table 1 lists eight irrigation projects that have been quite successful in obtaining high rates of fee collection with half reaching 100%, and three reporting substantial water savings. Using these projects as examples of successes in sustaining finances, what institutions were important in their success? Several key institutional arrangements have helped these systems move closer to financial sustainability. First, they needed to be able to legally establish water management entities that were, at least partially, financially autonomous from government. This helped establish an important set of incentives for management that fostered improved collection rates. A second important law was needed that allowed users to organize into water user associations (WUAs) and participate in the management of the irrigation system. If possible, they should have some authority to allocate water among users and authority to tax. Through active participation water users establish a closer working relationship with management, which helps increase system transparency and accountability which, in turn, improves users' willingness to pay water fees. Too many times WUAs have been established and given only responsibilities such as contributing "free" labor but no authority. This is why a number of WUAs have not improved the sustainability of their irrigation systems.

Table 1: Factors influencing fee collection rates.

Cases	Financial autonomy	Incentives to collect	Incentives to pay	User participation	Collection rate (percent)
Awati, China*	Yes	Yes	Yes	Yes	98
Bayi ID, China	Yes	Yes	Yes	Yes	100
Nanyao ID China	Yes	Yes	N.A.	Yes	95
Shangdong China*	N.A.	N.A.	Yes	N.A.	100
Yangtze Basin, China*	Yes	N.A.	Yes	Yes	N.A.
Gujarat, India	Yes	Yes	Yes	Yes	100
Haryana, India	Partly	N.A.	Yes	Yes	85–95
Mexico	Yes	N.A.	Yes	Yes	90
Alto Rio Lerma, Mexico	Yes	Yes	Yes	Yes	100

N.A., Not available. *These systems reported substantial water savings. Source: [7].

Why is it that establishing financial autonomy and active user participation improves the sustainability of irrigation projects both in terms of their operation but also their finances? Financial autonomy changes several important incentives. First, the fees collected from users are used for the project and do not go back to the state or federal treasury where they would be commingled with other tax returns. In other words, if you do not pay your water fees, it will have an impact on "your" project's ability to deliver water. That is not the case in many countries where the revenue arm of the government collects or tries to collect the water charges, which then go to the federal or state treasury. In such cases, farmers do not see any relationship between what they pay and the services they get from their irrigation system.

The Yangtze Basin Water Resource (Yangtze) project in China is a good example of an effective water management entity that is financially autonomous.

It also requires direct involvement of WUAs in water management decisions. This has increased crop yields and saved significant amounts of water, 1.2 million m^3 annually in each WUA. Yet they do not report how much fee collection rates have been improved [12].

Financial autonomy also means that government subsidies are not available and they must rely on fees collected from users to cover O&M costs. This is a second important incentive. Now the water management entity wants to create conditions that result in high collection rates from the water users. As a result management uses several different strategies. One is to use strictly enforced penalties for those who default on payment. For example, in the Bayi Irrigation District, irrigation water is denied those who defaulted [13]. A second strategy is for the management entity to give awards or penalties to encourage staff to achieve higher rates of collection. In Awati, China, staff salaries are dependent on water charges and collection rates have reached 98% [14]. The staff in Bayi Irrigation District receives rewards for turning in collected fees by a deadline, but is fined for late payments [13].

In terms of participation, it is beneficial to get users involved early in the process of project design and building. This is particularly true for rehabilitation projects that farmers are expected to repay the costs or contribute labor. This makes the decision making process more transparent and increases users' willingness to pay for the improvements. The Laur Project in the Philippines provides one example where the WUA were able to review the rehabilitation proposal before it was implemented. Coward [15] found that this improved the project design and the users' willingness to pay.

Finally, a water management entity that is financially autonomous has an incentive to provide users a good service. This will not only give users more reason to pay their fees, it should also increase their ability to pay. Better irrigation service should result in increased yields as it did in the Yangtze Project. This should increase farmers' incomes and ability to pay for the water.

Another more dramatic way to increase user participation and create management incentives has been irrigation management transfer (IMT) to the users. This strategy has had mixed results partly because of what management responsibilities were actually transferred and the condition of the infrastructure at the time of the transfer. Several of the transfers have gone well [16] while others have not. As Zekri and Easter [17] found, IMT tend to be "successful where farmers had their water rights established, farms are medium and large scale with good access to markets and the government had a strong willingness to empower users." In contrast, programs that emphasized only farmer participation were not very successful. Farmers need to perceive some clear benefits from participation and taking over system management.

6 Supporting institutions

Too many times we have focused only on the technical or engineering side of irrigation projects. Of course, the engineering aspects of a project are important and in a technical sense determine which farms can be irrigated. What has been left out are the institutional arrangements that are needed to determine who actually has access to the irrigation, who can claim income from the water, and who will bear the cost of water use [3]. Some of the institutions that will be critical in the sustainable financing of irrigation project were discussed above. These include the legal institutions that allow financially autonomous water management entities to be established and the organization of effective WUAs. It is also becoming clearer that water users need some type of water rights that will give them assurance regarding when and how much water they will receive during the crop season. These might be outright water rights that can be traded either permanently or temporarily, or water use rights that have a set lifespan, e.g., 30 to 99 years. Another option which establishes the right incentives is a water contract between the farmers and the water management entity that specifies the amount and timing of water to be delivered to farmers. This was one of the keys to the success of the improved irrigation project in Katepurna, India where they have saved 7.7 million m^3 of water annually and expanded the area irrigated by 80% [18]. Over-irrigation in the wet season was greatly reduced since farmers no longer had to try to store water in their soil for the dry season. They now have a contract assuring them water for the dry season.

As part of an institutional arrangement that establishes farmers' rights to water, it must be clear how the rights will be allocated. When they established water rights in Chile in 1981, the consumptive use rights were allocated based on past water use, which was fairly equitable since most irrigated land holdings were relatively small, 50 ha. or less [4]. It is a more difficult question for new projects. If the area to be irrigated is already farmed, then past land ownership will likely be a key factor in the water rights allocation. Another option would be for the state to reserve some of the water rights and auction off the rest for commercial use and irrigation. If the water rights to a new project are going to be given to farmers, as they were in Chile, then one option would be to limit the area one farmer can irrigate. For example, the U.S. Reclamation Act that authorized federal funds to develop irrigation in the western U.S., limited the area one individual could irrigate to just under 65 ha. However, the limit has proved difficult to enforce, particularly in California, even after it was raised six fold in the 1980s [19].

You also need an effective local system for enforcing water use rules and water rights. In more developed countries this might be the court system, a water agency, a WUA, or some combination of the three. In less developed countries the village leadership is likely to play a much larger role. In some cases the village leadership makes it difficult to establish other methods of managing water by essentially running any WUA.

Another important institutional arrangement that needs to be in place to improve fee collection is a mechanism to make the process of setting water charges more transparent. If the charges are based on the cost of O&M, as they are in many cases, then users need to know how costs are calculated and what costs are to be included. This knowledge can help reduce the fear among users that the fees will be used just to enrich water managers and their families.

Finally, an institutional arrangement is needed that allows a council or review board to be set up that can review the record of farmers who are not paying their water fees. This can serve two important purposes. First, it can determine if there are circumstances that make it difficult for some farmers to pay their fees, such as crop failure. In such cases the council can then decide to forgive some or all of the farmer's delinquent fees. Second, they can use the review as a time to make it public in the villages who is not paying their water fees.

7 Conclusion

This paper emphasizes the importance of institutional arrangements in sustainable irrigation and focuses on those institutions that are critical to financial sustainability. Williamson's [1] four-levels of institutions are used to show how institutional arrangements at these different levels can guide water use and facilitate the process of determining who should pay for the water. A well-constructed irrigation system only assures that the water can be delivered, not that it will be used effectively, or that it will be financially sustainable.

Based on past projects that have been successful in maintaining high rates of fee collection from their users, several institutional arrangements appear to be critical. One is for the management or operating entity to be financially autonomous from government. This creates a set of incentives that focuses management's attention on providing good service, accountability and the importance of fee collection. If the farmers are going to pay for the irrigation, they want assurance that they will receive dependable and timely service. One set of institutional arrangements that increase these assurances for farmers are those necessary to create water rights or water-use rights. There is also the possibility that similar assurances can be achieved through contracts between management and users regarding water delivery and payment schedules. The trick is to make these contracts binding on all parties.

A final set of institutional arrangements that will be important in maintaining high levels of fee collection are those that improve communication between management and users. Here WUAs that are given authority along with responsibility for management can play a key role. They can also be important in making the process of determining and setting water charges more transparent and in helping establish mechanisms to deal with farmers who default.

References

[1] Williamson, O.E., The new institutional economics: taking stock, looking ahead, *J. of Economic Literature*, **38 (September)**, pp. 595–613, 2000.

[2] North, D.C., *Institutions, Institutional Change and Economic Performance*, Cambridge University Press, 1990.

[3] Livingston, M.L., & Garrido, A., Entering the policy debate: the economic evaluation of groundwater policy in flux. *Water Resources Research*, **40**, 2004, W12S02, doi: 10.1029/2003WR002737

[4] Hearne, R.R. & Easter, K.W., Water allocation and water markets: an analysis of gains-from-trade in Chile. Washington, D.C.: World Bank Technical Paper 315, 1995.

[5] Birner, R. & Wittmer H., On the 'efficient boundaries of the state': the contribution of transaction-costs economics to the analysis of decentralization and devolution in natural resource management.

Environment and Planning C: Government and Policy, **22(5)**, pp. 667–685, 2004.

[6] Jones, W I., *The World Bank and Irrigation.* Washington, D.C.: World Bank Operations Evaluation Study, 1995.

[7] Easter, K.W. & Liu, Y., Who pays for irrigation: cost recovery and water pricing? *Water Policy,* **9**, pp. 285–303, 2007.

[8] Easter, K.W., Economic failure plagues developing countries' public irrigation: an assurance problem. *Water Resources Research,* **29(7)**, pp. 1913–22, 1993.

[9] Easter, K.W., Cost allocation and the distribution of costs for irrigation in Andhra Pradesh, India. Unpublished paper, University of Minnesota, Department of Applied Economics, 2003.

[10] Wang, X. & Lu, J., The application of IC (integrated circuit) cards in efficient irrigation management system. *China Water Resources,* **46(October),** 1999, (in Chinese).

[11] Easter, K.W. & Liu, Y., Cost recovery and water pricing for irrigation and drainage projects. Agriculture and Rural Development Discussion Paper 20, Washington, D. C.: World Bank, 2005.

[12] Lin, Z. *Participatory management by farmers—- Local incentives for Self-financing Irrigation and Drainage Districts in China.* Environment and Social Development East Asia and Pacific Region discussion paper, Washington, D.C.: World Bank, 2003.

[13] Johnson III, S. H.,Vermillion, D., Svendsen, M., Wang, X., Zhang, X. & Mao, X., *Management Reform and Performance Changes in Two Irrigation Districts in the North China Plain*. The Short Report Series on Locally Managed Irrigation, Colombo, Sri Lanka: Irrigation Water Management Institute, 1996.

[14] Awati County Government. *Report of Water Management Institutional Reform in Fengshou Irrigation District*. Awati County, Xinjiang, China, 2002, http://www.xjawt.gov.cn/xianqing/jinji/fsgq.htm (in Chinese).

[15] Coward, E.W., *Irrigation and Agricultural Development in Asia,* Ithaca, New York: Cornell University Press, 1980.

[16] Kloezen, W.H., Garces-Restrepo, C. & Johnson III, S. H., *Impact assessment of irrigation management transfer in the Alto Rio Lerma Irrigation District, Mexico.* Research report 15, Colombo, Sri Lanka: International Irrigation Management Institute, 1997.

[17] Zekri, S. & Easter, K.W., Water reforms in developing countries: management transfer, private operators and water markets. *Water Policy,* **96**, pp. 573–89, 2007.

[18] Belsare, E.S., Participatory irrigation management in Katepurna irrigation project: a success story. ICID Watsave Young Professional Award Winning Paper, New Delhi, India, 2001, http://www.icid.org/belsare_2001.pdf

[19] Wahl, R.W., *Markets for Federal Water (Subsidies, Property Rights and the Bureau of Reclamation)* Washington, DC: Resources for the Future, 1989.

Irrigation and water security: the role of economic instruments and governance

R. C. de Loë[1] & H. Bjornlund[2]
[1]*Guelph Water Management Group, Department of Geography, University of Guelph, Canada*
[2]*Department of Economics, University of Lethbridge, Canada*

Abstract

Irrigation is an important contributor to global food security. However, it also contributes to a number of serious water management problems in countries around the world, including groundwater subsidence, reduced water quality, salinization and degraded ecosystems. Steps are being taken to modernize irrigation, both in terms of technologies and institutions. Importantly, while improvements within the irrigation sector are critical, they alone will not ensure the continued sustainability of the sector because irrigation contributes to, and is affected by, the larger challenge of ensuring water security. In this paper, we explore the concept of water security and link it to irrigation through examining the link between water allocation and water security. Economic instruments, we argue, are an important part of strategies to improve allocative efficiency and thus they promote water security – but on their own they are not sufficient. Rather, we suggest that water security as a multi-dimensional challenge must be approached from the broader perspective of improved governance. In this context, attention is needed to considerations such as transparency in decision making; equity in stakeholder involvement; integration among related systems (e.g., land and water, water and economy); the scale of decision making; and the balance between state and non-state actors.

Keywords: water governance, irrigation, economic instruments, water security.

1 Introduction

Irrigation increases agricultural yields and outputs and helps to stabilize global food production. The roughly 17% of the world's cropland that is irrigated

WIT Transactions on Ecology and the Environment, Vol 112, © 2008 WIT Press
www.witpress.com, ISSN 1743-3541 (on-line)
doi:10.2495/SI080041

produces over one-third of the food and fibre grown globally (Oster and Wichelns [15]). Irrigation's importance in the global food system is expected to increase. Approximately 250 million hectares of land are irrigated worldwide, a five-fold increase relative to the beginning of the 20th century (Rosegrant et al. [16]). Expansion in the amount of land irrigated is likely in future, with the bulk occurring in developing countries where population growth and demand for increased food production are strongest (WWAP [18]).

Balanced against the benefits of irrigation for food security are its environmental costs. Agricultural production is by far the largest consumer of water, with up to 70% of water withdrawals being used in irrigation [18]. Globally, water withdrawals for irrigation are estimated at about 2,000 to 2,500 km^3 per year. However, on average only 40% of this water contributes directly to crop production, with the remainder lost to evaporation, infiltration or weed growth [18]. In many parts of the world irrigation practices have resulted in lowered groundwater levels, salinization of soils, reduced water quality, alteration and regulation of natural watercourses, and reduced surface water availability (Dougherty and Hall [7], Rosegrant et al. [16], WWAP [18]). Poor drainage and irrigation practices are estimated to have caused waterlogging and salinization in 10% of the world's irrigated land [18].

In lights of its importance for global food production and food security [16], and considering its implications for the quality and quantity of water resources, effective irrigation water management is a global priority. Whether or not current levels of irrigation are sustainable, and whether or not additional irrigation is possible, clearly depends strongly on the availability of water supplies. These supplies are not assured because demand for increasingly scarce fresh water supplies is growing rapidly around the world from a variety of human uses. Simultaneously, expectations regarding environmental quality are growing. Maintenance and enhancement of aquatic ecosystems has emerged as a priority in many parts of the world. As a result, demands for water resources in many regions where pressure already exists are increasing as environmental needs are addressed.

In this chapter, we explore links between irrigation and the concept of "water security". We argue that irrigation is an important determinant of water security and, at the same time, we suggest that broader water security concerns influence the long-term viability of the irrigation sector. Water allocation systems can be an important vehicle for enhancing water security. Economic instruments are a key part of any strategy to reform water allocation systems in ways that promote water security, but numerous other concerns relating to *governance* also must be addressed.

2 Water security as a governance challenge

"Water security" is a multi-dimensional concept that has widely differing interpretations. For example, in the United States, fears about terrorist attacks have spawned an industry focused on identifying vulnerabilities in drinking water systems (e.g., Haestad Methods [12]). A much broader perspective on water security is offered by the Global Water Partnership (GWP), which defines it as "access to adequate quantities of water, of acceptable quality, for human and environmental uses" (GWP [10]). It is the broader perspective of the GWP that informs this paper. Water security exists when sufficient water of good quality is available for social, economic and cultural uses while, at the same time, adequate water is available to sustain and enhance important ecosystem functions (de Loë et al. [5]). Achieving water security is challenging because it requires good governance.

Governance refers to the processes through which societies make decisions that affect water. Good water governance depends on broad participation by affected stakeholders, and is characterized by transparency, equity, accountability, coherence, responsiveness, ethical choices, and integration of water decision making with other pertinent concerns [18]. Increasingly, people concerned about governance explicitly recognize that decision making regarding water should involve not only governments, but also citizens, non-governmental organizations and businesses.

One specific area in which links between water security and good governance are strong is *water allocation*. Water allocation systems are the rules and procedures through which access to water for both consumptive and non-consumptive uses is determined. By establishing the availability and priority of access to water resources for consumptive uses such as irrigation, cities, and manufacturing, and for non-consumptive uses such as hydropower, recreation and environmental protection, water allocation systems influence economic productivity, social and cultural wellbeing and ecosystem quality (Gleick [9]; Ferreyra and Van Beek [8]; Warner et al. [17]).

The socioeconomic, cultural and ecological implications of water allocation are amplified when water resources become scarce due to population growth, climate change, and changes in societal preferences. In the context of scarcity – whether created by societal or natural processes – water allocation systems can increase or decrease water security. Thus, effective, efficient, and equitable water allocation systems are critical to maintaining and enhancing environmental quality, economic productivity, and social wellbeing. This is especially true in regions where local economies are strongly dependent on irrigated agriculture (Bjornlund [3]).

3 Irrigation, water security and governance

Modernization of irrigation systems is occurring throughout the world through upgrading of infrastructure and through improving water use efficiency and productivity in agriculture. This is being accomplished with technologies such as drip irrigation and laser-leveling, and through procedures such as improved irrigation scheduling; measures such as these permit water application in the optimum quantity and timing for crop development [15], [18].

Improvements in irrigation technologies and practices are not the sole focus of attempts to modernize the industry. Considerable attention also is being paid to adapting institutional arrangements and irrigation policies in order to change the focus from developing new water supplies to a more holistic approach which considers agricultural productivity and market conditions. Reform of water allocation systems and increased use of economic instruments is seen as a priority (Bjornlund [2], Dinar and Mody [6], WWAP [18], Young and McColl [19]).

Technological and institutional improvements within the irrigation sector are important contributors to its long-term viability. To the extent that they lead to more efficient water use and reduced impacts on land and water resources, they also may contribute to water security. However, improved practices and institutions within the sector will not be sufficient on their own to ensure water security, and thus the continued viability of the irrigation sector. As Bazzani et al. [1] have noted, "Irrigated agriculture is currently confronted with many issues relating to its role within the overall management of water at basin level." This suggests that what is also required is attention to the broader factors that shape water security in any particular place. These factors can be revealed through consideration of the links between water allocation systems and water security.

Considerations such as the ones outlined in Table 1 determine the extent to which sufficient water of good quality is available for social, economic and cultural uses while, at the same time, adequate water is available to sustain and enhance important ecosystem functions. For example, the need to find a balance between environmental and socio-economic concerns is strongly evident in concern for *ecosystem protection*, on the one hand, and *economic production*, on the other (Table 1).

- Water security is enhanced when water allocation systems recognize the need to provide water in appropriate amounts, and at appropriate times, to support environmental needs, and when they incorporate mechanisms for monitoring and enforcing environmental water allocations, and are sufficiently flexible to permit the use of new ecological knowledge.
- Equally, water security is promoted by allocation rules that are clear and stable, by systems that provide information needed to make economically sound decisions, and by mechanisms that promote allocative efficiency.

As is illustrated by Table 1, economic instruments should be part of any strategy to enhance water security. A well developed literature exists that explores approaches to improving efficiency in water allocation generally, and in the specific context of irrigation (e.g., Grimble [11], Kemper [13], Bjornlund [2], Dinar and Mody [6], Cantin et al. [4]). However, increase allocative efficiency in irrigation by itself is unlikely to ensure the sustainability of irrigation (Massarutto [14]), let alone water security. In light of the numerous ways in which agricultural water use affects communities and the natural environment [1], additional concerns are critical. These include the development of scientifically-sound bases for environmental water allocation; collection and dissemination of information needed to make economically sound decisions; creation of mechanisms to address conflicts at different scales, and to permit sustained in meaningful stakeholder and public participation; development and implementation of technologies and practices that promote water conservation;

consideration of the impacts of climate very ability to change on water resources and water allocation systems; incorporation of adaptation strategies in water allocation decision making; attention to state sovereignty; and, in areas where indigenous people exist, respect for indigenous customary water allocation systems.

Table 1: Linking water allocation and water security.

Broad Water Security Consideration	Specific Concerns Pertinent to Water Allocation
Ecosystem Protection	• Environmental water allocation • Monitoring and enforcement for ecosystem protection • Creation and incorporation of ecological knowledge
Economic Production	• Clear and stable allocation rules • Water allocation and related information to make economically sound decisions • Ability to re-allocate water between users, sectors and/or regions
Equity and Participation	• Equity • Sustained and meaningful stakeholder and public participation • Mechanisms to address potential conflicts at different scales
Integration	• Integration between groundwater and surface water resources occurs • Integration between water quality and water quantity • Integration between land use planning and water allocation
Water Conservation	• Conservation-related charges • Re-allocation of water to more efficient and less consumptive uses • Incorporation of water conservation practices
Climate Variability and Change	• Investments to understand impacts of climate variability and change • Development and application of adaptation strategies
Transboundary Sensitivity	• Coordination of water allocation systems across political boundaries • Respect for state sovereignty • Respect for indigenous customary allocation

Source: Adapted from de Loë et al. [5].

The extent to which many of these concerns can be addressed successfully depends strongly on the effectiveness of governance. Dinar and Mody [6] emphasize the importance of creating appropriate *institutional frameworks* within which policies relating to water pricing and cost recovery should be developed and implemented. This is an important consideration. However, from the broad perspective of water security, effective governance depends on much more than the institutional frameworks within which economic instruments are established and used. This was illustrated in a recent study of the links between water allocation and water security in Canada (de Loë et al. [5]), which revealed the following:

- Failing to involve stakeholders equitably can produce conflicts, as can failing to acknowledge and respect aboriginal water rights.
- Conflicts also can result when decision making relating to water allocation is not linked effectively with decision making around land use and economic development.
- A lack of transparency in decision making processes can create uncertainty, and can lead to poor investment decisions on the part of water users.
- Efforts to protect ecosystems through land-use planning, soil and water conservation programs, and water quality management initiatives can be undermined – or nullified – if water allocation systems ignore environmental water requirements.
- Carefully designed environmental water allocations can be undermined by a failure to recognize the impacts of decision making in other contexts, such as suburban development, waste water treatment, or aggregate development.
- The scale at which governance relating to water allocation should take place is a critical determinant of its success, and is highly context dependent. Equally context dependent is the appropriate role of non-state actors in water governance (e.g., citizens, industries, nongovernment organizations) relative to state (government) actors.

4 Conclusions

In this paper, we argue that that irrigation is an important determinant of water security because of the volumes of water used, and the impacts of this use on communities and the environment. However, we also argue that broader water security concerns influence the long-term viability of the irrigation sector. In other words, irrigation affects water security, and water security affects irrigation. Water allocation systems can be an important vehicle for enhancing water security. In that context, economic instruments are a critical concern in that they facilitate allocative efficiency, and thereby can contribute to increased water security. However, by themselves economic instruments cannot create water security. Numerous other concerns relating to *governance* also must be addressed. Effective governance, we argue, demands attention to considerations such as transparency in decision making; equity in stakeholder involvement; appropriate linkages and integration among related systems (e.g., land and water,

water and economy); the scale of decision making; and the balance between state and non-state actors.

Acknowledgements

This paper draws on a recent study of the relationship between water security and water allocation in Canada funded by the Walter and Duncan Gordon Foundation (de Loë et al. [5]). Becky Swainson (University of Guelph) conducted additional background research.

References

[1] Bazzani, G. M., Di Pasquale, S., Gallerani, V., Morganti, S., Raggi, M. and Viaggi, D. 2005. The sustainability of irrigated agricultural systems under the Water Framework Directive: first results. Environmental Modelling & Software 20, no. 2: 165–175.

[2] Bjornlund, H. 2003. Efficient water market mechanisms to cope with water scarcity. Water Resources Development, 19(4): 553–567.

[3] Bjornlund, H. 2004. Formal and informal water markets: drivers of sustainable rural communities? Water Resources Research 40: 1–12.

[4] Cantin, B., Shrubsole, D. and Ait-Ouyahia. M. 2005. Using economic instruments for water demand management: Introduction. Canadian Water Resources Journal 30(1): 1–10.

[5] de Loë, R.C., Varghese, J., Ferreyra, C. and Kreutzwiser, R.D. 2007. Water Allocation and Water Security in Canada: Initiating a Policy Dialogue for the 21st Century. Report prepared for the Walter and Duncan Gordon Foundation. Guelph, ON: Guelph Water Management Group, University of Guelph.

[6] Dinar, A., and Mody, J. 2004. Irrigation water management policies: allocation and pricing principles and implementation experience. Natural Resources Forum, 28: 112–122.

[7] Dougherty, T.C., and Hall, A.W. 1995. Environmental Impact Assessment of Irrigation and Drainage Projects. FAO Irrigation and Drainage Paper No. 53. FAO: Rome, Italy.

[8] Ferreyra, C., and Van Beek, C. 2006. Valuing and allocating water resources I: Setting the stage. In Freshwater for the Future: Policies for Sustainable Water Management in Canada – Conference Proceedings, Gatineau, Québec, May 8–10. ed. Policy Research Initiative, 19–36. Ottawa, Ontario: PRI, Government of Canada.

[9] Gleick, P.H. 1998. Water in crisis: Paths to sustainable water use. Ecological Applications, 8(3): 571–79.

[10] Global Water Partnership. 2000. Towards Water Security: A Framework for Action. Stockholm, Sweden: GWP.

[11] Grimble, R.J. 1999. Economic instruments for improving water use efficiency: Theory and practice. Agricultural Water Management, 40: 77–82.

[12] Haestad Methods. No Date. "Your Guide to Water Security" (Accessed 25 May 2007). www.watersecurity.org.
[13] Kemper, K.E. 2001. The role of institutional arrangements for more efficient water re-sources use and allocation. Water Science and Technology, 43(4): 111–17.
[14] Massarutto, A. 2003. Water pricing and irrigation water demand: economic efficiency versus environmental sustainability. European Environment 13: 100-119.
[15] Oster, J.D. and Wichelns, D. 2003. Economic and agronomic strategies to achieve sustainable irrigation. *Irrigation Science* 22, no. 3–4: 107–120.
[16] Rosegrant, M.W., Cai, X., and Cline, S.A. 2002. Global Water Outlook to 2025: Averting an Impending Crisis. International Food Policy Research Institute: Washington, D.C.
[17] Warner, J.F., Bindravab, P.S. and Van Keulen, H. 2006. Introduction: Water for food and ecosystems: How to cut which pie? Water Resources Development, 22(1): 3–13.
[18] World Water Assessment Programme. 2003. Water for People, Water for Life: The United Nations World Water Development Report. UNESCO: Barcelona, Spain.
[19] Young, M. and McColl, J. 2008. A Future-Proofed Basin: a New Water Management Regime for the Murray-Darling Basin, Adelaide, South Australia: University of Adelaide.

Irrigation to meet growing food demand with climate change, salinity and water trade

J. Connor[1], K. Schwabe[2] & D. King[1]
[1]*CSIRO Land and Water, Adelaide, Australia*
[2]*University of California, Riverside, California, USA*

Abstract

There is increasing demand for water due to rising world population and wealth. This, coupled with lower supplies of freshwater due to possible climate change, suggests further stress on an already over-allocated resource. Additionally, environmental concerns relating to low flow levels, and salinity may exacerbate the ability of irrigated agricultural regions to increase agricultural production. The objective of this research is to investigate the potential impacts of climate change induced water scarcity on irrigated agricultural productivity, water demand, and profitability. The extent to which output capacity is constrained by rising salinity levels and institutional regulating water trade are evaluated for a River Murray, Australia case study.

Keywords: irrigated agriculture, salinity, climate change, water scarcity.

1 Introduction

Recent trends in population growth, forecasting an average of nearly a billion more people every decade, suggest commensurate increases in food production will be required to meet future global demands. While the largest increases in food demand will occur in developing countries due to increases in both populations and daily calorie intake, the agricultural lands most capable of meeting these increases are likely to be located in developed countries, particularly the United States, with a sizeable increase in exports from Australia and Eastern Europe also likely (Anderson et al. [1]).

Given that the productivity of irrigated land is nearly three times greater than that of rain-fed land, significant increases in food production will most likely be met by expansion and intensification of irrigation, which currently produces over

www.witpress.com, ISSN 1743-3541 (on-line)
doi:10.2495/SI080051

40% of the world's food supply and uses approximately 60 to 80% of the world's freshwater supplies. Expansion and intensification of irrigated agriculture necessarily means large investments in irrigation infrastructure and, mostly likely, more water use. Increases in water use by irrigated agriculture for future food production will further stress a system that suffers from water scarcity presently. In addition, recent predictions from climate change models suggest further reductions in freshwater supplies in many of the already water-stressed semi-arid and arid regions worldwide.

Conceptually, allowing water to trade freely and independently of land should enhance the productivity of water in producing food and thus aid in meeting world food demand with irrigated crop production. In many parts of the world water rights are attached to land and not easily traded. Such property rights don't allow the flexibility or information signals required to guarantee allocation of water to highest value uses. The potential benefit of trade is that it allows dynamic re-allocation of water in response to changes in determinants of water value including evolving prices, technology and environmental conditions (Rosegrant [2]).

However, it has long been recognised that the economic case for free water trade is complicated by flow interdependencies. Trade not only reallocates water from low to high value uses, it can also reduce the reliability of third party water rights. This is the case in the Australian Murray Darling Basin (MDB), the focus of this case study. MDB water property rights holders can trade the entire quantity of water that they are entitled to divert. A result of this institutional rule and an increasingly active water market has been a tendency for irrigators to improve irrigation efficiency and sell water savings or use them to expand irrigated cropping. The net result is that less water flows down stream. In line with minimum flow maintenance rules, some of the lost flow is replaced with dam releases resulting in an erosion of the reliability of consumptive water rights. In addition the level of environmental flows in the MDB is reduced (Young and McColl [3]). In other part of the world (e.g. much of the western US) institutional rules to protect the reliability of third party water rights allow only the portion of a water property right that is consumptively used to be traded.

An additional flow interdependence issue arises because water trade may either improve or degrade water dependent environmental conditions. An important environmental concern in the River Murray case study considered here and many other major irrigated agricultural regions is salinity. As noted in Schwabe et al. [4], nearly one-third of the irrigated land worldwide is affected by salinization. Salinization inhibits the ability to intensify and expand irrigated lands through reducing crop yields. The fix—leaching salts out of the soil—leads to highly saline watertables and salt loading of water bodies. Both impacts can reduce crop yields. This research evaluates the impact of salinity on irrigated agriculture under alternative climate change scenarios.

The Murray-Darling Basin extends across one-seventh of the Australian continent, contains almost three-quarters of the irrigated land in Australia, and generates about 40% of national income derived from agriculture and grazing (MDBMC [5]). Extractions for irrigation, municipal and industrial use began in the late 1800's and grew dramatically since the mid-1950s. As a result median annual flow to the sea is now only 27% of the pre-development flow and the frequency of moderate and large volume floods has decreased. Adverse environmental effects of the altered flow regime include algal blooms, and salinity risks of further die off of ecologically significant floodplain forests (Overton [6]). Additionally, lack of flow and saline fresh water mixing threatens the ecological health of the Ramsar-listed Coorong and Lower lakes estuaries at the mouth of River Murray.

The particular empirical application is the lower portion of the Murray-Darling River Basin (LMDB). The LMDB is currently experiencing historically unprecedented water allocation shortages while continually confronting increasing salinity in source water. Three sub-regions within the LMDB are considered in this study to gain insight into how regionally varying salinity levels and water allocation security influence the economics and food security impacts of climate change adaptation. As shown in Figure 1.

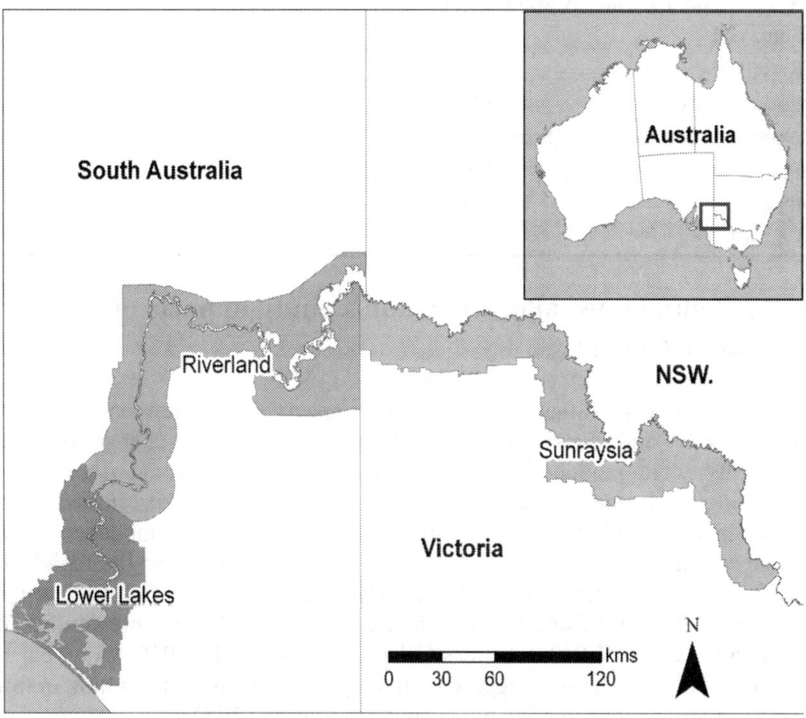

Figure 1: The Lower Murray Darling Basin river corridor, Australia.

The article reports on integrated hydro-economic optimisation modelling of:

1. Food supply and economic impacts of climate induced water scarcity with and without water markets
2. Food supply and economic impacts of salinity
3. Third party water right reliability and water quality (salinity) impacts

Table 1 summarises parameter values that were varied across climate change scenarios and regions.

Table 1: Scenario summary.

	Baseline climate	Mild Climate Change	Moderate Climate Change
Temperature	Historical mean	1 degree C	2 degrees C
Average Annual Rainfall	Historical mean	5% decrease	15% decrease
Average Annual Basin Inflow	1975 – 2000 25 year reference sequence levels	13% decrease	38% decrease
Average Water Price	\$53/ML	\$164/ML	\$250/ML
Average Annual Water Allocation (GL)			
Sunraysia	472	426	372
Riverland	342	272	210
Lower Lakes	95	76	59
Average River Salinity (μS/cm)			
Sunraysia	266	272	295
Riverland	419	440	502
Lower Lakes	1434	1618	2209

2 Modelling flow, allocation, and salinity impacts of climate change

The first step was estimation of potential impacts of climate change on river flow, salinity and water allocation. Impacts of climate change on water availability were modelled starting with a water use account developed to simulate rainfall-runoff partitioning, and basin in-flow for the Murray-Darling Basin. In-flows reductions were estimated for a mild and a moderate climate change scenario assuming average temperature increases across the region of 1 and 2 degrees Celsius, respectively. These climate scenarios are roughly consistent with the Australian Greenhouse Gas Office (AGO) predictions of a warming between 0.4 to 2.0 degrees Celsius in Australia by 2030. Water balance is maintained throughout the region with temperature increases resulting in lower rainfall and reductions in potential evapotranspiration (PET).

Basin inflow impacts consistent with these climate scenarios were predicted with the method developed by Kirby et al. [7] that involved partitioning rainfall between runoff and evapotranspiration (ET). The consequences of climate change on PET and rainfall are assumed to be felt uniformly across all rainfall events, both temporally and spatially.

The influence of climate change on water allocations to irrigation, river flow and salinity is determined by changes in in-flows given water allocation, dam storage rules, and flow management rules. For this analysis, water sharing rules consistent with the current Murray-Darling Basin Agreement are assumed and modelled with a basin river operation model known as Bigmod MSN. Bigmod MSN incorporates a daily flow and salinity mass balance model along with representations of all dam operating and water allocation rules governing the system.

The impacts of these climate changes scenarios on water allocation levels are presented in Table 1. For each climate change scenario, the variability of water availability is represented with four water allocation states of nature—low, moderately low, moderately high, and high allocation levels. Though the detail is not reported here, the level of allocation associated with each state of nature changes depending on the climate change scenario (e.g., low availability years become more frequent as the climate change scenario moves from the baseline to the severe). A key finding from the basin water-balance modelling is that annual salinity concentrations throughout the Murray-Darling Basin are estimated to increase with the degree of climate change relative to the baseline climate scenario. Similarly, salinity concentration increases are expected to be greater further downstream as flow volume reductions downstream are expected to be greater than reductions upstream.

3 Irrigation sector economic response model

Results of several irrigation sector adaptive response modelling exercises were used for this analysis. This includes a model developed to evaluate climate change impacts on the Lower Murray (Connor, et al. [8]). That analysis involved modelling adaptation to reduced and more variable water supply using a two-stage model of adjustment with recourse (Danzig, [9]; McCarl et al. [10]). The first stage models the choice of long-run water entitlement and irrigation cropping capital investments. The second stage models the short-run (annual) decisions regarding water application rates, area dewatered on a temporary annual basis, and annual water lease purchases and sales. Short-run decisions vary over states of nature characterising variation in annual water allocation and price. The short-run decisions are conditional on the fixed capital levels chosen in the first stage.

Part of the analysis reported on here involved an extension of Connor [8] in two directions. First, the crop-water production function was updated with a formulation that accounts for the effects of both water stress and salinity on yield, and is consistent with agronomic science and previous research into the economics of salinity management analysing these relationships (Kan et al. [11]; Schwabe et al. [4]). Second, the profit function formulation was extended to include a third stage which involves a weekly choice of how much water to apply based on yield response given weekly varying salinity levels. This stage incorporates weekly marginal economic decisions consisting of comparing the cost of additional water with the benefits of yield reductions from the salt-leaching as a result of applying the water.

Solutions are obtained through maximizing regional agricultural profits from irrigated agricultural production given constraints on land, water, and crop mix.

In the extended model, the following profit function was solved for three sub-regions within the LMDB:

$$\begin{aligned}\pi = & \sum_s prob_s * \left(\sum_w \sum_j \sum_h p_j^c y_{swjh} - p_s^m (w_{swjh}^c - w_s^a) - p^f w_s^a - vc_j\right) * ai_{sjh} \\ & - \sum_s prob_s * \sum_j \sum_h fyc_j * \left(ai_{sjh} - a_{jh}\right) \\ & - \left(\sum_j cec_j - \sum_j \sum_h iec_{jh}\right) * a_{jh}\end{aligned} \tag{1}$$

where the choice variables include:

a_{jh} ~ area (hectares) for crop *j* using irrigation system *h*;
ai_{sjh} ~ area (hectares) for crop *j* using irrigation system *h* irrigated in state of nature *s* (as opposed to being fallowed);
y_{swjh} ~ contribution to total yield (tonnes) from water applied in week *w* on crop *j* using irrigation system *h* for state of nature *s*;
w^c_{swjh} ~ water (ML) applied in week *w* to crop *j* using irrigation system *h* in state of nature *s*;

and the parameters in the objective function include:

p^c_j ~ price of crop *j*;
p^m_s ~ price of water on market in state of nature *s*;
p^f ~ delivery cost of initial water allocation;
w^a_s ~ initial water allocation in state of nature *s*;
vc_j ~ other variable costs to produce crop *j*;
fyc_j ~ future yield costs from under-irrigated perennial crop *j*;
cec_j ~ crop establishment costs for crop *j*;
iec_{jh} ~ irrigation establishment costs for crop *j* using irrigation system *h*;
$prob_s$ ~ probability of state of nature *s*.

4 Results

The climate change basin water balance model predicted 13% and 38% less flow into the basin for mild and moderate climate change scenarios. These inflow reductions result in even greater 36% and 47% average annual allocation reductions to the region as the result of basin water sharing rules. Consider first results of the scenario precluding water trade in which the water use is restricted to regional allocations.

Because there are a range of adaptive responses including improved irrigation efficiency and deficit irrigation, output reductions are only 9% and 13% for mild and moderate climate change, much less than the water supply allocation reductions. In the more realistic scenario consistent with Murray Darling Basin institutional rules it is assumed that water can be traded into the region independently of land. Despite high prices, significant amounts are predicted to be imported from lower value production regions. The economically optimal response still involves 3% and 5% less output, and 18% and 27% less water use than in the baseline climate scenario. Regional detail of water use, output, and profit impacts by region is provided in Table 2 for the scenario assuming water trade.

Table 2: Output, water use, and profit impacts of climate change.

Scenario	Region	Water use (gl)	Output (yield index)	Profit ($\$x10^6$)
baseline	Riverland	47.57	11.8	304.4
climate 1	Riverland	39.35	11.48	264.2
climate 2	Riverland	34.94	11.22	236.4
baseline	Sunraysia	59.12	11.87	385.7
climate 1	Sunraysia	48.43	11.56	333.1
climate 2	Sunraysia	43.3	11.35	298.8
baseline	lower lakes	14.43	11.31	59.7
climate 1	lower lakes	11.55	10.73	45.0
climate 2	lower lakes	10.38	9.87	26.7

Table 3 describes regional water use, output and profit impacts that precluding free water trade would have on the region. As can be seen the impact is estimated to be largest on Lower Lakes region where high salinity and low water allocations present the greatest constraints on regional production in absence of ability to import water

Table 3: Output, water use and profit impacts of precluding free water trade.

Scenario	Region	water use change	Output change	Profit change
climate 1	Riverland	-30.80%	-10.02%	-9.44%
climate 2	Riverland	-39.81%	-18.81%	-32.97%
climate 1	Sunraysia	-12.04%	-2.60%	-1.98%
climate 2	Sunraysia	-14.06%	-3.61%	-2.95%
climate 1	lower lakes	-34.55%	-9.97%	-12.42%
climate 2	lower lakes	-43.74%	-19.05%	-56.61%
climate 1	Region	-21.79%	-6.30%	-6.01%
climate 2	Region	-27.33%	-11.23%	-20.33%

An additional objective of this analysis was to assess how increasingly salinity may exacerbate the challenge of meeting growing food demand with reduced water supply. As summarised in Table 1, one impact of reduced flow in the Lower Murray is anticipated to be greater salt concentration in the river. To assess the impacts of this higher salinity, water use, output and profit were estimated with for climate scenarios with and without accounting for this higher salinity concentration. For the region as whole, elevated salinity is estimated to increase water use by 0.6% and 2.7% and decrease output by 0.2%, and 0.8% for the mild and moderate climate change scenarios respectively. This implies that 7% and 17% percent of the estimated total decline in regional output can be attributed to higher salinity in mild and moderate climate change scenarios. There clear regional differences, with much greater impact in the Lower Lakes region where very high concentrations as the result of reduced flow are anticipated.

Table 4: Output, water use and profit impacts of climate induced salinity.

Scenario	Region	Water use change	Output change	Profit change
climate 1	riverland	0.55%	-0.17%	-0.44%
climate 2	riverland	1.82%	-0.36%	-2.13%
climate 1	sunraysia	0.00%	0.00%	-0.14%
climate 2	sunraysia	1.72%	-0.09%	-0.74%
climate 1	lower lakes	3.96%	-1.29%	-6.66%
climate 2	lower lakes	11.36%	-6.45%	-35.63%
climate 1	Region	0.62%	-0.20%	-0.91%
climate 2	Region	2.72%	-0.83%	-4.78%

Table 5: Percolation, irrigation efficiency and salt leaching impact of climate change.

Scenario	Region	Percolation (GL)	Percolation less leaching (GL)	Irrigation efficiency & irrigation efficiency less leaching *	Irrigation efficiency less leaching
baseline	Riverland	84.3	84.3	82.3%	82.30%
climate 2	Riverland	48.5	40.1	86.1% (88.5%)	88.50%
baseline	Sunraysia	112.3	112.3	81.0%	81.00%
climate 2	Sunraysia	63.8	58.0	85.3% (86.6%)	86.60%
baseline	lower lakes	27.1	27.1	81.2%	81.20%
climate 2	lower lakes	21.0	12.7	79.8% (87.8%)	87.80%
baseline	Region	223.6	223.6	81.5%	81.5%
climate 2	Region	133.3	113.8	85.0% (87.5%)	87.5%

* Value in parenthesis are irrigation efficiency less leaching.

Climate change is anticipated to effect volumes of water percolating below irrigation in two potentially offsetting ways. The incentives created by greater water scarcity and higher water price should motivate increased efficiency and reduced percolation. The higher salinity concentrations associated with less flow should motivated increased leaching of salts and greater percolation. As shown in Table 5, the net result for the region as a whole is less percolation and greater irrigation efficiency. The effect is uneven across the region with significant increase in leaching volumes anticipated in the Lower Lakes where greatest salinity level increases are anticipated under climate change.

One significant third party impact of the improved irrigation efficiency is reduced reliability of third party water rights. Because regional rules allow trade in the full volume of water diverted, efficiency savings can be are often sold to third parties for consumptive. This reduces return flows that would have formed part of third party water rights or environmental flows. The efficiency savings for the moderate climate change scenario estimated here represents 4% of baseline water use. Given current allocation rules this amount would be shared (80%/20%) as reductions in environmental flow and consumptive use rights. A positive third party impact of reduced percolation is reduced river salt loading. Doble et al. [12] found that this effect is time delayed by decades and of lesser magnitude than the concentration effect of reduced flow.

5 Summary and conclusions

This article assessed the impact of climate change on capacity to supply food for the Lower Murray irrigation region in Australia. The region is a good case study as it is confronted with reduced water supply and rising salinity as a consequence of climate change. Results suggest that reduced water supply is likely to reduce the amount of food supplied from the region, but food supply reduction is likely to be less than proportional to water supply decline as a result of adaptation opportunities to use water more efficiently. Salinity impacts of reduced water supply exacerbate the impacts of increased water scarcity modestly, as more water is required to leach increased salinity associated with low flows in climate scenarios. The institutional arrangements in the region allowing water trade have a very significant impact on reducing adverse impacts of climate change. Externality impacts of water trade are found to have positive water quality and negative water right reliability, and environmental flow impacts.

References

[1] Anderson, K., Dimaranan, B., Hertel, T., Martin, W. Asia-Pacific food markets and trade in 2005: a global, economy-wide perspective The Australian Journal of Agricultural and Resource Economics 41 (1) , 19–44 doi:10.1111/1467-8489.00002 (1997).

[2] Rosegrant, M. Water resources in the twenty first century: challenges and implications for action. International Food Policy Research Institute. Food, Agriculture, and the Environment Discussion Paper 20 (March, 1997).

[3] Young, M. D. and J. C. McColl. Robust Reform: The Case for a New Water Entitlement System for Australia, *Australian Econ. Rev.*, 36(2), 225-234, doi:10.1111/1467-8462.00282 (2003).

[4] Schwabe, K.A., I. Kan, and K.C. Knapp. Drainwater Management for Salinity Mitigation in Irrigated Agriculture, Amer. J. Agr. Econ., 88(1), 135–149 (2006).

[5] MDBMC, Basin Salinity Management Strategy 2001–2015, Murray Darling Basin Ministerial Council, Canberra. (2001).

[6] Overton, I., Modelling Floodplain Inundation on a Regulated River: Integrating GIS, Remote Sensing and Hydrological Models, River Research and Applications (Forthcoming) (2004).

[7] Kirby, M, Evans, R, Walker, G. et al. The shared water resources of the Murray-Darling Basin. Murray-Darling Basin Commission Report, Canberra (2006).

[8] Connor, J. M., Kirby, K., Schwabe, A., Lukasiewicz and D. Kaczan. Impacts of reduced water availability on lower Murray irrigation, Australia. Socio-Economics and the Environment in Discussion (SEED). CSIRO Working Paper Series Number 2008-02. February 2008.

[9] Danzig G. B. Linear programming under uncertainty, Management Science 1:197–206 (1955).

[10] McCarl, B. A., Dillion, C. R., Keplinger, K. O. et al. Limiting pumping from the Edwards Aquifer: an economic investigation of proposal, water markets, and spring flow guarantees. Water Resources Research 34(4): 257–1268 (1999).

[11] Kan, I., K. A. Schwabe, and K. C. Knapp. Microeconomics of Irrigation with Saline Water, J. Agr. Resource Econ., 27(1), 16–39. (2002).

[12] Doble, R., J. Connor, M. Stenson, A. Elmahdi, I. Jolly, M. Miles and G. R. Walker. Future landscape scenarios in the lower River Murray Corridor and their implications for river salinity and floodplain health. Third International Salinity Forum. Adelaide, Australia. March 31–April 4, 2008.

Irrigated agriculture in an era of high energy prices

K. Schoengold[1], T. Sproul[2] & D. Zilberman[3]
[1]*Department of Agricultural Economics, University of Nebraska-Lincoln, USA*
[2]*Department of Agricultural and Resource Economics, University of California-Berkeley, USA*
[3]*Department of Agricultural and Resource Economics, University of California-Berkeley and the Giannini Foundation, USA*

Abstract

Rising energy prices will alter water allocation and distribution. Water extraction and conveyance will become more costly and demand for hydroelectric power will grow. The higher cost of energy will substantially increase the cost of groundwater, whereas increasing demand for hydroelectric power may reduce the price and increase supply of surface water. High energy prices and geopolitical considerations drive investment in land- and water-intensive biofuel technology, diverting land and water supplies to energy production at the expense of food production. Thus, rising energy prices will alter the allocation of water, increase the price of food and may have negative distributional effects. The impact of rising energy prices and the introduction of biofuels can be partly offset by the development and adoption of new technologies, including biotechnology. The models considered here can be used to determine the effects of rising energy prices on inputs, outputs, allocation decisions and impact on distribution.
Keywords: Biofuels, conveyance, groundwater, surface water, water price.

1 Introduction

At the outset of the 21st century, concern about global climate change seems likely to constrain the massive use of fossil fuels that characterized the preceding century. The 1900s witnessed significant human population growth and

www.witpress.com, ISSN 1743-3541 (on-line)
doi:10.2495/SI080061

increasing agricultural production, irrigation and energy use. It was a century of significant fossil fuel extraction that heralded the introduction of alternative forms of energy, such as nuclear power. The declining price of energy was crucial for growth and the development of technologies. Now, rapid economic growth in China and India are spurring rising energy demand. Energy prices are much higher than they were at the start of the millennium and are expected to remain above historical levels. These trends suggest the energy paradigm of the last century will need to adapt to the challenges of this century.

The increased cost of energy will impact water systems because of their reliance on energy. In many cases, water demand growth has been facilitated by energy-intensive systems. This paper discusses how increasing energy prices affect water systems, and, conversely, how water affects energy systems. Hydroelectric power, for instance, has been an important source of energy for millennia. New energy sources are emerging in the form of biofuels, which require water in production. We will consider the impact of the introduction of biofuels on water and on other agricultural outputs, including food.

The present paper proceeds with a short overview of the potential impacts of higher energy prices on agriculture. We will assess the impact of increased energy prices on the price of water, food, and other resources, and on the well-being of different water users and the end consumer. Then we will assess the impact of higher water prices on policies and institutes governing water use. Finally, we will consider how the introduction of biofuels will affect the water situation, the price of food, and the well-being of consumers and producers.

2 Overview of the potential impacts of higher energy prices on agriculture

Because more than 60 percent of water is used in agriculture, we will analyze the impact of high energy prices mostly on agricultural systems, and the benefits we consider are agricultural benefits defined as agricultural revenue. We also consider situations in which water first generates hydroelectric power and then provides agricultural benefits. The benefits of water in such situations are the sum of hydroelectric and agricultural benefits. The demand curve for water depicts the willingness of buyers to pay for an additional unit of water; the willingness to pay is equal to the marginal benefits of water, and thus represents the incremental benefits of water in agriculture and hydroelectric use. Water use entails several cost categories, and our analysis accounts for extraction and transfer costs. In some cases, it also includes the environmental costs associated with diversion of water to production from natural systems and the future costs associated with current use of water that affects water availability in the future. Supply curves depict the amount sellers require to produce an additional unit of the good. In our case, if the seller is extracting water, the supply curve is equal to the marginal extraction costs at a level associated with positive profits.

Energy and fuels are critical inputs for agriculture. Fuels are direct inputs used with farm machinery and beyond the farm gate, both upstream, by input suppliers, and downstream, by processors and marketers. Higher energy prices will raise the cost of production, thus reducing agricultural supplies and producing higher prices for agricultural outputs. We will denote this effect as the *food price effect*, though it also applies to agricultural outputs. Higher food prices may not translate to higher prices at the farm gate because the *transportation cost effect* will tend to reduce the price received by farmers. Higher transportation costs may cause farm production to locate closer to market and may reduce trade. Higher energy prices have a third effect on resource allocation in agricultural production, which we call the *other inputs effect*. It embodies the impacts of higher energy prices on the prices and use of agricultural inputs other than water. For example, with higher energy prices, fertilizers and pesticides may become more expensive, and their use may decline. Also, the amount of land used in production is subject to change.

3 The impacts of higher energy prices on water systems without conveyance

In a simple water system, water is used near the extraction site. The optimal water use is determined by the condition that the marginal benefit of water is equal to the marginal extraction cost of water (assuming there are no externalities or dynamics). This establishes the optimal price of water. Both marginal benefits and extraction cost depend on the price of energy. The marginal extraction cost is equal to the incremental energy required for extracting an additional unit of water times the price of energy. Therefore, the marginal extraction cost grows proportionately with the price of energy. The marginal benefit of water is equal to the price of output produced by water times the marginal productivity of water in production. Since we consider water systems that are small relative to the output market, farmers are price takers and unable to influence the output price. Thus, a higher energy price will affect the marginal benefit of water in the short run mostly through its impact on the price of output. Consequently, the direction of the shift in the marginal benefit after an increase in energy prices depends on the relative strength of the *food price effect* and the *transportation effect*. In some locations far from markets, transportation costs dominate, but in most locations, the food price effect will dominate.

We can analyze how an increase in energy prices will affect water use and the optimal water price for a region that starts with traditional irrigation and extraction technologies. *The likely outcome of higher energy prices is to increase water price and reduce water use*, but impacts vary among regions. In regions where farmers produce for the local market only (the region is an importer of food) and have relatively low transportation costs for their products, the increase in the price of water will be larger and the reduction in water use smaller than in regions where farmers primarily produce for a more distant export market. *The higher price of water that is likely to result from an increase in energy prices may actually lead to an increase in the economic surplus from water use*—the area between both curves. The higher price of output may more than compensate for the increase in energy costs. This is likely to occur in regions with low transportation costs and with smaller increases in marginal energy requirements as water use increases.

3.1 Future and environmental considerations

The analysis thus far has ignored some of the future costs and environmental side effects associated with water use. Water may be a non-renewable resource. In addition, water extraction may impose environmental costs as water is diverted away from its de facto use of providing ecosystem services. At the optimum, the price of water, is equal to the total marginal cost, where the total marginal cost is the sum of the externality cost, the marginal future cost of water extracted today, reflecting the discounted, reduced benefits of water use in the future, and the marginal externality cost of water use. The higher energy price will increase extraction costs and may or may not increase marginal benefits.

In the likely case that higher energy costs reduce the gap between future marginal benefits and the sum of future extraction and environmental costs, they will reduce the future cost of water extracted today. It is not clear, *a priori*, how an increase in energy prices will affect marginal externality costs, but it is plausible to assume that marginal externality costs will not be affected by changes in energy prices. Thus, the difference between marginal extraction costs and the social marginal cost, which is the sum of the extraction, future, and externality costs, declines as the energy price increases. Marginal benefits may not change with higher energy prices because the food price effect is countered by the transportation and other input effect. In this case, the higher energy cost increases the socially optimal price of water and reduces the level of optimal water use. If the marginal benefit of water increases with higher energy cost because of a dominant food price effect, the social optimum results in a higher water price and higher water extraction than when the food price effect is not dominant.

In the case of an open access groundwater aquifer, the future costs will not be considered by individual users, which will lead to over-extraction. *However, the distortion associated with ignoring the future costs will decline as energy costs increase and the open-access outcome may become closer to the social optimum.* The environmental externalities will be completely ignored unless water providers and users are provided incentives or regulated. In many cases, the sub-optimal water allocations resulting from ignorance of marginal future and marginal externality costs are trivial compared to the misallocations of water resulting from policy, such as the subsidization of water that make the price of water lower than the cost of extraction.

3.2 Market and policy constraints

An increase in water pricing may require adjustment to accommodate the poor. A preferable solution response to poverty is the development of a welfare payment system that offers the poor income support, but does not distort production decisions. When governments are unable to raise the funds (by taxation) to support the poor, alternative policies may be considered, such as the introduction of tiered pricing to allow access to a minimal amount of water at a low price, or the introduction of a system of transferable rights that will distribute water rights broadly (perhaps proportionally to historical use) and allow poorer and less-efficient farmers to gain income by selling, renting, or trading their water rights. The transition to economic water pricing will be easier if the higher energy pricing in question has significant food price effects that dominate the transportation and other input effects. The domination of food price effects will increase the income of farmers, which will make water pricing reform more feasible because of reduced demand for adjustments to address distributional considerations.

4 Water systems with extraction, conveyance, and use

Frequently, water extraction and use occur at different locations, and a key element of water systems is the transport of water from the source to its use in irrigation or other activities. The transfer requires construction of canals and aqueducts and sometimes investment in energy to lift water. On the other hand, frequently water moves downhill and, as it moves, it is a source of hydroelectric power. Thus, increasing energy prices may impose extra cost or provide extra earning that may affect the amount of water available and its productivity. A related issue is the maintenance and upkeep of the conveyance canals that affects conveyance losses and water allocation in terms of availability over different regions.

As was mentioned earlier, in assessing the optimal management of water systems, one must consider the benefits, the extraction costs, the future and externality costs, and another element—the conveyance net benefit. For simplicity, we will ignore the future and environmental costs by assuming water is a renewable resource with no externalities. We assume there are no transmission losses and that all of the water extracted will go through the system. Net conveyance benefit is a net energy benefit associated with the transfer of water. This function is positive when the transfer of water generates energy, and negative when it consumes energy.

If the transfer of water produces more hydroelectric power than it uses, higher energy prices will increase the demand for water and increase extraction of water. If, on the other hand, transfer of water requires energy, increases in energy price will reduce the extraction of water. There is a gap between the price of extracted water, and the price of water to the final users. The optimal price of water for the final users is the price of the extracted water minus the marginal net benefits from energy.

If there is a net gain from the production of hydroelectric power, the addition of the net conveyance benefit consideration reduces the optimal price of water to the farmer, but if there is a net energy loss it will increase the optimal price of water to the farmer. If higher energy prices have a positive effect on marginal benefits that surpasses the extra marginal cost of extraction and other marginal input costs, an increase in the price of energy may actually lead to an increase in water use in farming. If an increase in the price of energy results in a bigger increase in the marginal net benefits from transmission than the increase in the marginal extraction cost, the price of water to farmers will decline. These results suggest geography matters. Water systems where end users are downhill from the source, and which use water to produce hydroelectric power, may actually be expanded after energy prices increase, while water systems where end users are uphill from the source, which require energy for the conveyance of water, will decline.

5 Biofuels and alternative energy

The rise in energy prices in recent years reflects a growing demand spurred by high economic growth rates in Asia and physical and political constraints on the supply of fuels. The growth in energy demand is likely to continue, and it is anticipated that world energy demand will increase 70 percent in the next quarter-century. This coupled with the concern about global warming and the depletion of oil reserves has fueled a scramble for alternative fuels, and biofuels appear to be highly competitive among alternatives, especially in the shorter run. Thus, provision of crops for fuel production is likely to become an increasingly important agricultural activity. Furthermore, significant water resources will be diverted for the generation of biofuels and alternative fuels, including oils from coal, gas, tar sands, and from deep, partially depleted wells. So, assessing the impacts of increasing energy prices and scarcity on the water sector has to incorporate the growing demand for water in energy production, in addition to the changes in demand for foods and the changes in the cost of water extraction and transfer.

We will analyze the impact of biofuels within large systems that affect food prices (national or international water systems). Because the demand for food is inelastic [1], the derived demand for water for food production is inelastic and has a steeper slope than the demand for water for energy production, since the demand for energy has a relatively more elastic demand, reflecting that a small change in price may lead to a relatively large increase in quantity demanded.

As the price of energy increases, biofuels can compete with food production for the use of water. The higher marginal cost of water and the increase in demand, mostly due the introduction of biofuels, will increase the optimal price of water and the quantity of water demanded may increase, but the amount of water going to agriculture declines. This decline in the amount of water going to agriculture will reduce the production of food from irrigated agriculture and contributed to reduction of food availability, resulting in higher food prices. Lower availability of food and the resulting higher costs are likely to disproportionately affect the poor, especially in developing countries. Thus, the introduction of biofuels with the aim of providing cheaper fuel may harm the poor.

As a result of increased demand for ethanol, corn prices in 2007 are at historic highs and the stock-consumption ratio for corn is at the lowest level it has been for a long time. Unlike urban poor and the rural landless poor, higher prices for agricultural commodities can be beneficial to the landholding poor. The negative effects of biofuels on the poor can be mitigated by research that increases the productivity of conventional crops and bio-fuels.

The introduction of conservation technologies, and in particular, yield-enhancing technologies that increase the productivity of traditional crops, may reduce the impact of biofuels on the food sector. Specifically, given the demand for food is inelastic, introduction of technologies that will enhance yields will reduce food pricing, and most importantly, will reduce the amount of water needed for food production for a given set of water and energy prices. The downward shift in the energy (biofuel) demand for water may also shift water demand downwards, but less so than the shift in food demand for water. Thus, the net effect will be a reduction in both water use and water prices, and the lower water prices will further reduce the price of food and increase food production (relative to the case without the new technology). If the improvements in productivity are very drastic, the new technologies may even result in an outcome that increases water price (because of the increased cost of extraction) but may not increase water use relative to the initial solution before biofuels.

Since the expansion of water use for agricultural production has environmental costs, the introduction of new technologies that increase the productivity both of biofuels and of traditional food crops not only reduces the pressure on consumers, it also contributes to reduced depletion of water resources and addresses environmental problems. The analysis in this section assumes that water pricing is determined by the intersection of extraction costs and demand, and we do not consider environmental or future costs. Our analysis provides qualitative insight that fits the existing reality, but the reality is worse since water is subsidized. Continued subsidization of water as energy costs increase, along with increased demand for biofuels, may result in depletion of water resources, very high subsidy costs for government, and inefficiency. It also delays innovation for improved water-efficiency technologies. On the other hand, considering the externality and future costs would reduce the optimal level of water consumption and production of biofuels, while raising food prices for consumers.

The results suggest that policies that increase efficiency of production of biofuels and traditional crops will help to alleviate some of the pressures introduced by rising energy costs. One particular area that can make a difference is a more open-minded approach to transgenic crops. Already there are several hundred million acres grown with these crops, but pressure that is not scientifically supported to restrict the use of these technologies is crippling their growth and restricting their potential to reduce demand for water while increasing yields. The work of Graff and Zilberman [2] shows that the evolution of second generation transgenics that include new drought resistant varieties has been suffocating following the heavy registration requirements the EU introduced in 1999. The requirements serve as a practical ban on transgenic crops. Investors are not likely to invest in technologies that are excessively regulated, and the impact as energy and water become scarcer are likely to become substantial.

Investment in research and development for a new generation of biofuels can reduce the pressures on food crops and some of the inefficiencies associated with the production of biofuels from corn and traditional crops. A regulatory environment that would reduce transaction costs and provide the incentive to convert vehicles and transportation infrastructure in a way that would allow for a smooth transition to biofuels would also increase investment in the technology and in it efficient utilization.

Improved technologies both for traditional crops, pumping and biofuels would not be a complete solution to the pressures imposed on water systems by high energy costs. Incentives to reduce demand for energy that take into account the environmental costs of various energy forms and provide incentives to introduce more fuel efficient vehicles, alternative modes of transportation, and better management of energy and fuels will likely reduce the demand for biofuels crops and water for biofuels, which would help to alleviate the pressure on water systems. At the same time, increased pressure on water systems because of high energy demand may lead to a new round of investment in water projects and transfer of water from water-abundant pristine ecosystems to water-scarce regions, costing resources and causing significant environmental problems.

6 Conclusion

As we have considered here, rising energy prices will affect water use in a variety of ways. Higher energy prices will make the extraction and conveyance of water more costly and may produce increased demand for hydroelectric energy. High energy prices and political considerations are producing greater demand for alternative fuels, including biofuels, which are both land and water intensive. Food prices have increased and are expected to continue increasing because of the land and water supplies diverted to biofuels and because food production becomes more costly with higher energy prices. These trends will alter the allocation of water and may induce distributional problems for poor and dry regions of the world. The development of technologies to more cheaply extract water and more efficiently transport water can offset the effects of rising energy prices, which are expected to induce adoption of new and better technologies. In addition, technologies that improve agricultural productivity can mitigate the demand effects stemming from the new energy paradigm. Efficient water allocation mechanisms will become increasingly important. The results suggest that higher energy prices increase the cost of flawed institutions and of flawed distributional policies. Water rights that limit trading result in insufficient investment in modern irrigation technologies and overproduction of water-intensive crops. The cost of such inefficiencies is likely to increase as the price of energy increases, except for the case of water systems that are integrated with expandable hydroelectric power production. The literature on water reform suggests that the transition from systems of water rights to water trading is costly, and the gain from reform increases when water is scarcer. Increased energy prices will likely encourage such transitions.

Economic models can be used to determine the effects of rising energy prices on inputs, outputs, allocation decisions, and distributional impacts. This analysis suggests that the development and optimal utilization of a variety of technologies will be important as energy becomes increasingly costly. Public sector research will be important in driving innovation and the fruits of research should be developed, marketed and rapidly diffused without hindrance by political speculation.

References

[1] Gardner, B. *The Economics of Agricultural Policies*. Macmillan, New York, 1988.

[2] Graff, G.D., and D. Zilberman. "Explaining Europe's Resistance to Agricultural Biotechnology." ARE Update 7:5 (May/June 2004).

References

Factors influencing water allocation and entitlement prices in the Greater Goulburn area of Australia

S. Wheeler[1], H. Bjornlund[1,2], M. Shanahan[1] & A. Zuo[1]
[1]*CRMA, University of South Australia, Australia*
[2]*Economics Department, University of Lethbridge, Alberta, Canada*

Abstract

This paper examines the factors that influence the price of water allocations and entitlements by using actual monthly prices paid for water allocations and entitlements from 1993 to 2007. Water allocation prices are influenced by seasonal factors, current water allocations received, some rural commodity prices and policy impacts. The price of water entitlements are most significantly influenced by current prices of water allocations, allocations currently received, seasonal factors and government water policy. Such an analysis highlights the importance of government policy in influencing water markets.
Keywords: water markets, allocation water prices, entitlement water prices.

1 Introduction

Australia has increasingly promoted markets in water allocations and entitlements as an integral part of agricultural water management. Water markets have been functioning in Australia since 1984 and by now farmers in all states can buy and sell water on a temporary basis (allocation trading), or on a permanent basis (entitlement trading), as required. Farmers may sell their 'allocation' each year (the amount the government announces each may receive each year, based on rainfall and water storage conditions) or their 'entitlement' (the amount of water that 'by right' is linked to their ownership of property and is permanently associated with that land (unless sold).

Within the Goulburn-Murray Irrigation District (GMID) in Victoria market prices for both entitlements and allocations have increased considerably since

WIT Transactions on Ecology and the Environment, Vol 112, © 2008 WIT Press
www.witpress.com, ISSN 1743-3541 (on-line)
doi:10.2495/SI080071

1993 with a mean annual growth of allocation prices of 20.2% and 12.3% for entitlement prices. There is also clear evidence that the two prices are closely linked, following the same cyclical pattern but with allocation prices fluctuating more than twice as much as entitlement prices [1, 2].

Comprehension of the various factors underlying price determination in water markets is essential in order to understand the effects of climate, policy changes and other shifts in market factors. Prices are determined by the interaction between supply and demand of water, which are in turn influenced by a range of other policy and market factors. Generally, economic analysis of agricultural prices considers factors such as stocks, imports, production, usage and exports. However, modeling water prices is slightly different and despite agricultural water markets being in existence for some twenty years in Australia, there have only been a few attempts to estimate the factors impacting on prices paid for water in the markets for water allocations and entitlements; mainly because of the difficulty in obtaining data and relatively thin trading in these markets.

The GMID is Australia's largest irrigation district, and is the area upon which this study is based (Figure 1). The GMID is located in Northern Victoria along the River Murray which itself forms the border with New South Wales. Irrigation within the district is mainly supplied by two major sources: the Goulburn and the Murray Rivers.

Initially, trade in this region in both the markets for water allocations and entitlements were low. Trading by irrigators has increased considerably in the area, and by July 2004, more than eighty per cent of farm businesses within the GMID had traded some type of water at some time. During very dry seasons, the percent of farm businesses active in buying or selling allocations or entitlements increases considerably with more than 60% of farm businesses trading during the season of 2002/03 [3].

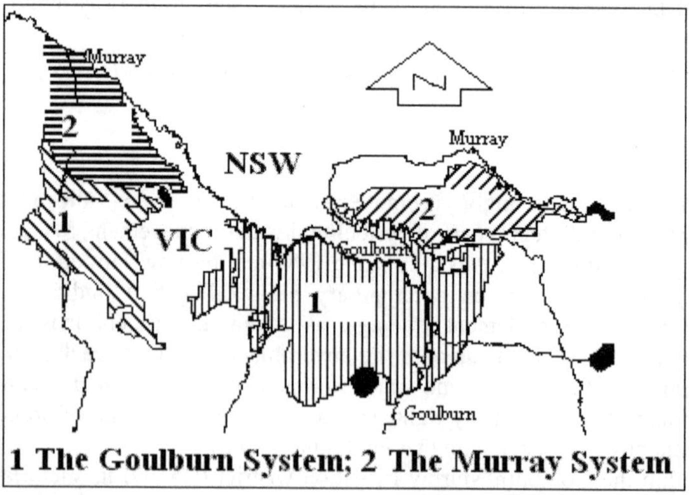

Figure 1: The Goulburn-Murray Irrigation District.

2 Influences on prices – overview and literature review

Figure 2 displays the allocation and entitlement water prices paid in the GMID from 1993 to 2007.

Although allocation and entitlement prices have increased steadily since water trading began, they have been very variable, and seem to be strongly influenced by seasonal conditions such as the drought in 2002–03 and 2006–07.

Most economic work in water markets has concentrated on modeling water demand rather than water prices (i.e. Scheierling *et al.* [4] present a meta-analysis of 24 studies on allocation water demand in the United States). Wheeler *et al.* [2] analyzed demand for water allocations while Wheeler *et al.* [5] analyzed demand for water entitlements in the GMID.

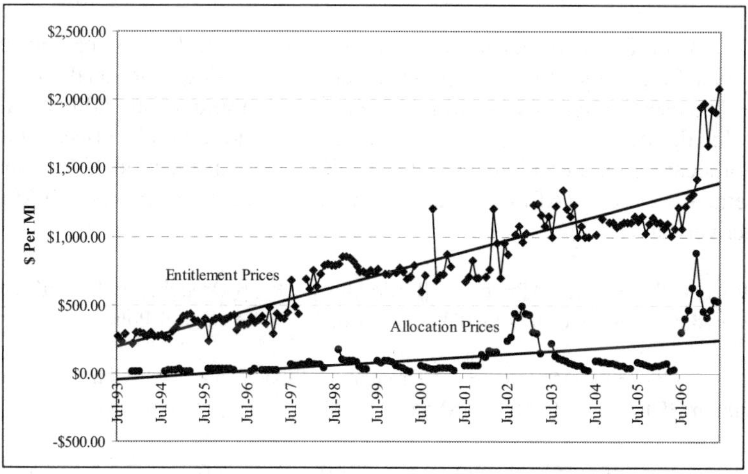

Figure 2: Monthly average prices paid for water allocations and entitlements in the Greater Goulburn from 1993 to 2007.

There are few studies of the factors influencing water market prices. Two exist for the US; one analyzes individual water transactions [6] and one analyzes mean annual prices [7]. It was found that the factors that most influence entitlement prices in the United States were urban and industrial activity in the market, the priority of the water right that is traded, the volume of water traded, and prior to urban expansions in Colorado, agricultural commodity prices. Two studies exist for Australia. These found that the major influences identified in the US studies - priority of water rights and competition from urban and industrial users, are not relevant for Australia. A recent analysis by Bjornlund and Rossini [8] analysing average monthly entitlement prices in the GMID from 1993 to 2003 found that entitlement prices were most influenced by water scarcity, price of water allocations, interest rates and some commodity prices. An earlier study analysing individual transactions of water entitlements found that the major determinants of entitlement prices are the level of restrictions on trade in the area, the water use efficiency of buyers and sellers, the value of commodities produced and the relative bargaining strength of the buyer and seller [9].

The single study on factors influencing the market price of water allocations of which we are aware is by Bjornlund and Rossini [10]. They studied water allocation prices in the GMID from 1993 to 2003 and found prices were most influenced by water scarcity, rainfall and evaporation. The present study builds on the earlier analyses by Bjornlund and Rossini [8, 10] and provides a more structurally rigorous model and longer dataset over a period of severe drought in the GMID.

3 Methodology and results

3.1 Influences on average monthly allocation and entitlement prices

Price demand functions have traditionally used linear, log-log or log-linear functional forms, with the most popular being the log-log functional form (i.e. [11]). This section calculates influences on prices of allocation and entitlement water for the period of 1993 to 2007 utilizing average monthly prices paid for water allocations and entitlements. The specifications for the models for average entitlement prices (AEPrice) and average allocation prices (AAPrice) respectively are:

$$\ln AEPrice_t = \beta_0 + \beta_1 \ln AAPrice_t + \beta_2 \ln MAlloc_t + \beta_3 Drought_t + \beta_4 NDKyab_t + \beta_5 CommodityPrices_t + \beta_6 \ln FarmGDP_t + \beta_7 Month_t + \beta_8 Year_t + \beta_9 Policy_t \quad (1)$$

$$\ln AAPrice_t = \beta_0 + \beta_1 \ln MAlloc_t + \beta_2 Drought_t + \beta_3 NDKyab_t + \beta_4 CommodityPrices_t + \beta_5 \ln FarmGDP_t + \beta_6 Month_t + \beta_7 Year_t + \beta_8 Policy_t \quad (2)$$

where t is the time period, and the monthly average prices paid/received for water entitlements and allocations by farmers in the GMID are the dependent variables. Water allocation prices are used as an independent variable in the entitlement price model, as it is expected that they should have a positive influence on entitlement prices (as when water allocation prices go up, then it becomes more rational to invest in water entitlements). Bjornlund and Rossini [1, 8] found such an effect.

The next independent variable $MAlloc_t$ is the current allocation level of irrigation water for that month, and it is expected to have a negative influence on both types of prices, as when farmers receive higher allocations it decreases their need/incentive to buy more water, hence decreasing the willingness to pay for additional water. It is expected that the allocation level should have a greater influence on allocation prices than entitlement prices. Drought is a dummy for the two severe drought years of 2002–03 and 2006–07, and it is hypothesised that prices will be positively influenced during this time. $NDKyab_t$ (calculated by subtracting monthly rainfall from monthly evaporation rates obtained form the Bureau of Meteorology for the Kyabram station) is the net monthly water deficit in the region. It is expected that an increase in the net water deficit (that is, more water is evaporating from the soil than is being replaced by rainfall), will lead to an increased demand (and hence price) for water, as suggested by some results in Brennan [12] and Bjornlund and Rossini [9]. Again, theory would suggest that the relationship between $NDKyab_t$ and allocation prices should be stronger than the relationship with entitlement prices.

The fourth independent variable in the allocation price model is commodity prices, namely using the price of feed barley ($PriceFB_t$) and the price of red wine: the fifth independent variable in the entitlement price model. (Note too that a wide variety of agricultural prices were collected and tested but no others displayed any significance in the models suggesting that during this period of high scarcity levels the main driver of price is scarcity rather than commodity prices. Farmers are buying to minimize their losses rather than maximizing their profits. They buy water to protect their investment in diary herd and equipment or permanent plantings).

Prices for barley and wine are the unit export commodity prices received (provided by the Australian Bureau of Agricultural Resource Economics). Previous studies have found some relationship between water prices and commodity prices [8, 10]. One of the largest buyers of water allocations over this time period has been dairy farmers, who can choose to water their grass or buy feed barley to feed their cows. Hence, feed barley can be a substitute for buying water, suggesting a positive relationship between water allocations and feed barley, though there is no expected relationship between water entitlements and feed barley. Another significant influence on prices may have been the price of red wine (the price of red wine has been significantly higher than the price of white wine over the time period in question). Permanent plantings of viticulture over the course of the past decade have significantly increased the demand for water downstream. It is expected that this would have more of an effect on water entitlements than water allocations, because farmers would need an ongoing water source for their vines. Commodity prices, apart from wine grapes, are not expected to have an influence on entitlement water prices.

$FarmGDP_t$ is the average monthly farm GDP as estimated by the Reserve Bank of Australia (RBA). It is hypothesised that farm GDP (which is a proxy for income of farmers) will be positively related to water prices.

Month is a seasonal dummy for the three months of highest trading in the year (January, February and March – high summer). It is expected that there will be a positive influence on prices in this period, though the relationship is theorized to be stronger with allocation prices than entitlement prices. Year is a continuous variable for the trading year in the database to detect if there is a time trend or some effect associated with a developing regional water market over time. The last independent variable is $Policy_t$, which is a dummy variable for 1998, reflecting two main water policy changes. First, this was the year that the Northern Victoria Water Exchange was introduced, providing fast, cheap and secure trading in water. Second, from 1998 onwards the water authority changed its allocation policy, and only incorporated minimum expectations to inflows during the seasons (resulting in lower opening allocations) [3]. Such policy changes are expected to have a positive influence on water demand and prices.

Table 1: Monthly allocation and entitlement actual prices in the GMID - 1993 to 2007.

	Allocation Prices			*Entitlement Prices*			
	Coef.	*t*	*P>|t|*		*Coef.*	*t*	*P>|t|*
lnalloc	-0.24	-2.02	0.05	lnaaprice	0.09	2.39	0.02
drought	0.43	1.20	0.23	lnalloc	0.02	0.32	0.75
ndkyab	0.00	1.02	0.31	drought	-0.02	-0.20	0.85
lnpricefb	0.64	1.60	0.11	ndkyab	0.00	-0.74	0.46
lnpricerw	0.31	0.65	0.52	lnpricerw	0.29	1.62	0.11
lnfarmgdp	-0.57	-1.01	0.31	lnfarmgdp	0.01	0.04	0.97
month	0.04	0.48	0.63	month	0.06	1.93	0.06
yeartrend	0.16	4.11	0.00	yeartrend	0.09	7.91	0.00
policy	1.13	1.86	0.07	policy	0.29	2.25	0.03
cons	4.46	0.84	0.41	cons	4.86	2.79	0.01
n	128			n	115		
Adj R^2	0.74			Adj R^2	0.99		
F Test	40.21			F Test	868.56		
Prob > F	0.00			Prob > F	0.00		
DurbinWatson Stat	1.60			DurbinWatson Stat	1.97		

Water volume was not used in either price model specification because of the problem of endogeneity. Lagged allocation and entitlement prices could also not be used in either model because of collinearity problems. Real prices (base year of 1996–97) were used in all models.

Table A1 in Appendix 1 provides a summary of the descriptive statistics (in terms of means, minimums, maximums and standard deviations) for the average price monthly model from July 1993 to June 2007.

Table 1 presents the results of the two models (the allocation price model has 128 monthly observations and the entitlement price model has 115). These final specifications showed no problems with heteroscedasticity or multicollinearity, however they did have a problem with serial correlation. We addressed this problem by using Prais-Winsten AR(1) regressions (iterated estimates) for both models.

The analysis shows that the price of water allocations are significantly and positively influenced by a year trend and government policy introduced in 1998 (the price of feed barley was very weakly significant at 11 per cent). The price of water allocations in the current month was negatively and significantly related to the current allocation level. In further testing of the model, we tested for significance of lags of key variables. When lagged by one month, allocation levels are still significant and net deficit of water also becomes significant with both having the correct sign as hypothesized. This suggests that farmers may be reacting to weather in the previous month in the market for water allocations, although the current weather does not seem to influence significantly allocation prices.

The price of water entitlements is significantly and positively influenced by allocation prices, the month, a year trend, and policy introduced in 1998 (the price of red wine was also very weakly significant at 11 per cent.

Overall, more variation in entitlement prices was explained than the variation in allocation prices. The most significant influence on entitlement prices was the year trend, highlighting the fact that over time investing in water entitlements has proved much more profitable for irrigators. The other most significant influence on entitlement prices was allocation prices, which are correlated to a degree because they have both increased in value over the time period in question (but not to the extent to cause problems with collinearity in the model). As allocations become more expensive, it is only rational that irrigators are willing to pay higher prices for water entitlements in order to reduce the need to buy allocations and to ensure a more secure supply during periods of scarcity.

The price of water allocations over the time period from 1993 to 2007 has been much more influenced by short-term factors (such as the drought, the current allocation of water received and the net deficit of water) than the prices of water entitlements – with the exception of the seasonal month influence – the dummy for January, February and March. Permanent plantings of grapes, and hence the prices received, seemed to have had an influence on the prices of entitlement water in the GMID, but not so much water allocations. The key change in water policy in 1998 had a significant and large positive influence on both allocation and entitlement prices, indicating that government policy can make a significant difference to water prices through administrative changes. Time itself has been the most significant factor in influencing entitlement prices. Over time the demand and prices paid for water entitlements has increased significantly as irrigators recognise the need to have long-term secure rights to a resource that is proving to be more and more finite.

There are a number of key differences between the results of this paper and Bjornlund and Rossini [8, 10]. Such differences arise because of the different functional and theoretical form of the model/s, differences in the data of key variables (such as commodity prices), and a longer time-series of data used.

4 Conclusion and policy recommendations

This paper has provided an estimate of the factors influencing the prices of water entitlements and water allocations within one region along the Murray River in Australia over the period 1993 to 2007. It was found that there were key differences between the models for water allocations and water entitlements. Water allocations are much more influenced by short-term water factors (such as drought and allocation received), while the price of water entitlements is much more likely to be influenced by variables such as a time trend and the price of water allocations. Water policy changes in 1998 significantly influenced the prices paid in both markets for water allocations and water entitlements, indicating that governments can indeed play a role in regulating the market with administrative changes.

Acknowledgements

This research is part of a larger project funded by the Australian Research Council and six industry partners: Murray-Darling Basin Commission, Department of Natural Resources; Department of Sustainability and Environment, Goulburn-Murray Water, Department of Water, Land and Biodiversity Conservation and UpMarket Software Services.

Appendix 1

Table A1: Descriptive statistics of variables used.

	Mean	*Standard Deviation*	*Range*	*Minimum*	*Maximum*
Water Entitlement Model					
lnaeprice	6.57	0.44	2.00	5.49	7.48
lnaaprice	3.99	1.05	4.51	2.06	6.56
lnalloc	4.52	0.61	3.35	1.95	5.30
drought	0.25	0.44	1.00	0.00	1.00
ndkyab	118.76	86.28	339.60	-57.40	282.20
lnfarmgdp	8.52	0.20	0.85	7.99	8.84
month	0.35	0.48	1.00	0.00	1.00
lnpricerw	1.55	0.19	0.80	1.13	1.93
yeartrend	8.11	3.94	13.00	1.00	14.00
policy	0.08	0.27	1.00	0.00	1.00
Water Allocation Model					
lnaaprice	4.00	1.03	4.16	2.06	6.21
lnalloc	4.48	0.65	3.69	1.61	5.30
drought	0.24	0.43	1.00	0.00	1.00
ndkyab	113.02	86.54	346.10	-57.40	288.70
lnpricefb	5.18	0.21	0.95	4.70	5.64
lnpricerw	1.54	0.18	0.79	1.13	1.93
lnfarmgdp	8.52	0.19	0.85	7.99	8.84
month	0.32	0.47	1.00	0.00	1.00
yeartrend	8.21	3.80	13.00	1.00	14.00
policy	0.07	0.26	1.00	0.00	1.00

References

[1] Bjornlund, H. & Rossini, P. Are the fundamentals emerging for more sophisticated water market instruments? *Proceedings from the 14th Annual Conference of the Pacific Rim Real Estate Society*, Kuala Lumpur, Malaysia, January. Available at www.prres.net, 2008.

[2] Wheeler, S., Bjornlund H., Shanahan M. & Zuo, A., Price Elasticity of Allocation Water Demand in the Goulburn-Murray Irrigation District, *Australian Journal of Agricultural and Resource Economics*, 52, pp. 37-55, 2008.

[3] Bjornlund, H. Increased participation in Australian Water Markets. In G. Lorenzini, and C. Brebbia Eds. *Sustainable Irrigation Management, Technologies and Policies*, Southampton: WIT Press, pp. 289-302, 2006.

[4] Scheierling S., Loomis, J. & Young R., Irrigation water demand: A meta-analysis of price elasticities, *Water Resources Research,* 42, W01411, 2006.

[5] Wheeler, S., Bjornlund H., Shanahan M. & Zuo, A., Price Elasticity and the Demand for Water Entitlements in the Goulburn-Murray Irrigation District of Australia, paper accepted for presentation at Water Down Under conference, 2008, Adelaide.

[6] Colby, B., Crandall, K. & Bush D., Water right transactions—market values and price dispersion, *Water Resources Research,* 29, pp.1565-1572, 1993.

[7] Person, P. & Michelsen A., Determinants and Trends in Water Right Prices: An Econometric Analysis, Wyoming Water Resources Center, Laramie, Wyoming, 1994.

[8] Bjornlund, H. & Rossini, P., Fundamentals determining prices in the market for water entitlements – an Australian Case Study, *International Journal of Water Resources Development,* 23, pp. 537-553, 2007.

[9] Bjornlund, H. (2002): Signs of Maturity in Australian Water Markets *New Zealand Property Journal*, July, 31–46

[10] Bjornlund, H. & Rossini, P. Fundamentals determining prices and activities in the market for water allocations, *International Journal of Water Resource Development,* 21, pp. 355-69, 2005.

[11] Goodwin, B., Schnepf R. & Dohlman, E., Modelling Soybean Prices in a Changing Policy Environment, *Applied Economics*, 37, pp. 253-263, 2005.

[12] Brennan, D. Water policy reform in Australia: lessons from the Victorian seasonal water market, *The Australian Journal of Agricultural and Resource Economics*, 50, pp. 403-423, 2006.

Sustainability of groundwater resources in India: challenges and scope for economic instruments and policy

S. Msangi
International Food Policy Research Institute, Washington DC, USA

Abstract

In this paper we address the groundwater resource management issues in selected regions of India, within the context of the policy and natural environments that exist. We illustrate the underlying nature of the resource management problem, and suggest possible alternative interventions, with the use of an economic model that is linked to a simplified representation of the characteristic hydrology. By illustrating the policy problem in this way, the paper explores the scope that alternative economic instruments could have in correcting the perverse incentives that exist for groundwater conservation, in some regions, while improving human welfare. In doing so, we hope to better clarify the role of market (and non-market based) instruments in addressing common pool resource management problems in India, and provide guidance to researchers and policy makers on how they can best study these cases, and further refine their policy recommendations

Keywords: groundwater, common pool resources, policy and institutions, market-based instruments, natural resource management.

1 Introduction

The importance of irrigation in maintaining the necessary productivity within the Indian agricultural sector has long been recognized by researchers and policy makers, and is a major driver behind the growth in output that was observed during the period of the 'Green Revolution'. While the agro-ecological conditions vary widely across the Indian sub-continent, there is a sizable share of agricultural production that relies on irrigation, as is shown in Table 1, below.

www.witpress.com, ISSN 1743-3541 (on-line)
doi:10.2495/SI080081

Table 1: India Irrigation, 2003–2004.

State	Total Cultivable Area ('000 ha)	Net Irrigated Area ('000 ha)	Irrigated as % of Total Cultivable
East	13,077	2,416	18.5%
West	76,758	16,862	22.0%
North Central	36,529	18,977	52.0%
North East	6,129	439	7.2%
North West	11,477	7,818	68.1%
South	39,498	8,592	21.8%
All India Total	183,468	55,104	30.0%

Source: CWC, Delhi, India – Water Related Statistics.

The deepening scarcity problems that have been observed in many parts of India, as a result of increasing demands on limited water resources for both agricultural and non-agricultural uses, has also been a source of concern for both policy makers and analysts who seek to promote the sustainability of water resources through improved management on both the demand and the supply side.

Much of the literature that deals with the management of water resources in Indian agriculture has focused, in particular, on groundwater management, which represents a significant share of the water resources that are available to the majority of Indian farmers. Table 2, below, shows the agricultural area that is supported by tubewells and other wells, as opposed to other kinds of water withdrawals from surface sources.

Table 2: Indian Irrigation by Source ('000 ha).

State	Tube Wells	Other Wells	Other Sources	Total All Sources
East	233	98	265	2,416
West	3,865	8,220	893	16,867
North Central	12,898	569	622	18,977
North East	2	2	193	437
North West	4,865	20	183	7,818
South	2,304	2,186	597	8,592
All India Total	24,167	11,095	2,753	55,107

Source: CWC, Delhi, India – Water Related Statistics.

The deepening scarcity problems that have been observed in many parts of India, as a result of increasing demands on limited water resources for both agricultural and non-agricultural uses, has also been a source of concern for both policy makers and analysts who seek to promote the sustainability of water resources through improved management on both the demand and the supply side. A number of authors have focused upon the electricity subsidies to agriculture, which might be giving perverse incentives to farmers using electric pumps to withdraw more groundwater than would otherwise be socially optimal [3,6]. Others have suggested alternative policy instruments that might help groundwater users to internalize the external effects that they impose upon other users of the common-pool groundwater resource, through such instruments as taxes on pumping, based either on the volumetric quantity of water used or electricity that is consumed [4].

These types of instruments, however, are difficult to impose, due to problems of observability and measurability of actual volumes of water that are used by individual groundwater users. The recommendations that might apply to regions which have groundwater pumpers with very large landholdings, like Kern County, California [1], may not be as applicable to developing country regions where there are numerous small-holder farmers, like in East and South Asia. A number of other studies have tried to examine the role of markets for water, and how they can potentially improve the allocation of water among users, by using market-based incentives and trading mechanisms, as well as the possible welfare losses that can result from the exercise of monopoly power by wealthier and larger landowners on their poorer and smaller neighbours [2,5].

In this paper, however, we will focus on the choice of instruments that might be employed by policy makers to encourage more efficient use of water resources, and promote the sustainability of limited groundwater resources, within the context of India. We will consider the efficacy of tax- or tarrif-based approaches to control, as well as that of quantity-focused instruments – and discuss the situations in which the efficacy of one might outweigh that of the alternative methods. We will discuss these instruments within the context of a theoretical model of economic behaviour, which addresses both groundwater use as well as inter-agent re-allocation and trade of permits. Our conclusions and recommendations will then close the paper.

2 Economic behaviour in water usage

The economics of water usage is typically based upon the behavioural economics of a profit-maximizing agent, who seeks to maximize the net revenue that accrues from irrigated agricultural production, and faces a trade-off in terms of costs of inputs (including water), or constraints in water or land use. Even without considering the dynamics of groundwater usage, and how the stock of water held in the underground aquifer evolves over time, the individual profit-seeking agent can be hypothesized to behave according to the following, simple maximization problem.

$$\max_{x,A} \quad pAf(x) - cx - \alpha A^2 \qquad s.t. \quad A \leq \overline{A} \tag{1}$$

where the decision-maker's problem is defined in terms of choosing the optimal level of input (i.e. water) which enters into the agricultural yield function $f(x)$, while also choosing the optimal land area over which to farm, A, which is available up to a particular limit $\bar{A}$. The price of the agricultural output is denoted by p, whereas the cost of the productive input is given as c. The quadratic term in land area, α, captures the decreasing returns to adding land area which is due to limited management and labour, as well as variable land quality over the available area. The maximization problem written in eqn (1) can be stated in terms of the full Lagrangian function, shown in eqn (2) below, which has the shadow value of the constraint included as a choice variable

$$\max_{x,A,\lambda} \quad L(x,A,\lambda) = pAf(x) - cx - \alpha A^2 - \lambda\left[A - \bar{A}\right] \tag{2}$$

The necessary conditions for profit-maximizing choice over the variables x and A, lead to the following set of equations

$$\begin{array}{lll} pAf'(x) - c \le 0 & x \ge 0 & x\left[pAf'(x) - c\right] = 0 \\ pf(x) - 2\alpha A - \lambda \le 0 & A \ge 0 & A\left[pf(x) - 2\alpha A - \lambda\right] = 0 \\ A - \bar{A} \le 0 & \lambda \ge 0 & \lambda\left[A - \bar{A}\right] = 0 \end{array} \tag{3}$$

which we can use to define the optimal choice of water and land usage by the representative farmer. If we were to assume that the decision-maker always chooses to use as much land as is available, then we can reduce the set of conditions in eqn (3) to the following pair of equations

$$\begin{array}{l} p\bar{A}f'(x) - c = 0 \\ pf(x) - 2\alpha\bar{A} - \lambda = 0 \end{array} \tag{4}$$

which we can use to examine the sensitivity of the optimal choices to the key economic parameter values. By totally differentiating the pair of first-order conditions in eqn (3), with respect to each variable and parameter, we obtain the following linear system

$$\begin{bmatrix} p\bar{A}f''(x) & 0 \\ pf'(x) & -1 \end{bmatrix}\begin{bmatrix} dx \\ d\lambda \end{bmatrix} = \begin{bmatrix} -\bar{A}f'(x) & -pf'(x) & +1 \\ -f(x) & +2\alpha & 0 \end{bmatrix}\begin{bmatrix} dp \\ d\bar{A} \\ dc \end{bmatrix} \tag{5}$$

which allows us to relate changes in the vector of decision variables, on the left hand side, to the vector of the key parameters, on the right hand side. The sign of the principle determinant, D, shown below

$$D = \begin{vmatrix} p\bar{A}f''(x) & 0 \\ pf'(x) & -1 \end{vmatrix} > 0 \tag{6}$$

conforms to our expectations of a well-behaved maximization problem, in which the set of production possibilities can be circumscribed by a convex hull. Applying Cramer's rule to the linear system in eqn (5), allows us to perform the following comparative static calculations, in which we examine the impact of parameter changes on the key decision variable of interest – that of water use

(x). In order to look at the impact of changes in the volumetric cost of water (c) on water usage, we can compute the following differential

$$\frac{\partial x}{\partial c} = \frac{\begin{vmatrix} +1 & 0 \\ 0 & -1 \end{vmatrix}}{D} = \frac{-1}{D} < 0 \qquad (7)$$

in which the marginal effect of increasing the cost, serves to decrease the level of water usage, as we would expect. Similarly, we can examine the impact of changing the constraint on land, such that we allow for a marginal change in the binding quantity $\overline{A}$.

$$\frac{\partial x}{\partial \overline{A}} = \frac{\begin{vmatrix} -pf'(x) & 0 \\ +2\alpha & -1 \end{vmatrix}}{D} = \frac{pf'(x)}{D} > 0 \qquad (8)$$

while these effects are opposite in sign, we can see that their relative magnitude depends on the magnitude of the value marginal product of water $pf'(x)$ relative to unity. If it turns out that the value marginal product exceeds unity, then a change in allowable land area might be more effective in reducing water usage, than a change in the unit variable cost of water. The converse would be implied by the case where $pf'(x) < 1$. While the comparison of marginal impacts on water use, from the point of view of comparative statics is useful in judging the efficacy of alternative instruments for managing water demand – it does not describe the relative efficacy of the institutions that might be used to apply those instruments. We will now take up this discussion, in the following section.

3 Institutional efficacy of economic instruments

Now that we have discussed the technical aspects of groundwater user behaviour, with respect to key parameters of economic behaviour, we can now turn to the institutional aspects of implementing groundwater management policy. The unit cost of water usage, which we described by the single parameter c, in eqns (1) and (2), can be conceptualized as a function of several variables, within the context of groundwater usage. Typically, the marginal cost of groundwater pumping is thought to vary according to the hydrological conditions under which water is withdrawn from the aquifer. In particular, the distance over which water must be lifted from the groundwater table to the surface (i.e. the 'lift') is a key determinant to the marginal cost of pumping a single volumetric unit of water, as well as the energy costs that are associated with the action of the pump. Therefore, we can describe the marginal cost of water as a function of the 'state' of the system, which we can describe by the state variable h, which denotes the height of the groundwater table with respect to a reference level, and the distance it lies below the ground surface $\overline{S}$, shown in Figure 1 below.

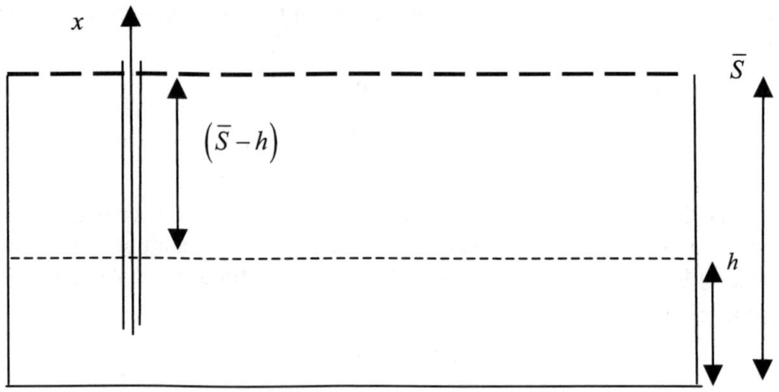

Figure 1: Simplified representation of pumping from an aquifer.

Combining the 'lift' ($\bar{S}-h$) with the energy cost of pumping (e), we can express the marginal cost of water usage as

$$c(h,e) = \gamma e \cdot \left(\bar{S} - h\right) \tag{9}$$

where γ is a conversion factor.

3.1 Alternative economic instruments for demand management

If we were to introduce a volumetric charge for pumping, then this would act as an additive factor that raises the marginal cost of pumping above that which is determined by energy and the hydrological state of the system. Such a charge or tax, t, would change the marginal cost in eqn (9) to equal

$$\tilde{c}(h,e,t) = \gamma e \cdot \left(\bar{S} - h\right) + t \tag{10}$$

If we were to use the upper limit on available land as a policy instrument that limits the area of a heavily water-consumptive crop – we could then specify a limit that either coincides with or falls below the 'natural' limit that the farmer would otherwise face in the absence of policy intervention. If we denote this limit as a quota ($\bar{Q}$), which would be allocated to each irrigator, then we would have

$$\max_{x,A} \quad pAf(x) - c(h,e,t)x - \alpha A^2 \qquad s.t. \quad A \le \bar{Q} \le \bar{A} \tag{11}$$

which represents the modified maximization problem of the irrigator who faces two possible policy instruments. Applying the same assumptions as those used to derive eqn (4), we would have a modified set of first-order necessary conditions

$$\begin{aligned} &p\bar{Q}f'(x) - \gamma e \cdot \left(\bar{S} - h\right) + t = 0 \\ &pf(x) - 2\alpha\bar{Q} - \lambda = 0 \end{aligned} \tag{12}$$

which yields a similar linear equation system, when we take the total differentials with respect to all the decision variables and parameters of the problem. The magnitude of the differential, with respect to the tax, is identical to that shown in eqn (7), and is only a function of the parameters in the determinant. We can derive an additional expression that illustrates the impact of the quota on the implicit shadow value λ , which can be written as

$$\frac{\partial \lambda}{\partial \bar{Q}} = \frac{2\alpha p\bar{Q}f''(x) + \left(\tilde{c} / \bar{Q} \right)^2}{D} < 0 \qquad (13)$$

and is negative in sign, if the marginal pumping cost is sufficiently small compared to the allocated quota. The sign of this expression conforms to what we would expect from a downward sloping demand curve for allocated quota, and denotes that the willingness to pay for additional quota goes up as the allocation gets smaller.

In the hands of a well-informed agent, who is forward-looking with respect to the state of the groundwater table, and the implications that a lower water table has for the future pumping costs and producer welfare – we could construct an optimal path of extraction that would ensure the long-term efficiency of water usage, and maximize the sustainability of groundwater usage. Such a path would be obtained by solving the following social planner's problem.

$$V^{CP}(h) = \max_{x,A} \begin{Bmatrix} pAf(x) - \gamma e \cdot (\bar{S} - h)x - \alpha A^2 + \beta V^{CP}(h + \varphi x - \tilde{r}) \\ s.t. \quad A \le \bar{A} \end{Bmatrix} \qquad (11)$$

where the inter-temporal optimization is carried out with respect to the pumping of all players in each period, and where β is the discount rate that captures the social planners inter-temporal preferences. The function $V(h)$ is the maximized value of the dynamic game problem, for each player, beginning with the current level of groundwater lift (h), and proceeding under the assumption that actions taken in subsequent periods are done optimally with respect to the groundwater lift in each period. This recursive relationship linking the implied optimality of behavior from period-to-period captures the essence of Bellman's "Principle of Optimality" (Bellman, 1957).

The solution to the social planner's problem would give a 'benchmark' outcome for water and land use that would maximize long-run benefits over time, and enhance groundwater resource sustainability over time. This benchmark would guide the policy maker as to the best choice of tax ($\hat{t}$) or quota ($\hat{Q}$) to impose as an economic instrument on the less-informed, myopic irrigator who would not otherwise consider the long-run benefit. Where numerous irrigators are involved, the social planners problem can be generalized to yield a vector of decisions over all agents that prescribe their optimal pumping $\{\hat{x}_i\}_{i=1}^N$ and land use $\{\hat{A}_i\}_{i=1}^N$ behaviour, over time. The well-informed policy-maker would choose the optimal quota allocations as $\hat{Q}_i = \hat{A}_i$ and the optimal pumping tax ($\hat{t}$) equal to $\beta V'(h+\varphi x-\tilde{r})$, which represents the extra marginal cost that the social planner imputes to each unit of water withdrawn from the aquifer, and is proportional to the derivative of the optimal value function.

The information that would be needed by the policy maker to set these policy instruments to their optimal levels is daunting, when dealing with many individual agents, and is beyond the capacity of typical institutions of groundwater management. The fact that the individual irrigators know their own technologies and productive possibilities much better than the central administrator is a persuasive argument behind using market-based instruments – such that the individuals can interact within a decentralized framework, on the basis of their privately-held information.

3.2 Market-based instruments for demand allocation

Allowing the quotas for land area to be tradeable, allows each player to engage in decentralized transactions that allow them to trade their initial allocations up to the point that their private benefits are maximized, and their individual shadow values for quota allocations (λ_i), are equalized across agents. Taking the case of just two agents into consideration, for simplicity, we can depict such an equilibrium outcome in terms of Figure 2, below. This figure shows the equilibrium outcome where transaction costs exist (τ), which cause the agent transactions ($\hat{z}$) to deviate from those levels that would be realized in the absence of transaction costs (z^*). In the presence of transactions costs, the volume of trade is less than that which is otherwise, achievable, and the aggregate benefits of both agents fall below that which the central planner could achieve in the benchmark case.

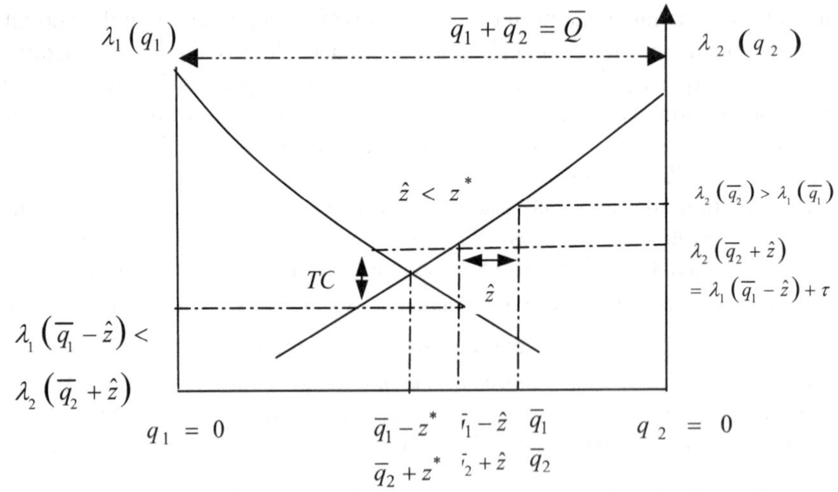

Figure 2: Decentralized allocation of tradable quota between agents.

4 Tradeoffs between alternative policy instruments

This represents a common trade-off that is faced by policy makers, when implementing market-based instruments to improve the efficiency of decentralized schemes for resource allocation. The excessive information requirements of implementing an optimal tax on volumetric extraction – both in terms of monitoring the volumes of water extracted by each individual and deciding on the welfare-maximizing level across all agents – has to be balanced against the likely transaction costs that would be incurred by the individual agents who try and enact trades within a decentralized framework.

It remains an empirical matter to determine how large the loss of welfare due to transaction costs are, within a decentralized allocation scheme, relative to the administrative 'errors' that would be incurred by implementing a centralized allocation of quota with imperfect information. The advantage of allocating land areas, from an administrative point of view, lies in the fact that it is much easier to observe land area and cropping patterns, than it is to observe individual volumetric quantities of water withdrawal. So by replacing the taxing of groundwater pumping with the restriction of land area, based on the type of crop that is grown – such that more water-consuming crops can be limited to a maximum area – we can overcome some of the information problems that would otherwise face the regulator in implementing a demand management scheme over many irrigators.

5 Conclusions

In this paper, we have discussed the relative effectiveness of price-based instruments that increase the volumetric cost of water, compared to a quantity-based limit on available land, within the context of a profit-maximizing irrigator. We have argued for the relative efficacy of these instruments, on the basis of the transaction and administrative costs that might be incurred in their implementation, either within a centralized or decentralized scheme of water demand management. We conclude that quantity-based instruments are more easily observable, and might impact total consumptive use of water within a groundwater basis to a greater degree than price-based instruments that only act to reduce water withdrawals from the common pool resource. Despite the welfare losses that are inevitable, when implementing market-based schemes for decentralized demand management, these might still serve to overcome problems of asymmetric information that will inevitably occur when a less-informed central administrator faces a large number of individual irrigators. The degree to which these transaction costs might reduce the efficacy below that of a 'naïve' implementation scheme by a central regulator, remains a question that can only be answered by further empirical work. The importance of irrigation to Indian agriculture, warrants such research, as it will undoubtedly enhance the knowledge base that policy makers need to have to better promote the sustainability of India's limited and increasingly stressed water resources.

References

[1] Feinerman, E. & Knapp, K.C., Benefits from groundwater management: Magnitude, sensitivity and distribution. *American Journal of Agricultural Economics*, **65**, pp. 703–710, 1983.

[2] Kajisa, K. & Sakurai, T., Efficiency and Equity in groundwater markets: The case of Madhya Pradesh, India. *Environment and Development Economics*, **10(6)**, pp. 801–819, 2005.

[3] Kumar, M.D., Impact of electricity prices and volumetric water allocation on energy and groundwater demand management: Analysis from western India. *Energy Policy*, **33(1)**, pp. 39–51, 2005.

[4] Kumar, M.D. & Singh, O.P., Market instruments for demand management in the face of scarcity and overuse of water and Gujarat, western India. *Water Policy*, **3**, pp. 387–403, 2001.

[5] Prakash, A., *The dark zone*. Orient Longman: New Delhi, pp. 75–100, 2005.

[6] Scott, C.A. & Shah, T., Groundwater overdraft reduction through agricultural energy policy: Insights from India and Mexico. *Water Resources Development*, **20(2)**, pp. 149–164, 2004.

Water options contracts to facilitate intersectoral trade

J. Byrnes[1], L. Crase[2] & B. Dollery[3]
[1]*Economist at KPMG Australia*
[2]*Associate Professor and Executive Director, La Trobe University, Wodonga Campus, Australia*
[3]*Professor and Director of the Centre for Local Government, University of New England, Australia*

Abstract

While much faith has been placed in the ability of market based solutions to allocate water entitlements efficiently, relatively little effort has been made in fostering trade between urban and rural sectors. One barrier has been concerns regarding the decline of rural communities following the trade of water out of rural areas. In this paper we demonstrate how the use of options contracts to facilitate temporary trade between the sectors may benefit both urban water utilities and water entitlement holders in irrigation districts.
Keywords: water options, intersectoral trade, water reform.

1 Introduction

Creating the conditions for deep and liquid markets for the trade in water entitlements has been an enduring motif in the development of Australian water resource policy for at least the last ten years (Crase et al. [4]). Yet in terms of implementation, the focus has almost solely been on trade in rural water allocations. Trade from rural to other sectors has received relatively little attention due to at least two factors. First, in some locations considerable geographic barriers necessitate elaborate engineering solutions and considerable investment. Second, politicians and some rural communities have raised objections on the grounds that access to water is the lifeblood of rural Australia (Crase et al. [4]).

WIT Transactions on Ecology and the Environment, Vol 112, © 2008 WIT Press
www.witpress.com, ISSN 1743-3541 (on-line)
doi:10.2495/SI080091

In this paper we develop water options contracts previously proposed by Leroux and Crase [7] and Page and Hafi [12] for the temporary trade of water between rural and urban sectors without requiring the transfer of ownership over the allocation. We show that when purchased in relatively benign markets conditions, the instrument is an economically rationale alternative to purchasing permanent entitlements.

The paper itself consists of five additional parts. The background to the study area is provided in section 2 in order to give the context for this paper. In section 3 a description of the general nature of options contracts is given, and we outline the applicability of this instrument to water markets. The two existing papers to have investigated the potential for water markets in Australia are also reviewed in section 3. The methodology followed, the data analysed and the results obtained are presented in section 4. A discussion of some policy implications is given in section 5. The paper itself ends with brief concluding remarks in section 6.

2 Potential for intersectoral trade in the Murrumbidgee valley

Residential consumers in the Australian Capital Territory (ACT) have been subject to water use restrictions since 2002 (Pagan and Crase [11]), and current government policy is to make this regime a permanent feature of life for the foreseeable future (Byrnes et al. [3]). The fact that the Canberra water storage network suffers from a distinct deficit of capacity (Pagan and Crase [11]) explains why a sizable portion of its water allocation is sent down the Murrumbidgee River as environmental flow (Edwards [5]) while restrictions are simultaneously imposed on urban water use. It shouldn't surprise that the ACT government has been one of the most aggressive in investigating the gamut of solutions to solve the urban water supply 'crisis' that has prevailed in most of Australia's capital cities since the first few years of this decade. One of the 'solutions' identified was the purchase of permanent water entitlements in the Tantangara Dam, located some 100km upstream from Canberra on the Murrumbidgee River. Other options included the upgrade of existing dams, a more punative set of water restrictions and the recycling of wastewater for potable use. In comparison the 'Tantangara option' has a number of advantages. First, Tantangara represents a sizeable storage capable of meeting the water demands of Canberra in most years. Second, the capital cost of the infrastructure to allow the transfer of water from Tantangara to one of Canberra's existing water storages is relatively cheap (estimated to be \$38 million). Finally, the transfer of water from Tantangara would entail beneficial environmental flows (ACTEW [2]).

Notwithstanding the advantages outlined above, the Tantangara option was rejected when first formally considered in 2005. This was essentially due to concerns surrounding less than certain property rights over allocations to be held by the ACT urban water utility, stemming from the fact that Tantangara is located in the Snowy Mountains region of NSW, and subject to the control of the

NSW government water bureaucracy (ACTEW [2]). In essence, the ACT government feared that the transfer of water from Tantangara to meet the needs of urban water consumers in Canberra would be vetoed in times of extremely low storage levels, precisely when the water would be needed (Pagan and Crase [11]).

However, a review of water supply options for the ACT in 2007 saw the Tantangara option revived. The plan would see high security water allocations purchased from those willing to sell (most likely irrigators from the Murrumbidgee and/or Coleambally irrigation districts) on a permanent basis. When required the water attached to the allocation would be transported to a storage dam in Canberra, either via the Murrumbidgee River (a journey of approximately 100km) or a newly constructed 20km pipeline connecting the Tantangara to the Corin Reservoir. The plan calls for the purchase of rights to approximately 20GL of water at an estimated cost of $30 million dollars (ACTEW [2]; ACT Government [1]).

Implementation appears likely to fall flat for at least two reasons. First, there are non-trivial differences between the stipulation of water allocation rights in NSW and the ACT (Pagan and Crase [11]). Second, political objections to the permanent transfer of water rights from rural to urban areas have been rapidly raised in other contexts, and seem just as likely to float to the surface in this instance (see Crase et al. [4] for a detailed discussion of this barrier to trade). Water options contracts potentially alleviate the second of these concerns.

3 The nature of options contracts and their applicability to water markets

An options contract gives the holder the right, but not the obligation, to undertake a defined action at a pre-determined future time, and at an agreed rate of exchange. When the holder of the contract opts to invoke the right, the option contract is said to have been exercised. There are two parties to an options contract: the writer and the holder. The writer of the contract is obligated to undertake an action (such as delivering an asset) in the event that the holder invokes her rights under the contract. Options contracts can be further classified into puts and calls. A call option gives the holder the right (but not the obligation) to purchase an asset, while the holder of a put option has secured the right (but not the obligation) to sell an asset (Jones et al. [6]).

Options contracts have become an efficient means of transferring risk in financial markets, and although the basic premise of an options contract holds in their application to water markets, they differ in one key respect. Ownership is not transferred upon exercise. Rather, the contract gives the right to temporary use of the water allocation. This brings two additional advantages. First, ownership remains with the writer, which may dilute some of the objections to intersectoral transfers. Second, the option can be exercised multiple times over a contract period, delivering some supply security for the urban water utility.

Given the special nature of water options contracts, Michelson and Young [8], who were the first to propose options as a means of transferring water between urban and rural sectors, identified a number of other features that must be incorporated. First, an option must take into account the possibility of allocations being varied in drought conditions and market conditions being varied by governments. Second, since ownership is not transferred, provisions must be included to allow the writer to sell her water allocation, even after having entered into an options contract.

3.1 Valuing water options contracts

There are two steps to determining whether a water options contract has value to the holder. In the first, the capital cost of obtaining the next least cost alternative supply source is compared to the cost of exercising the option. In the case that exercising the contract is the cheapest of the two alternatives, the option has economic value. In the second step the cost of purchasing the contract is compared to the value of holding the contract. Should the premium payable not extinguish the value of holding the contract, the urban water utility would benefit from purchasing the option.

3.2 Previous efforts at exploring water options in Australia

Page and Hafi [12] examined the potential for the use of options contracts to facilitate trade between irrigators in the Murrumbidgee Valley and the urban water utility in Canberra. Following the methodology of Michelson and Young [8], they found that there was net present value from holding an options contract, when compared with a range of other supply augmenting alternatives. In a similar vein, and LeRoux and Crase [7] specified an options contract following Michelsen and Young's [8] model in the context of the Victorian town of Wangaratta. They also found a net economic benefit to the urban water utility from entering into an options contract rather than purchasing permanent allocations.

4 Methodology, data and results

In this section we contribute to the literature by both estimating the Present Value Of Benefit (PVOB) to the ACT water authority from holding the option and the premium that would be required by the writer of the contract, via the Black-Scholes pricing model.

4.1 Estimation of the PVOB from holding a water option

Following Michelson and Young (1993), Crase and LeRoux (2007), Page and Hafi (2007) and Williamson et al. (2007), the PVOB from holding the option contract is estimated as follows.

$$PVOB = \sum^{T}_{t=0} [(K_{t=0}*r + M_t - B(1-P)) - (E*P)_t]d_t + [K_{t=0} - K_{t=0}(1+a)^T]d_t \quad (1)$$

where:

PVOB = net present value of option benefits

t = year

T = contract termination year
$K_{t=0}$ = capital cost of alternative supply at the beginning of the option term
r = annual interest rate
M= annual maintenance cost of the alternative
B = price per ML of temporary water sales
E = exercise cost of option
P = annual probability of option exercise ($0 \leq P \leq 1$)
d_t = discount factor for present value, $1/(1 + r)^t$
a = annual rate of appreciation of alternative supply.

In essence, equation 1 compares the capital cost of obtaining the alternative supply and any annual costs and/or benefits associated with obtaining the alternative supply ($K_{t=0}*r + M_t - B(1-P)$), to the cost of exercising the option contract (E*P). The present value of future transactions is obtained by applying a standard discounting factor (d_t). A PVOB is estimated for each of the contracts bundled into the multiple exercise contract. Thus, in the case of a 10 year contract, ten PVOBs are estimated and summed to provide the PVOB of holding the contract over 10 years. The second term, $[K_{t=0} - K_{t=0}(1+ a)^T]d_t$, allows for inclusion of appreciation or depreciation in the value of the alternative supply, discounted to allow valuation at present value. It is important to note that the second term is only considered in the final year of the contract.

4.1.1 Data

In this example we estimate the value of contracts spanning both 5 and 10 year periods. The alternative supply to be secured is the purchase of 20 GL of high security water entitlements on the permanent market. As outlined in section 2, investigations by the ACTEW suggested the cost of this purchase would be approximately $30 million, which equates to $1,500 per ML (ACTEW, 2007). Since data regarding trades on the permanent market in the relevant district is scarce, the estimate by ACTEW was taken on face value.

The annual risk-free interest rate was arbitrarily set at 5% and left constant, however it is acknowledged that this need not be the case. The annual maintenance cost in this instance consisted of payments that would be required by the Snowy River Scheme for lost hydro electricity generation, estimated at $270 per ML. We assume that unused portions of any alternative supply could be sold into the temporary water market at $300ML, and that the probability of making temporary sales is given by (1- p).

Following Michelson and Young [8], Le Roux and Crase [8] and Williamson et al. [13], the exercise price was taken as the gross margin per ML of water consumed in the irrigation of three separate crops in the Murrumbidgee Valley: Barley, faba beans and lucerne. Data was sourced from NSW Department of Primary Industries gross margin budgets for 2007/08 (NSW DPI [10]), and acts as the 'base' scenario 1. Two further scenarios were modelled in which gross margins were deliberately increased.

The annual probability of exercise is based upon modelling by ACTEW (ACTEW [2]) that suggests water restrictions are likely to be imposed every three years out of 10. We leave the rate of appreciation at zero for the sake of simplicity.

Table 1: Gross margin assumptions.

Crop	*Gross margin: 1st scenario*	*Gross margin: 2nd scenario*	*Gross margin: 3rd scenario*
Barley	48.75	100	150
Faba beans	127	300	350
Lucerne	163	370	450

4.1.2 PVOB from holding water option contract

Table 2 presents the estimates of the PVOB to the urban water utility from holding the water options contract written by the three different irrigators, under each of the three scenarios and for contract lengths of five and 10 years.

Table 2: PVOB from holding option contract under three scenarios.

PVOB	*Barley*		*Faba Beans*		*Lucerne*	
	5yr	*10yr*	*5yr*	*10yr*	*5yr*	*10yr*
Scenario 1	$521	$929	$420	$748	$373	$665
Scenario 2	$455	$811	$195	$347	$104	$185
Scenario 3	$390	$695	$130	$232	$32	$58

A number of patterns are apparent. First, the value of holding a 10 year contract is universally higher than that of holding the 5 year contract, regardless of who the contract is purchased from. Second, the higher the exercise price of the contract, the lower the present value of holding the contract. This is intuitively appealing, since the economic principle underpinning water options is that water will move from a relatively lower value use to a higher value. Combined, the results suggest an economic benefit (in present terms) to the Canberra urban water utility of entering into either of the two contracts. This is consistent with the findings of Page and Hafi [12].

4.2 Estimation of premium payable to writer of option contract

The Black Scholes (BS) approach to determining the 'fair value' premium payable to the writer of an options contract is now almost universally employed in the pricing of financial options (Jones et al. [6]). Although the underlying nature of water prices violates a basic assumption (normality), we make use of this approach because of its general acceptance. The results are obviously limited by this modelling choice.

In essence, the BS model prices an options contract on the likelihood that the contract will be exercised, with higher probabilities requiring relatively higher premiums. It follows that the relative values of spot and exercise prices is of crucial importance. The BS approach is specified as follows:

$$c = S_0N(d_1) - Ke^{-rT}N(d_2) \tag{2}$$

and

$$d_1 = (\ln(S_0/K) + (r+\sigma^2/2)T)/\sigma\sqrt{T} \tag{3}$$

$$d_2 = d_1 - \sigma\sqrt{T} \tag{4}$$

where c is the European call price, S_0 is the stock value at time zero, N is the cumulative normal distribution, K is the strike/exercise price, r is the risk-free rate of interest, σ measures price volatility and T is the time to maturity of the option in years.

4.2.1 Data

In the case of water option contracts the stock value (S_0) is the price of temporary trades in high security water, expressed in dollars per ML. We make use of a data set supplied by the Murrumbidgee Horticultural Council (MHC, [9]) over the period 2006 to 2007. A feature of the data set is the relative lack of depth in trades and the extreme volatility from mid 2006 to late 2007. The series is presented in Figure 1, with both volume and spot price displayed.

K and r are as previously defined. Volatility in spot prices is clearly greater than 1, evident from the range of prices ($29 to $1401 per ML). Under the bold assumption that a degree of stability will return to the market as the current drought breaks we have imposed a volatility term of 0.5. As outlined above contracts of two maturities (five and ten years) are specified.

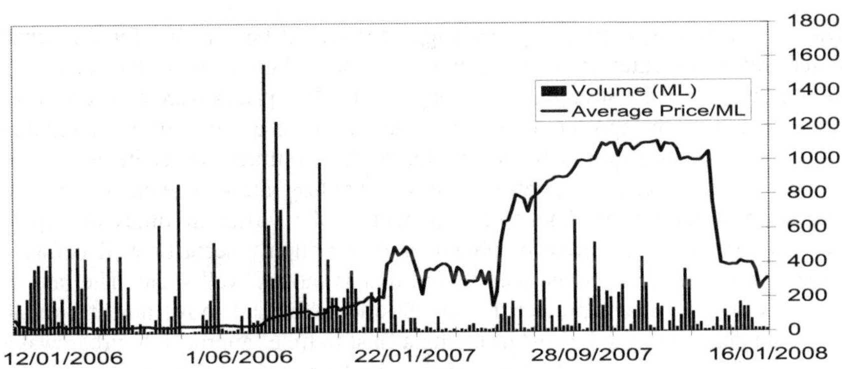

Figure 1: Average price and volume of trade in Murrumbidgee valley. (Source: Murrumbidgee Horticultural Council [9].)

If the option was in the money, (when the contract is out of the money, it is left unexercised, even though the urban water utility foregoes the premium, since water can be sourced from the spot market at a price less than the exercise price) the urban water utility would compare the value of holding the option with the cost of purchase, represented by the premium. In Table 3 the range of spot water prices at which there is still a surplus PVOB after having deducted the premium are reported. The lower value is by definition the spot price at which the contract becomes in the money. This is reported for each of the three crops and each of the three gross margin scenarios.

4.3 Results

A generally applicable principle is evident in Table 3. Those producing crops that add the least economic value per ML of water are most likely to trade with the urban water authority via an options contract. This stems from the fact these farmers require a relatively lower exercise price, resulting in relatively higher PVOB surpluses after the payment of premiums.

Table 3: Spot prices at which PVOB is greater than premium payable.

Scenario	Barley	Faba Beans	Lucerne
		5 year contract	
1	$70.31 to $525	$130 to $450	$175 to $416
2	$100 to $525	Premium > PVOB	Premium > PVOB
3	$151.5 to $475	Premium > PVOB	Premium > PVOB
		10 year contract	
1	$70.31 to $955	$130 to $777.50	$175 to $416
2	$100 to $843	$300 to $402.00	Premium > PVOB
3	$151.5 to $750	Premium > PVOB	Premium > PVOB

5 Policy implications

The results presented in this paper suggest it would be sensible for the urban water utility to enter into an options contract when it was not facing an immediate supply constraint due to drought, but when prices were relatively low, since the higher the spot price of water relative to the exercise price, the higher the compensation required by the writer of the contract. Thus, in periods of relatively high water prices, options contracts are less likely to be entered into by the urban water utility. The premium paid to the writer in times of supply security represents the price of obtaining future supply security well before a 'crisis' is encountered. Some may regard this as money well spent. The rash of expensive infrastructure projects being fast-tracked by state governments around Australia demonstrates the cost of taking a 'just in time' approach to urban water planning. This would also be beneficial to the farmer since she receives income from selling the option at time when the option is unlikely to be exercised.

6 Conclusions

In this context at least, there would appear to be value for the urban water utility in the purchase of water options contracts. However, the results suggest that the transaction is likely to be with those irrigators adding relatively little value per ML of water. This is intuitively appealing, since it sits well with the basic premise of water trading: trade will move water from relatively low to higher value uses. Yet as is so often the case in water policy, options contracts do not represent a universal or perfect solution to the thorny issue of intersectoral trade. Furthermore, while the results presented here are promising, a number of important caveats must be considered. Property rights must be certain. Planning must be such that there is a high degree of certainty surrounding the proportion of the applicable resource available. Knowledge of interactions between basins and the consequences of moving allocations from one part of the system to another must be well advanced so that the parties can be reasonably certain of the quantity of water required to make the delivery.

Acknowledgements

The views and opinions expressed herein are those of the authors and do not necessarily represent the views and opinions of KPMG, an Australian partnership.

The information contained herein is of a general nature and is not intended to address the circumstances of any particular individual or entity. Although we endeavour to provide accurate and timely information, there can be no guarantee that such information is accurate as of the date it is received or that it will continue to be accurate in the future. No one should act upon such information without appropriate professional advice after a thorough examination of the particular situation.

References

[1] ACT Government (2007). 'Water supply security fact sheets.' Retrieved 21 December, 2007, from http://www.thinkwater.act.gov.au/more_information/publications.shtml#water_security.

[2] ACTEW (2007). Water security for the ACT and region: Recommendations to the ACT Government. Canberra, ACTEW.ACTEW, 2005

[3] Byrnes, J., Crase, L. and Dollery, B. (2006). 'Regulation versus pricing in urban water policy: The case of the Australian National Water Initiative', *The Australian Journal of Agricultural and Resource Economics*, Vol. 50, pp. 437–49.

[4] Crase, L., Byrnes, J. and Dollery, B. (2007). 'The Political Economy of Urban-Rural Water Trade', *Public Policy*, Vol. 2, No. 2, pp. 130–40.

[5] Edwards, G. (2007). Urban water management. Water policy in Australia: The impact of change and uncertainty. L. Crase. Washington, RFF press: 144–165.

[6] Jones, C., Shamsuddin, A, and Naumann, K. (2007). *Investments Analysis and Management*, 2nd Edition, John Wiley & Sons, China.

[7] Leroux and Crase (2007). 'Options Contracts for Managing Inter-Sectoral Water Trade: A Preliminary Investigation of the Feasibility of Urban-Irrigation Options Contracts in the Ovens Basin, Victoria, Australia', proceeding from *Australian Agricultural and Resource Economics Society Annual Conference*, February, Queenstown.

[8] Michelsen, A. and Young, R. (1993). 'Optioning agricultural water rights for urban water supplies during drought', *American Journal of Agricultural Economics* Vol. 75, 1010–1020.

[9] Murrumbidgee Horticultural Council (2007), 'High Security Water Entitlement'. Retrieved 21 January, 2008, from http://www.murrumbidgeewater.com.au/index.php?id=6&no_cache=1

[10] NSW Department of Primary Industries (2007), Farm enterprise budget series: Gross margin budgets. Retrieved 21 December, 2007, from www.agric.nsw.gov.au.

[11] Pagan, P. and Crase, L. (2008). 'Institutional Constraints and Inter-Jurisdictional Water Trade: The Case of the Australian Capital Territory and New South Wales', proceedings of *The 51st Annual Conference of the Australian Agricultural and Resource Economics Society*, February, Canberra.

[12] Page, S. and Hafi, A. (2007). *Urban Water Options Contracts: Rural to Urban Water Trade*, Australian Bureau of Agricultural and Resource Economics, Canberra.

[13] Williamson, B., Villano, R. and Fleming, E. (2007). 'Structuring Exotic Options Contracts on Water to Improve the Efficiency of Resource Allocation in the Water Spot Market', proceedings of *The 51st Annual Conference of the Australian Agricultural and Resource Economics Society*, February, Canberra.

Section 2
Irrigation management

Why do two-thirds of Australian irrigators use no objective irrigation scheduling methods?

K. D. Montagu[1] & R. J. Stirzaker[1,2]
[1]Cooperative Research Centre for Irrigation Futures, PO Box 56, Darling Heights, Queensland 4350, Australia
[2]CSIRO Land and Water, PO Box 1666, ACT 2601, Australia

Abstract

There is significant pressure on irrigators to improve on-farm water use. A range of objective irrigation scheduling methods, such as soil monitoring, evaporation and decision support tools, have been developed to address this need. We examined how these tools have been adopted by Australian irrigators using data from an Australian Bureau of Statistics water survey of 7,280 irrigators in 2003. A total of 2.2 million hectare's were irrigated in 2002–2003 with irrigated pasture accounting for 39% of the area irrigated and 36% of the water consumed by agriculture. Cotton, grape and fruit irrigators, who account for 27% of the water used for irrigation, are the biggest users of objective scheduling methods. But still only one quarter to a third of growers are using these methods. For most irrigated crops the use of objective irrigation scheduling methods increases with farm size. The exceptions being pasture and sugar, where the use of objective irrigation scheduling methods remains low irrespective of the farm size. The major users of objective irrigation scheduling tools are where enterprise profitability is directly linked to improved crop water management, such as cotton, grape and fruit production. Other industries, particularly pasture, will lag in the use of these tools because the profitability of their enterprisers is not as sensitive to water management. Until new drivers emerge then it is unlikely that these enterprises will invest in tools to improve irrigation management. With drought and increased competition water reducing allocations, and the increased focus on river and ecosystem health, we may be seeing some of the new drivers emerge.

Keywords: objective irrigation scheduling methods, Australia, survey, soil water monitoring tools, evaporation.

WIT Transactions on Ecology and the Environment, Vol 112, © 2008 WIT Press
www.witpress.com, ISSN 1743-3541 (on-line)
doi:10.2495/SI080101

1 Introduction

In many countries irrigated agriculture accounts for typically 70% of the water consumed. The decision on when and how much water to apply to the crop is arguably the most important decision, in terms of both determining the fate of applied water and the financial return (Meyer and Montagu [1]). This is a decision made many times over a season by irrigators. To assist in managing the timing and amount of irrigation applied, science has provided a range of soil monitoring (e.g. Charlesworth [2]), evaporation and decision support tools (e.g. Inman-Bamber and Attard [3]) and more recently remote sensing products (e.g. Johnson *et al* [4]).

With many regions of the world needing to improve on-farm water management for water conservation, sustainable food production, farm profitability and environmental quality, what is the use of these tools? The limited evidence available suggests that less than one in four irrigators make use of some objective irrigation scheduling tool or service. In Washington, USA, between 18–28% of producers directly, or indirectly via consultants or extension services, utilise some form of crop evaporation data or soil moisture sensors to determine when and how much to irrigate (Leib *et al* [5]). In Australia, 13% of irrigators used soil moisture monitoring products (Stirzaker [6]).

In this paper we look at what tools different irrigation sectors are using and the impacts of farm size. With less than 40% of irrigators using some form of objective irrigation scheduling tools we propose some explanations for the low uptake.

2 Methods and materials

Survey data was obtained from the Australian Bureau of Statistics (ABS [7]) first detailed irrigation Water Survey undertaken in November 2003. The survey was comprised of 25 questions which covered the areas of:

1. Water sources, entitlement and trade,
2. Crops irrigated and irrigation methods,
3. Storages,
4. Irrigation tools,
5. Irrigation practices and barriers to change.

This paper uses data from areas 2, 4 and 5.

To examine changes in irrigation scheduling methods over time, earlier data from ABS Agricultural Census undertaken in 1996 and 2001 was obtained. Farm size categories varied for each irrigation sector. The actual size (ha) of farms for the very small, small, medium, large and very large, respectively, are as follows, ; pasture 0-75, 75-150, 150-250, 250-500, 500+; cotton 0-500, 500-1,000, 1,000-2,000, 2000-4,000, 4,000+; cereals 0-100, 100-250, 250-500, 500-1,000, 1,000+; sugar 0-50, 50-100, 100-150, 150-250, 250+; fruit 0-10, 10-20, 20-40, 40-80, 80+; rice 0-250, 250-500, 500-1,000; 1,000-2,000; grapes 0-8, 8-15, 15-30, 30-80, 80+; vegetable 0-10, 10-50, 50-100, 100-200, 200+.

2.1 Surveyed population

The Water Survey was undertaken on approximately 8,000 farm establishments. Farms were selected from a previous Agricultural Survey undertaken in 2001-02 which answered yes to the question, 'Did you irrigate between 1 July 2002 and 30 June 2003?'

A response rate of 91% was achieved for the Water Survey resulting in 7,280 responses.

3 Results

In 2002–2003 2.2 million hectare's were irrigated in Australia. Pasture accounted for 39% of the area irrigated and 36% of the water consumed by agriculture (Table 1).

Table 1: Irrigated area and volume of water applied in 2002–2003 to the major irrigated agricultural sectors in Australia.

Crop	Area irrigated ('000 ha)	Water volume applied (GL)	% of agriculture water use
Pasture	872	3,648	36
Cotton	234	1,525	15
Cereals	478	1,374	14
Sugarcane	238	1,293	13
Fruit	138	659	6
Rice	44	615	6
Grapes	150	588	6
vegetables	72	447	4

Since 1996, the proportion of irrigators who use some form of soil water monitoring equipment to decide when or how much to irrigate has risen from 13% to 22% (Table 2). Evaporation values are the only other objective tool used by a reasonable proportion of growers to decide when to irrigate or how much to apply.

Most irrigators relied on their knowledge/observation to schedule irrigations. Almost half the growers surveyed, exclusively use knowledge (i.e. personal experience on the farm) or observations only to determine when and how much irrigation to apply. Approximately 40% of growers combine local knowledge with more objective measures to make their irrigation decision.

Cotton, grapes and fruit irrigators, which account for 27% of the water used for irrigation in 2002/03, are the biggest users of objective methods (Figure 1). But still only one quarter to one third of growers are using these methods. Pasture, the sector using the most water, have the lowest use of objective irrigation scheduling methods.

Table 2: The percentage of irrigation scheduling methods used in Australian irrigation enterprises based on 1996 and 2001 Agricultural Census and the 2003 Water Survey. Data for 1996 and 2001 n=28,000; 2003 n=7,280.

Irrigation scheduling method	1996	2001	2003
	%		
Tensiometers		8	9
Soil probes[a]	13[b]	8	13
Government/commercial scheduling service		2	3
Evaporation figures/graphs	14[c]	7	10
Calendar/rotational scheduling		12	13
Knowledge/observation	93	81	91
Other methods	-	6	4
Total[d]	120	124	143

[a]neutron and capacitance probes.
[b]The 1996 census grouped tensiometers, soil probes and scheduling services together.
[c]The 1996 census grouped Evaporation figures/graphs and calendar/rotational scheduling together.
[d]The data adds to more than 100% because multiple answers were permitted.

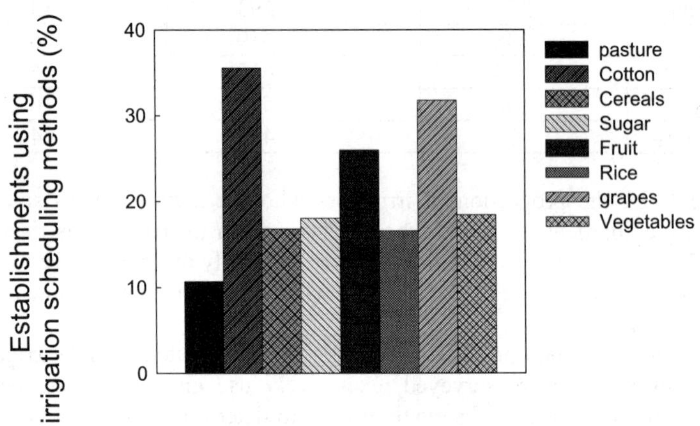

Figure 1: Proportion of irrigation establishments using objective irrigation scheduling techniques (evaporation values, tensiometers, soil probes and scheduling services). Commodities are listed in decreasing order of percentage of water consumed (Table 1). Values are maximums as some establishments maybe using two or more methods, e.g. soil probes and evaporation values).

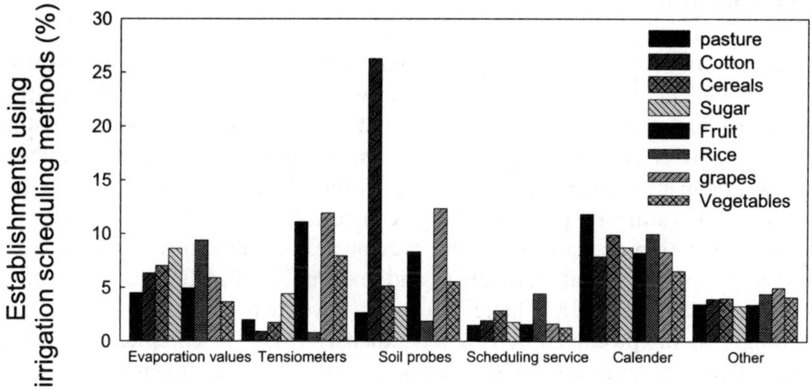

Figure 2: Proportion of irrigation establishments using irrigation scheduling tools. Commodities are listed in decreasing order of percentage of water consumed (Table 1). Values are maximums as some establishments maybe using two or more methods.

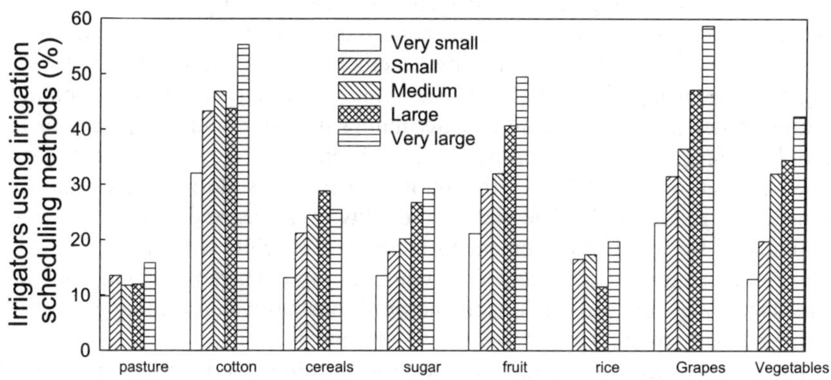

Figure 3: Impact of farm size on the proportion of irrigators using objective irrigation scheduling methods. See methods for actual farm size (ha).

Clearly, the irrigated cotton, grape and fruit sectors are leading the use of soil water monitoring tools (Figure 2). By contrast, pasture irrigators were most reliant on calendar and rotation scheduling methods. Fewer then 20% of cereal, sugar, rice, or vegetable irrigators reported using some objective method for scheduling irrigation.

In all irrigation sectors the use of objective irrigation scheduling methods increases with farm size. The exceptions being pasture and rice sugar (some trend), where the use of objective irrigation scheduling methods remains low irrespective of the farm size.

4 Discussion

The survey of over 7,280 Australia irrigators clearly indicates that objective irrigation scheduling tools, such as soil water monitoring and evaporation techniques, have not captured the hearts and minds of two thirds of irrigators (Table 1). This is despite increases in the use of these techniques, particularly those using capacitance and neutron probes, which rose from 8 to 13% between 2001 and 2003 (almost all capacitance). As a result, around half of our growers base their irrigation decision solely on previous experience, observations of the soil or crop, or local irrigation practises and make no farm scale measurements.

Low adoption of the available tools can have two root causes. First, the technologies being promoted may be too expensive, risky or complex to confer a relative advantage over what the irrigator is already doing, given that water is often a small proportion of the input costs (Batz *et al* [8]; Stirzaker [9]; Pannel *et al* [10]). Second, the technologies may not be compatible with the broader goals of the farm family or farm manager, which are influenced by social, cultural and historical factors (Vanclay [11]; Kaine *et al* [12]; Lineham *et al* [13]; Pannel *et al* [10], Montagu *et al* [14]). To understand why, we need to look closer at who are using objective irrigation scheduling methods.

Let's start by looking at which irrigation sectors are using objective irrigation scheduling techniques (Figure 1). The cotton, fruit and grape sectors standout as users of objective irrigation scheduling methods. But at best only 25–35% of irrigators are using these tools. These three industries all share a common feature. Water plays a key role in determining the yield and /or quality of the crop. Mismanagement of the crops water requirements at key periods can have a major impact on profitability. In cotton, water stress during flowering will dramatically reduce yield. The yield and quality of grapes and fruit crops can be manipulated by water management to increase profitability. In these industries irrigation management has become an integral crop management tool. The need to manage water for crop production and profit has motivated leading growers to overcome many of the barriers to using these objective tools.

The pasture sector is the standout non-user of objective irrigation scheduling methods. For crops such as pasture (and lucerne), the irrigation decision has less impact on yield and profitability, compared to cotton, grapes or fruit crops. Mis-timing an irrigation decision may temporarily reduce grass growth but doesn't dramatically reduce overall yield. Furthermore, the irrigated pasture industry essentially provides input into a larger production system, such as dairy or meat production. As a consequence the focus is on the animal production system, with irrigation being one input into this system. This may be the underlying reason for the low adoption of irrigation scheduling tools and the greater reliance on calendar and rotational schedules.

The vegetable sector deserves a special mention. Despite water playing a key role in determining yield and/or quality, the use of objective tools is low at less than 20%. We suggest application of water far in excess of that needed by the crop effectively reduces the need to use any objective scheduling methods. For vegetable producers, water limitations are one of the easiest and cheapest constraints to remove. Because the value of production is very high, it is common practice to apply inputs to excess as a risk minimisation strategy Stirzaker [9]. For example, even when vegetable growers are using town water and paying over $1,000 per ML, there is little use of objective scheduling methods. In this situation the cost of getting it wrong is so high that the risk is mimimised by applying luxury levels and thereby ensuring that there are no yield reductions.

The bright spot in the data is that in most irrigation sectors the larger farms are make greater use of objective scheduling methods. This is most noticeable in the fruit, grape, vegetable and sugar sectors where the proportion of farms using objective scheduling methods more than doubles with increasing farm size. But despite this only around half of irrigators use some form of objective measure for scheduling irrigation.

Influence of the corporate farm sector

Looking to the future we can expect to see increased growth in the soil probe category, which is now predominantly made up of logging or manual capacitance probes. Recent data shows the adoption curve is still rising. Adoption is likely to continue as the word spreads, the technology gets cheaper and reliability and confidence in using the equipment grows among irrigators (Table 2).

There is a clear preference of irrigators to invest in new irrigation systems rather than using objective scheduling practices to optimise existing systems. Whether improvements in irrigation are realised by irrigation system upgrades without instigating monitoring has not been tested. Many farmers claim to have improved the efficiency of irrigation scheduling without investing in soil water monitoring (or other scientific methods). Here, increasing irrigation efficiency must comprise a diverse array of activities, including benchmarking activities, implementation of new equipment and on-going training.

Table 3: Percentage of irrigators who have made changes in irrigation practice, 1998–2003, and who intend to make changes in the future.

	Percentage change in irrigation practice	
	1998-2003	Future
No change	30	56
One or more changes	70	44
Type of change		
More efficient application system	46	22
More efficient scheduling	37	17
On-farm soil water monitoring	15	9

5 Conclusion

Industries where enterprise profitability is directly linked to improved crop water management, such as cotton, grapes and fruit, are the major users of objective irrigation scheduling tools. Other industries, particularly pasture, will lag in the use of these tools because the profitability of their enterprisers is not as sensitive to water management. Until new drivers emerge then it is unlikely that these enterprises will invest in tools to improve irrigation management. With the ongoing reductions in water allocations due to drought and increased competition for water, and the increased focus on detrimental impacts of poor irrigation management on river and ecosystem health, we may be seeing some of the new drivers emerge.

References

[1] Meyer, W. & Montagu, K. The irrigation industry in the Murray and Murrumbidgee basins. *Water*, **33(5)**, pp. 53–56. 2006.

[2] Charlesworth, P.B. *Soil Water Monitoring 2nd edition.* Irrigation Insights No. 1. Land & Water Australia, Canberra, 2005.

[3] Inman-Bamber, G. & Attard, S. Inventory of Australian Software Tools for On Farm Water Management. Technical Report No. 02/05. CRC for Irrigation Futures, Toowoomba, Australia. 2005.

[4] Johnson, L. Pierce, A. Michaelis, T. Scholasch, and Nemani, R. Remote Sensing and Water Balance Modeling in California Drip-Irrigated Vineyards. Proc ASCE World Environmental & Water Resources Congress, Omaha NE, 2006.

[5] Leib, B.G., Hattendorf, M., Elliot, T., & Matthews, G. Adoption and adoption of scientific irrigation scheduling: trends from Washington, USA as of 1998. *Agricultural Water Management* **55**: 105–120. 2002.

[6] Stirzaker, R.J. Soil moisture monitoring: state of play and barriers to adoption. Irrigation Matters Report No. 01/06. CRC for Irrigation Futures, Toowoomba, Australia. 2006.

[7] Australian Bureau of Statistics. Water Use on Australian Farms. Commonwealth of Australia, Canberra. 2005.

[8] Batz, F.J., Peters, K.J., & Janssen, W. The influence of technology characteristics on the rate and speed of adoption. *Agricultural Economics* **21**: 121–130. 1999.

[9] Stirzaker, R.J. The problem of horticulture matching the biophysical efficiency with the economic efficiency. *Agroforestry Systems* **45**: 187–202. 1999.

[10] Pannell, D.J., Marshal, G.R., Barr, N., Curtis, A., Vanclay, F., & Wilkinson, R. Understanding and promoting adoption of conservation practices by rural landholders *Australian Journal of Experimental Agriculture* **46**: 1407–1424. 2006.

[11] Vanclay, F. Social principles for agricultural extension to assist in the promotion of natural resource management. *Australian journal of experimental agriculture* **44**: 213–224. 2004.

[12] Kaine, G., Bewsell, D., Boland, A., & Linehan, C. Using market research to understand the adoption of irrigation management strategies in the stone and pome fruit industry. *Australian Journal of Experimental Agriculture* **45**: 1181–1187. 2005.

[13] Linehan, C.J., Krstic, M.P., & Kaine, G. Combining Biophysical and social science to achieve change through targeted extension. *ACTA Horticulturae* **672**: 211–216. 2005.

[14] Montagu, K., Thomas, B., Thomas, G., Christen, E., Hornbuckle, J., Baillie, C., Linehan, C., Smith, P., Gallo, F., North, S., Meyer, W., Stirzaker, R. & Cornish, P. From Enterprise Planning to the Paddock**.** Irrigation Insights No. 6. Land & Water Australia and IF Technologies Pty Ltd. 2006.

Salt tolerance classification of winter cereal varieties according to grain yield performance and water use efficiency

N. Katerji[1], M. Mastrorilli[2], M. M. Nachit[3], J. W. van Hoorn[4], A. Hamdy[5], T. Oweis[3], O. Abdalla[3] & S. Grando[3]
[1]*INRA, Grignon, France*
[2]*CRA-SCA, Bari, Italy*
[3]*ICARDA, Aleppo, Syria*
[4]*Sub-department Water Resources, Wageningen University, The Netherlands*
[5]*CIHEAM-IAM, Valenzano, Italy*

Abstract

Irrigation management with nonconventional waters (saline water, reused drainage-water, waste-water) require the identification of varieties which are adapted to saline conditions. The study aims to identify the varieties which combine high yield with the efficiency in using irrigation waters of different qualities. Durum wheat, barley and winter wheat showed an ascendant curvilinear relationship between grain yield and water use efficiency. The durum wheat varieties showed large differences in grain yield that increased at increasing salinity. Barley also showed large differences between the varieties, even more pronounced than durum wheat but not increasing with salinity. Among the bread wheat varieties only one variety was less suitable under saline conditions. The durum wheat varieties (Cham.1 and Belikh.2) and the barley varieties (California Mariout) and Melusine/A) present a combination of high yield and high water use efficiency in a saline environment, whereas the bread wheat variety (Johara.14) is less suitable under saline conditions than the other varieties. The varietal selection combining high yield and high water use efficiency constitutes an important point with respect to managing irrigation with saline waters.

Keywords: barley, durum wheat, bread wheat, water salinity, drought, water use efficiency.

WIT Transactions on Ecology and the Environment, Vol 112, © 2008 WIT Press
www.witpress.com, ISSN 1743-3541 (on-line)
doi:10.2495/SI080111

1 Introduction

When assessing the suitability of saline irrigation water, the choice of varieties adapted to saline conditions plays an important role (Rhoades et al. [1], Minhas [2]). Agronomists prefer varieties with the highest yield in a saline environment (Reitz [3]). Irrigation specialists often do not take yield as a leading indicator (Pereira et al. [4]), but they prefer the water use efficiency to identify the best irrigation scheduling strategies (Shideed et al. [5]) and to analyse the water saving performance of irrigation systems (Ayars et al. [6]).

The water use efficiency of cereals is defined as the ratio between grain yield and total evapotranspiration (Sinclair et al. [7]). The choice of the water use efficiency is well justified, because a high valorization of saline water means at the same time water economy and less salt input (Burt et al. [8]). Sinclair and Muchow [9] underline the complementariness of both indicators and suggest to select varieties presenting high yield together with high water use efficiency.

Research on links between yield and water use efficiency is a new approach in the selection of cereal varieties tolerant to drought or salinity. Actually, no information is available on the relationship between yield and water use efficiency of cereal varieties under saline conditions. The comparison of cereal varieties in a saline environment of Kingsbury and Epstein [10] and the most recent one of El-Hendawy [11] do not take into account evapotranspiration and water use efficiency. Data, however, are available on the relationship between yield and water use efficiency under non-saline conditions for trials conducted on winter cereals growing irrigated or non-irrigated conditions in the Mediterranean region. The results for winter cereals vary between the species. Bread wheat shows a curvilinear relationship without a maximum (Zhang and Oweis [12]). Whereas durum wheat shows a maximum in its curvilinear relationship, indicating a decrease of the water use efficiency at very high yields (Oweis and Zhang [13]). The approach based on multi-location trials for drought and salinity studies is not recommended, as other interacting factors are involved in the experiment. According to Maas and Grattan [14] it is not always clear if the varietal differences reflect differences in salt tolerance or differences in adaptation to the particular climatic or nutritional conditions under which the crops were tested.

This study comprises three experiments with varieties of durum and bread wheat and barley at three salinity levels. The trials were conducted in greenhouse, which permits to measure simultaneously evapotranspiration on a large number of cultivars and repetitions. The aim of this study was as follows: 1) to relate yield and water use efficiency in saline water for the three Mediterranean winter cereals (durum wheat, bread wheat, and barley); and 2) within each species, to study the relationship between salt tolerance rankings and water use efficiency.

2 Material and methods

The experiments were done in a greenhouse (inside temperature 20°C) at Bari, southern Italy. The set-up consisted of micro-lysimeters with a diameter of 0.4m,

Table 1: Composition of irrigation water (mmol l^{-1}).

EC (dS/m)	Ca^{2+}	Mg^{2+}	Na^{+}	K^{+}	Cl^{-}	HCO_3^{-}	SO_4^{2-}
0.9	4.0	3.4	1.6	0.5	3.2	6.0	0.3
4	6.5	14.0	22.0	1.3	36.0	5.8	2.0
5	6.6	8.4	40.0	0.8	45.0	8.0	7.0
8	7.8	20.0	48.0	2.7	70.0	6.0	2.5
10	8.2	19.4	78.6	1.4	80.0	8.0	12.0

Table 2: The cultivars.

		DURUM WHEAT
V1	Omrabi.5	ICARDA breeding line, high yielding, widely adapted, drought tolerant; released in Jordan, Iran, Turkey, and Iraq
V2	Hagla	ICARDA breeding line, salt tolerant
V3	Haurani	Syrian Landrace was grown on large scale in Syria, Lebanon, Jordan, and Turkey
V4	Gidara 2	ICARDA breeding line, high yielding in continental areas, cold tolerant; released in Turkey
V5	Cham.1	ICARDA breeding line, high yielding, good performance under higher rainfall and supplementary irrigation; released in several Mediterranean countries.
V6	Jennah Khetifa	Landrace
V7	Belikh.2	ICARDA breeding line, high yielding, some salt tolerance; released in Lebanon and Syria
		BARLEY
V1	Arar	ICARDA breeding line,, drought resistant, 2 rowed
V2	Arta	ICARDA breeding line, drought resistant, 2 rowed
V3	California Mariout	6 rowed, resistant to drought, salinity and diseases
V4	Zanbaka	2 rowed
V5	WI2737	ICARDA breeding line, selected for high yield potential, 2 rowed
V6	Melusine/A	ICARDA/CIMMYT breeding line, selected for high yield potential, 2 rowed
		BREAD WHEAT
V1	Sakha.8	Egyptian breeding line, irrigated crop, salt tolerant
V2	Cham.8	ICARDA breeding line, irrigated crop, salt sensitive
V3	Cham.6	ICARDA breeding line, dry land crop
V4	Haamam.4	ICARDA breeding line, dry land crop
V5	Qafzah.8	ICARDA breeding line, dry land crop
V6	Qimma.5	ICARDA breeding line, dry land crop
V7	Johara.14	ICARDA breeding line, dry land crop

a height of 0.6m. The lysimeter was filled with clay soil. Table 1 presents the chemical composition of the irrigation waters. The used crop cultivars (table 2) were developed for their high productivity and drought tolerance (Nachit and Elouafi [15] for durum wheat, Ceccarelli [16] for barley, and Ortiz-Ferrara and Abdalla [17] for beard wheat). No data are available about the salt tolerance, except for durum wheat where preliminary studies were made on limited cultivars (Almansouri et al. [18]).

Evaporation (class A pan) was used to schedule irrigations. Evapotranspiration (ET of the irrigation interval) was calculated as the difference between the amounts of irrigation and drainage water. Table 2 presents the studied varieties.

Fertilization doses were 150, 100 and 120 Kg/ha of P_2O_5, K_2O and N.

The durum wheat experiment comprised 63 lysimeters (7 varieties x 3 water qualities x 3 replicates), the barley 72 (6 varieties x 3 waters x 4 replicates), and the bread wheat 84 (7 varieties x 3 waters x 4 replicates).

The variables understudy in the experiments were analyzed with two-way ANOVA followed by multiple pair wise comparisons (Student-Neuman-Keuls test).

3 Results and discussion

Figure 1 presents the relationship between yield and WUE. The curves of barley and bread wheat are similar to those observed for crops (Zhang and Oweis [12]; Zhang et al. [19]; Oweis et al. [20]) cultivated in fields with or without irrigation. A difference, however, exists between our observations and those of Zhang and Oweis of durum wheat. The latter showed a maximum of water use efficiency at a yield of 0.6 Kg/m^2, after which the water use efficiency declined.

The decline of the WUE in the study of Zhang and Oweis could be attributed to the evapotranspiration not being measured, but calculated. In case of heavy rainfall and irrigation runoff and drainage cannot be neglected (Katerji et al. [21]) and a simplified water balance leads to overestimating the evapotranspiration and underestimating the water use efficiency (Rana and Katerji [22]).

The WUE data found in the present study conducted under greenhouse conditions are similar to those generated under field conditions. In the case of durum wheat, the average value for WUE on 7 cultivars is 1.37 Kg/m^3 (1.19 Kg/m^3 ± 0.2 according Zhang and Oweis [12]) and 1.9 Kg/m^3 for bread wheat (from 1 to 2.5 Kg/m^3, according Oweis [23]). The high WUE for bread wheat in comparison with durum wheat is in agreement with breeders observations (Nachit et al. [24]). As irrigation is not practiced in barley, therefore no WUE studies were undertaken in this present study.

The ranking of the durum wheat varieties in Table 3 shows almost the same order for yield and for water use efficiency, as it was expected from the strong relationship between yield and water use efficiency in Figure 1.

Salinity has little effect on the order of ranking. However, some slight changes have occurred. For example, in the case of fresh water the varieties V1, V2 and V3 show a significantly lower yield than variety V6, whereas only the

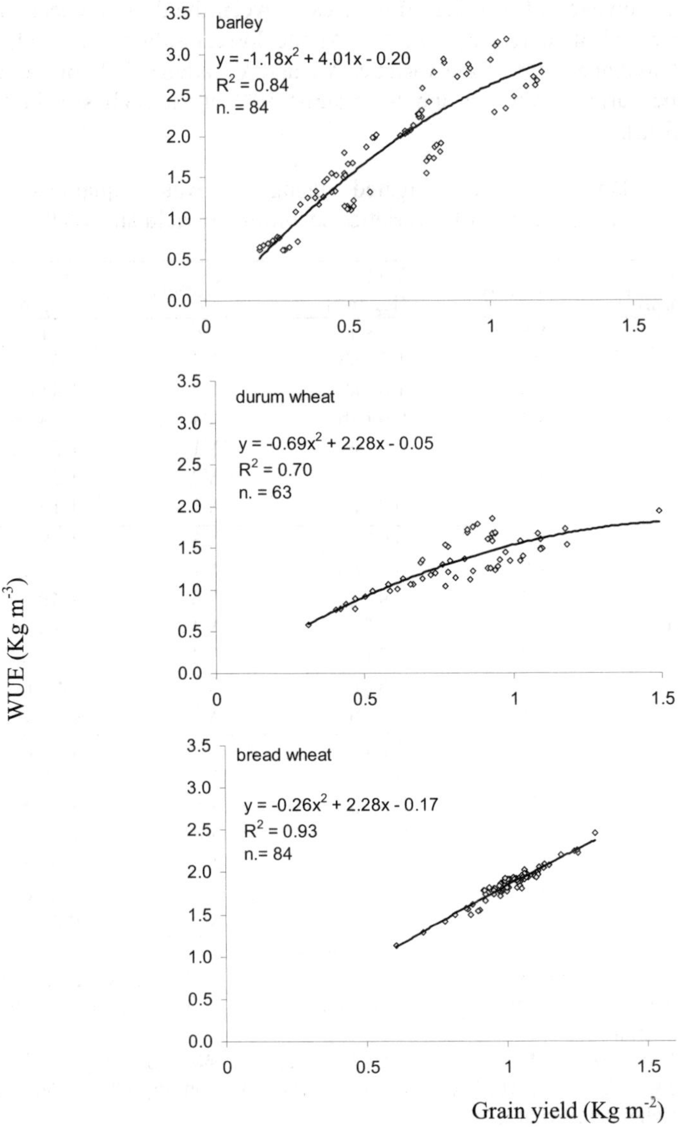

Figure 1: Water use efficiency (WUE) and crop yield for barley, bread and durum wheat.

water use efficiency of variety V3, is significantly lower than that of variety V6. In the case of the most saline treatment, the varieties V1, V2 and V3 present the lowest water use efficiency, whereas only the varieties V1 and V3 show the lowest yield.

Salinity, however, affects the differences between the durum wheat varieties, which increased at increasing salinity. At the lowest salinity level (ECw 0.9 dS/m), the average yield of the varieties V1 and V3 attains 77% of the average yield of the varieties V5, V6 and V7 against 50% at the highest salinity level (ECw 8 dS/m).

Table 3: Durum wheat irrigated using 3 water qualities (ECw): classifications of 7 varieties according to yield and WUE.

ECw (dSm^{-1})	Variety	Yield ($Kg\ m^{-2}$)	Variety	WUE ($Kg\ m^{-3}$)
	V6	1.23 a	V6	1.61 a
	V7	1.08 ab	V7	1.59 a
	V5	1.07 ab	V5	1.45 ab
0.9	V4	0.99 ab	V4	1.34 ab
	V2	0.89 b	V1	1.24 ab
	V3	0.88 b	V2	1.21 ab
	V1	0.87 b	V3	1.16 b
	V6	0.96 a	V5	1.68 a
	V5	0.94 ab	V7	1.61 a
	V7	0.92 ab	V6	1.49 a
4.0	V4	0.77 bc	V4	1.26 b
	V2	0.74 c	V2	1.21 b
	V3	0.67 cd	V3	1.09 bc
	V1	0.56 d	V1	0.93 c
	V7	0.86 a	V7	1.76 a
	V5	0.85 a	V5	1.69 a
	V6	0.83 a	V6	1.41 b
8.0	V4	0.72 a	V4	1.40 b
	V2	0.57 b	V2	1.04 c
	V1	0.44 c	V1	0.83 c
	V3	0.42 c	V3	0.80 c

F-values (ANOVA: 7 varieties, 3 salinity levels, 3 repetitions).
Dependent variable yield: variety, $31.3 > 3.29 = (6,40; 0.01)$, highly significant; ECw, $87.0 > 5.18 = (2,40; 0.01)$, highly significant; interaction, $1.47 < 1.71 = (12,40; 0.10)$, not significant.
Dependent variable WUE: variety, $40.9 > 3.29 = (6,40; 0.01)$, highly significant; ECw, $2.9 > 2.44 = (2,40; 0.10)$, low significant; interaction, $3.3 > 2.66 = (12,40; 0.01)$, highly significant.
Values followed by the same letter are not significantly different at $P < 0.05$ according to the Student-Neuman-Keuls test.

Barley in Table 4 shows the same picture as durum wheat: similar order of ranking according to yield and water use efficiency and little effect of salinity on the varietal ranking. Differences between the varieties do not increase at increasing salinity. At the lowest salinity level the average yield of the varieties V1 and V4 attains 48% of the average yield of the varieties V3 and V6 against

Table 4: Barley irrigated using 3 water qualities (ECw): classification of 6 varieties according to yield and WUE.

ECw (dSm^{-1})	Variety	Yield ($Kg\ m^{-2}$)	Variety	WUE ($Kg\ m^{-3}$)
0.9	V3	1.15 a	V3	2.68 a
	V6	1.07 b	V6	2.44 b
	V5	0.81 c	V5	1.90 c
	V2	0.81 c	V2	1.71 d
	V1	0.54 d	V4	1.23 e
	V4	0.52 d	V1	1.20 e
5.0	V3	1.03 a	V3	3.09 a
	V6	0.91 b	V6	2.80 b
	V5	0.76 c	V5	2.32 c
	V2	0.73 c	V2	2.15 d
	V2	0.46 d	V4	1.41 e
	V1	0.46 d	V1	1.39 e
10.0	V6	0.81 a	V6	2.80 a
	V3	0.71 b	V3	2.06 b
	V2	0.59 c	V2	1.96 b
	V5	0.49 d	V5	1.66 c
	V4	0.42 e	V4	1.45 d
	V1	0.35 f	V1	1.19 e

F-values (ANOVA: 6 varieties, 3 salinity levels, 4 repetitions).

Dependent variable yield: variety, 761.5 > 3.51 = (5,51; 0.01), highly significant; ECw, 492.2 > 5.18 = (2,51; 0.01), highly significant; interaction, 25.6 > 2.80 = (10,51; 0.01), highly significant.

Dependent variable WUE: variety, 442.2 > 3.51 = (5,51; 0.01), highly significant; ECw, 95.8 > 5.18 = (2,51; 0.01), highly significant; interaction, 22.5 > 2.80 = (10,51; 0.01), highly significant.

Values followed by the same letter are not significantly different at $P < 0.05$ according to the Student-Neuman-Keuls test.

50% at the highest salinity level. The differences between the varieties are more pronounced than for durum wheat.

Table 5 presents the results of bread wheat. Salinity affects the varictal ranking by decreasing the differences between the varieties; the largest difference was observed at the lowest salinity level. Only variety V7 differs significantly from the other varieties at the highest salinity level. Its yield reduction compared to the yield of variety V5 remains stable: 83% at the lowestand 82% at the highest salinity level. Among the varieties selected by ICARDA the durum wheat varieties V5 (Cham.1) and V7 (Belikh.2) and the barley varieties V3 (California Mariout) and V6 (Melusine/A) present a combination of high yield potential and high water use efficiency in a saline environment, whereas the bread wheat variety V7 (Johara.14) is less suitable under saline conditions than the other varieties.

All the 3 cereal species were bred for drought tolerance and high yield under Mediterranean dryland conditions. Interestingly, the varieties with relatively good salt tolerance show also good yield stability. The two varieties of durum wheat, e.g.; Cham.1 and Belikh.2 are among the cultivars with high yield performance and yield stability across environments in the Mediterranean region (Nachit [25]). The additional salt tolerance to drought tolerance and yield potential may have increased the yield stability of Cham.1 and Belikh.2.

Table 5: Bread wheat irrigated using 3 water qualities (ECw): classification of 7 varieties according to yield and WUE.

ECw (dSm^{-1})	Variety	Yield ($Kg\ m^{-2}$)	Variety	WUE ($Kg\ m^{-3}$)
	V2	1.16 a	V2	2.10 a
	V5	1.12 ab	V1	2.05 ab
	V1	1.11 ab	V5	1.98 ac
0.9	V3	1.08 ab	V3	1.94 ac
	V6	1.04 ac	V6	1.83 bc
	V4	0.99 bc	V4	1.78 cd
	V7	0.93 c	V7	1.60 d
	V5	1.07 a	V3	1.94 a
	V3	1.04 a	V5	1.93 a
	V1	1.01 a	V1	1.91a
4.0	V2	1.01 a	V2	1.88 a
	V6	1.00 a	V6	1.82 a
	V4	0.95 a	V4	1.80 a
	V7	0.92 a	V7	1.67 a
	V5	1.02 a	V5	1.91 a
	V3	1.00 a	V3	1.91a
	V1	0.99 a	V2	1.88 a
8.0	V2	0.99 a	V1	1.86 a
	V6	0.98 a	V4	1.83 a
	V4	0.94 a	V6	1.81 a
	V7	0.84 b	V7	1.54 b

F-values (ANOVA: 7 varieties, 3 salinity levels, 4 repetitions).

Dependent variable yield: variety, 5.3 > 3.12 = (6,60; 0.01), highly significant; ECw, 7.7 > 5.18 = (2,60; 0.01), highly significant; interaction, 0.39 < 1.66 = (12,60; 0.10), not significant.

Dependent variable WUE: variety, 6.8 > 3.12 = (6,60; 0.01), highly significant; ECw, 1.5 < 1.87 = (6,60; 0.10), not significant; interaction, 0.52 < 1.66 = (12,60; 0.10), not significant.

Values followed by the same letter are not significantly different at P < 0.05 according to the Student-Neuman-Keuls test.

4 Conclusion

Durum wheat, barley and bread wheat showed an ascendant curvilinear relationship between yield and water use efficiency, which explains the similarity between the ranking order according to grain yield and that according to water use efficiency.

The durum wheat varieties showed large differences in grain yield that increased at increasing salinity. Barley also showed large differences between the varieties, even more pronounced than durum wheat but not increasing with salinity. Among the bread wheat varieties, only one variety was less suitable under saline conditions.

A varietal selection combining high yield and high water use efficiency constitutes an important point with respect to saline water irrigation management.

References

[1] Rhoades, J.D., Kandiah, A. & Mashali, A.M., The use of saline waters for crop production, Irrigation Drainage Paper 48, FAO: Rome, 1992.

[2] Minhas, P.S., Saline water management for irrigation in India. Agricultural Water Management, **30**, pp. 1–24, 1996.

[3] Reitz, L.L., Breeding for more efficient water use, is it real or a mirage? Agricultural Meteorology, **14**, pp. 3–11, 1974.

[4] Pereira, L.S., Oweis, T. & Zairi, A., Irrigation management under water scarcity. Agricultural Water Management. **57**, pp. 175–206, 2002.

[5] Shideed, K., Oweis, T., Gabr, M. & Osman, M., Assessing on-farm water use efficiency: A new approach, ICARDA publishing: Aleppo, 2005.

[6] Ayars, J.S., Phene, C.J., Hutmacher, R.B., Davis, K.R., Schoneman, R.A., Vail, S.S. & Mead, R.M., Subsurface drip irrigation of row crops: review of 15 years research in the Water Management Research Laboratory. Agricultural Water Management, **42**, pp. 1–27, 1999.

[7] Sinclair, T.R., Tanner, C.B. & Bennett, J.M., Water use efficiency in crop production. Bioscience, **34**, pp. 36–40, 1984.

[8] Burt, C.M., Clemmens, A.J., Strelkoff, T.S., Solomon, K.H., Bliesner, R.D., Hardy, L.A., Howell, T.A. & Einsenhauer, D.E. Irrigation performance measures: efficiency and uniformity. Journal of Irrigation and Drainage Engineering, **123**, pp. 423–442, 1997.

[9] Sinclair, T.R. & Muchow, R.C., System analysis of plant traits to increase grain yield on limited water supplies. Agronomy Journal, **93**, pp. 263–270, 2001.

[10] Kingsbury, R.W. & Epstein, E., Selection for salt resistant spring wheat. Crop Science, **24**, pp. 310–315, 1984.

[11] El-Hendawy, S.E., Hu, J., Yakout, G.M., Awad, A.M., Hafez & S.E., Schmidhalter, U., Evaluating salt tolerance of wheat genotypes using multiple parameters. European Journal of Agronomy, **22**, pp. 243–253, 2005.

[12] Zhang, H. & Oweis, T., Water yield relations and optimal irrigation scheduling of wheat in the Mediterranean region. Agricultural Water Management, **38**, pp. 195–211, 1999.

[13] Oweis, T. & Zhang, H., Water use efficiency: index for optimizing supplemental irrigation of wheat in water-scarce areas. Journal of Applied Irrigation Science, **33**, pp. 321–336, 1998.

[14] Maas, E.V. & Grattan, S.R., Crop yields as affected by salinity. Agricultural drainage, eds. R.W. Skaggs & J. van Schilfgaarde, ASA-CSSA-SSSA publishing: Madison, pp. 55–108, 1999.

[15] Nachit M.M. & Elouafi I. Durum wheat adaptation in the Mediterranean dryland: Breeding, stress physiology, and molecular markers. Challenges and Strategies for dryland Agriculture. CSSA special publishing: 32, pp. 203–218, 2004.

[16] Ceccarelli, S., Grando, S., Baum, M. & Udupa, S.M., Breeding for drought resistance in a changing climate. Drought resistance in cereals, ed. F.W.G. Backer, CAB Publishing int: Wallingford, pp. 167–190, 2004.

[17] Ortiz-Ferrara G. & Abdalla, O.S., Spring bread wheat improvement. Germoplasm improvement program. Annual report 1997. ICARDA publishing: Aleppo, pp. 153–168, 1997.

[18] Almsansouri M., Kinet J.M. & Lutts S., Physiological analysis of salinity resistance in Triticum turgidum var durum Desf.: Callus versus whole plant responses. Durum wheat Improvement in the Mediterranean Region: New challenges, eds. C. Royo, M.M. Nachit & N. Di Fonzo, Options Mediterraneenes 40, CIHEAM: Zaragosa, pp. 263–265, 2000.

[19] Zhang, H., Pala, M., Oweis, T. & Harris, H., Water use and water use efficiency of chickpea and lentil in a Mediterranean environment. Australian Journal Agricultural Research, **51**, pp. 295–304, 2000.

[20] Oweis, T., Hachum, A. & Pala, M., Faba bean productivity under rainfed and supplemental irrigation in northern Syria. Agricultural Water Management, **73**, pp. 57–72, 2005.

[21] Katerji, N., Daudet, F.A. & Valacogne, Ch., Contribution des réreserves profondes du sol au bilan hydrique. Détermination et importance. Agronomie, **4**, pp. 779–787, 1984.

[22] Rana, G. & Katerji, N., Measurements and estimation of actual evapotranspiration in the field under Mediterranean climate: A review. European Journal of Agronomy, **13**, pp. 125–153, 2000.

[23] Oweis, T., Supplemental irrigation: A highly efficient water-use practice, ICARDA: Aleppo, 1997.

[24] Nachit, M., Picard, E., Monneveux, P., Labhilili, M., Baum, M. & Rivoal, R., Présentation d'un programme international d'amélioration du blé dur pour le basin méditerranéen. Cahiers Agricultures, **7**, pp. 510–515, 1998.

[25] Nachit, M.M., Sorrells, M.E., Zobel, R.W., Gauch, H.G., Fischer, R.A., & Coffman, W.R., Association of morpho-physiological traits with grain yield and components of genotype-environment interaction in durum wheat. I. Journal of Genetics & Breeding, **46**, pp. 363–368, 1992.

Midsummer deficit irrigation of alfalfa as a strategy for providing water for water-short areas

B. Hanson[1], K. Bali[2], S. Orloff[3], H. Carlson[3], B. Sanden[4] & D. Putnam[5]
[1]*Dept. of Land, Air and Water Resources, University of California, Davis, USA*
[2]*University of California Cooperative Extension, Imperial County, USA*
[3]*University of California Cooperative Extension, Siskiyou County, USA*
[4]*University of California Cooperative Extension, Kern County, USA*
[5]*Dept. of Plant Sciences, University of California, Davis, USA*

Abstract

Alfalfa is California's single largest agricultural water user due to its large acreage and long growing season. As a result, interest exists in midsummer deficit irrigation (no irrigation in July, August, and September) of alfalfa in water-rich areas to provide water for water-short areas with the amount of transferred water equal to the difference in the evapotranspiration (ET) between fully-irrigated and deficit irrigated alfalfa. However, little data exists on the ET of midsummer deficit-irrigated alfalfa. Commercial fields were selected for fully-irrigated and deficit-irrigated irrigation treatments of alfalfa. The fully-irrigated alfalfa was irrigated according to the irrigator's normal practices. The deficit irrigation treatments were no irrigation in July through to September. Alfalfa ET was measured using the eddy covariance and surface renewal energy balance methods. Deficit irrigation of alfalfa during the midsummer reduced both ET and yield. The amount of reduction, however, was very site-specific. The Davis site showed the largest reduction in ET (198 mm), while the other sites showed much smaller reductions (5 to 62 mm).

Keywords: alfalfa, evapotranspiration, irrigation.

WIT Transactions on Ecology and the Environment, Vol 112, © 2008 WIT Press
www.witpress.com, ISSN 1743-3541 (on-line)
doi:10.2495/SI080121

1 Introduction

Alfalfa is California's single largest agricultural water user due to the amount grown, typically about 405,000 ha, and its long growing season. Thus, interest exists in midsummer deficit irrigation (no irrigation in July, August, and September) of alfalfa to provide water for transfer from water-rich areas such as the Sacramento Valley to water-short areas such as the San Joaquin Valley and southern California. In theory, the amount of water available for transfer would equal the difference in the evapotranspiration (ET) between fully-irrigated and deficit irrigated alfalfa. This midsummer deficit irrigation strategy maintains the relatively high yields of the first part of the year and eliminates irrigations during the summer when yields are small and quality is poor.

Studies have shown that midsummer deficit irrigation reduces crop yield [1–5]. Different alfalfa yield responses to deficit irrigation were found with the main factors contributing to the yield response to be soil texture and climate. These studies also showed that while no irrigation during the midsummer reduced the alfalfa yield, it generally did not stop all plant growth, and as a result, some level of evapotranspiration occurred. Little information exists on differences in midsummer ET between fully-and deficit-irrigated alfalfa. This study investigated the effect of midsummer deficit irrigation on yield and evapotranspiration in commercial alfalfa fields at sites throughout California, a major alfalfa production state in the USA.

2 Methods and materials

2.1 Experiment design

Sites were selected near Davis (southern end of the Sacramento Valley), Tulelake (Intermountain area of northern California), El Centro (Imperial Valley of southern California), Scott Valley (Intermountain area of northern California), and in Kern County (southern end of the San Joaquin Valley) for fully-irrigated and deficit-irrigated treatments of alfalfa. These locations were selected to obtain a wide range of climatic and field conditions for this experiment. Field elevations ranged from –2 m to 1,220 m; maximum midsummer air temperatures ranged from 22 to 45 °C; minimum midsummer air temperatures were 5 to 25 °C; minimum relative humidity ranged from 10 to 40 percent; and soil types were loam to clay. All sites were in commercial fields except the Tulelake site, which was located at the University of California Intermountain Research and Extension Center. The irrigation method was border (flood) irrigation at the Davis, El Centro, and Kern County sites and was sprinkle irrigation at the Tulelake and Scott Valley sites. The fully-irrigated alfalfa was irrigated according to the irrigator's normal practices. Deficit irrigation treatments at each site consisted of a dedicating a section of the field for which no irrigation occurred during the midsummer.

The field-scale approach was used instead of a randomized replicated experiment to obtain the field-wide conditions experienced by commercial agriculture and because a randomized replicated experimental design was not practical in the commercial fields.

2.2 Measurements

Alfalfa evapotranspiration (ET) was determined using the surface renewal (SR) energy balance method [6] in both fully irrigated and deficit irrigated treatments at all sites. Eddy covariance (EC) energy balance systems (Campbell Scientifc, Inc.) [7] were also installed at the Scott Valley and El Centro fully irrigated sites and used to calculate the ET of the fully irrigated treatments at those sites. Sensible heat data from the EC systems were used to calibrate the SR method. The uncalibrated sensible heat flux determined by the SR system must be corrected for unequal heating between the ground and the measurement height using independent measurements of sensible heat flux such as the EC data.

Alfalfa yield and quality were determined by machine-harvesting a predefined area at all sites except the El Centro site. At this site, hand-harvested plots were used to determine yield in both fully and deficit irrigated treatments. Soil water potential was measured with Watermark® electrical resistance blocks (Irrometer, Inc.) installed in both fully and deficit irrigated treatments.

2.3 Instrumentation

Sensors of the eddy covariance system were a CSAT3 sonic anemometer (Campbell Scientific, Inc.), four Hukseflux self-calibrating soil heat flux plates installed at 8 cm deep, two sets of spatially averaging soil temperature probes installed at 2 and 6 cm deep (Campbell Scientific, Inc.), two CS616 soil water sensors (Campbell Scientific, Inc.) installed 2.5 cm deep, and a Q7.1 net radiometer (Radiation and Energy Balance Systems, Inc.). The sensors were connected to a CR5000 data logger (Campbell Scientific, Inc.).

Sensors of the surface renewal system were a Q7.1 net radiometer, two soil heat flux plates (Radiation and Energy Balance Systems, Inc.), one set of spatially averaging soil temperature probes, and two fine wire thermocouples (0.076 mm diameter) (Campbell Scientific, Inc.) mounted about 1.5 m above the ground surface to measure air temperature. The sensors were connected to either a CR23X, a CR10X, or a CR1000 data logger (Campbell Scientific, Inc.), depending on the site, which recorded high frequency temperature data and low frequency data of other variables. Data archiving and processing to determine the uncalibrated sensible heat flux density is described in Snyder et al. (1996).

Sensible heat is transported in the vertical direction by upward and downward wind velocities, called eddies. The eddy covariance energy balance method used in this study calculates the sensible heat fluxes by measuring both vertical wind velocities and air temperatures using the CSAT3 sonic anemometer. The instantaneous deviations of wind velocity and temperature from the mean are determined, and the covariance of these deviations is used to calculate the sensible heat flux over the desired averaging period. At the same time, measurements of net radiation and soil heat flux are also made. The latent heat flux is calculated as net radiation minus the sum of the sensible heat and soil heat fluxes. The daily cumulative latent heat flux, converted to mm d^{-1}, is the ET.

The surface renewal method is a less expensive method for measuring sensible heat fluxes. The theory is that high frequency temperature data above and within a plant canopy has a ramp-like structure over time that occurs when an air parcel sweeps from above to the surface. Energy transfer between the air and canopy causes heating or cooling of the air while at the surface. The air parcel is then ejected from the surface and replaced by a new air parcel sweeping down from above. The sensible heat flux is directly related to the temperature change rate with time. Information of the surface renewal method is available on the website biomet.ucdavis.edu/SR.zip.

3 Results and discussion

3.1 El Centro

Daily ET of the fully-irrigated treatment (determined from the EC system) increased from about 1 mm d^{-1} at the beginning of 2007 to maximum values of 10 to 12 mm d^{-1} in June, and then decreased to about 1 mm d^{-1} at the end of the year (fig. 1A). Just after a harvest, ET decreased to values of 1 to 2 mm d^{-1}. Seasonal ET was 1,379 mm.

Deficit irrigation started at the end of June. A trend of decreasing ET over time occurred for the deficit treatment with ET values smaller than those of the fully-irrigated treatment. After DOY286, ET of both treatments was similar. ET of the fully irrigated and deficit irrigate treatments between DOY167 and 271 (considered to be the period of deficit irrigation) were 514 and 452 mm, respectively. The difference was 62 mm.

During the midsummer growth periods, daily cumulative net radiation values, expressed as an equivalent evaporation rate in mm d^{-1}, were smaller than the daily ET. At those times, the sensible heat flux was towards the ground surface, indicating an advective heat source contribution to the evapotranspiration which caused the daily ET to exceed the daily cumulative net radiation. The sensible heat flux direction was away from the ground surface just after a harvest due to heating of the ground surface as a result of the reduced canopy coverage.

Three harvests (July 24, September 24, and December 12) occurred after the start of deficit irrigation. Yields of the fully-irrigated alfalfa, based on hay samples, were 2.78, 1.39, and 1.81 Mg ha^{-1} for the respective harvest dates, while the deficit-irrigated respective yields were 2.17, 0, and 1.66 Mg ha^{-1}.

3.2 Kern County

ET values ranged from about 1 mm d^{-1} at the start of the measurement period to generally between 6 to 8 mm d^{-1} during the midsummer (data not shown). However, values between 8 and 10 mm d^{-1} occurred in May and June. Cumulative ET as of October 12 was 1,346 mm.

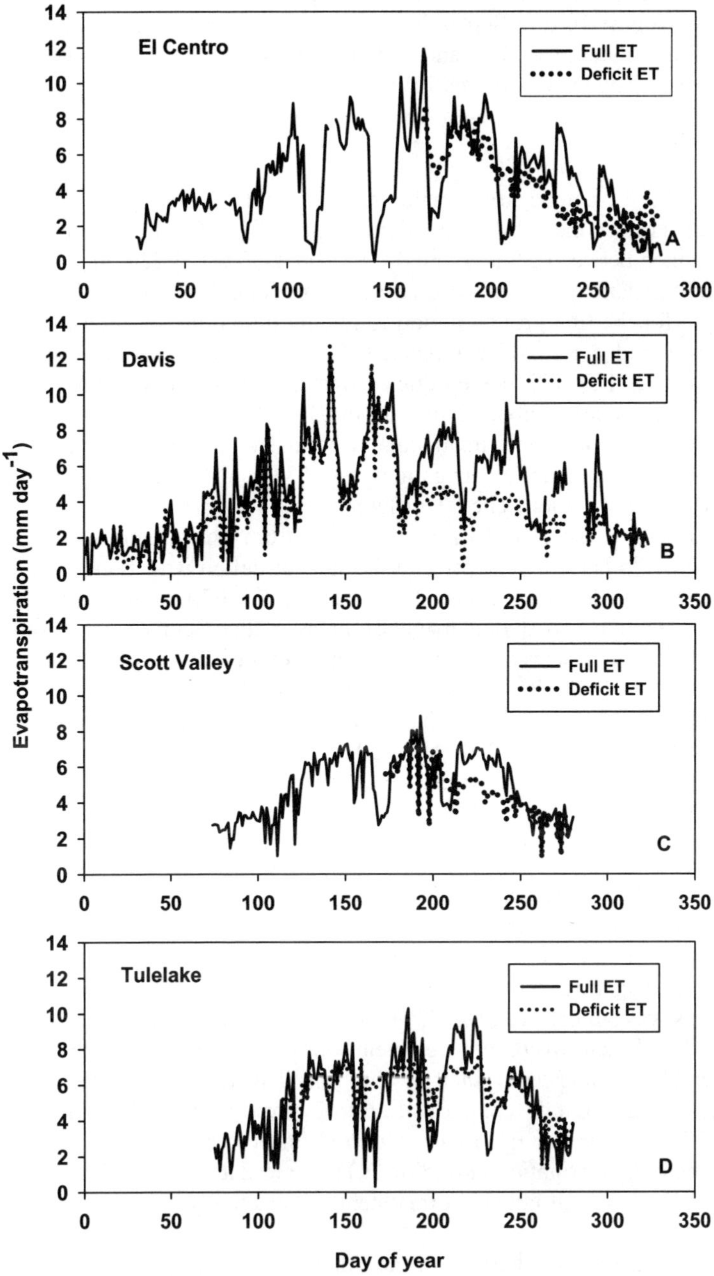

Figure 1: Evapotranspiration of fully-irrigated and deficit-irrigated alfalfa at four sites in California.

Deficit irrigation, which occurred in August, had a slight effect on daily ET. Between August 10 and September 14, cumulative ET of the fully and deficit irrigated treatments was 191 and 155 mm, respectively, and the ET difference was 36 mm. The yield loss was 1.90 Mg ha^{-1}.

3.3 Davis

Daily alfalfa ET ranged from smaller than 2 mm d^{-1} for the first 50 days of 2007 to values greater than 8 mm d^{-1} during June (fig. 1B). Just after harvest, ET levels ranged between 2 and 4 mm d^{-1}. Seasonal ET was 1,520 mm.

Deficit irrigation was imposed at the end of June. Maximum ET of the deficit treatment for the first growth period following initiation of deficit irrigation was between 4 and 5 mm d^{-1} compared to 7 to 8 mm d^{-1} for the full irrigation treatment (fig. 1B). ET of the deficit irrigation treatment continued to decrease over time to values of about 3 mm d^{-1} by DOY270. After DOY300, similar ET values occurred for both treatments. Cumulative ET between DOY182 and DOY300 (period of deficit irrigation) was 534 mm for full irrigation and 335 mm for deficit irrigation. The ET difference was 198 mm.

Yield per harvest of full irrigation varied from a maximum of 4.64 Mg ha^{-1} for the June 26 harvest (just prior to the start of deficit irrigation) to a minimum of 2.68 Mg ha^{-1} for the October 26 harvest (last 2007 harvest). Yields between the full irrigation and deficit irrigation areas of the field were not statistically different for each harvest prior to the start of deficit irrigation at a level of significance of 0.05 ($P = 0.279$ to 0.598). Yields during the period of deficit irrigation were 0.45 to 0.91 Mg ha^{-1}, statistically different from those of the full irrigation treatment ($P = 0.0001$ to 0.0008).

3.4 Scott valley

ET values at the start of the measurement period were between 2 and 3 mm d^{-1}, but increased to between 3 and 4 mm d^{-1} by DOY89 (fig. 1C), and reached maximum values of nearly 8 mm d^{-1} on DOY188 to DOY181. ET just after harvest ranged between 3 and 4 mm d^{-1}. The seasonal ET was 990 mm.

Deficit irrigation started after the first harvest (DOY174). ET of both treatments was similar until DOY203, at which time the second harvest occurred. ET of the deficit irrigation treatment during the third growth period was considerably smaller than that of the full irrigation treatment, with values between 4 and 5 mm d^{-1} compared to 6 to 7 mm d^{-1} for the full irrigation. After about DOY246, similar ET values occurred for both treatments. Cumulative ET during the deficit irrigation period of DOY210 to 246 was 227 mm and 170 mm for the deficit and full irrigation treatments, respectively. The ET difference was 57 mm.

Yields of the second harvest of the full and deficit irrigation treatments were 3.90 and 3.05 Mg ha^{-1}, respectively, not statistically different at a level of significance = 0.05 ($P = 0.119$). Yields of the third harvest were 2.15 and 0.65 Mg ha^{-1} for the respective treatments, which were statistically different ($P = 0.0207$).

3.5 Tulelake

ET values at the start of the measurement period were between 2 and 3 mm d^{-1} and increased over time to values between 8 and 10 mm d^{-1} during the second and third growth periods, after which ET decreased to values between 1 and 4 mm d^{-1} at the end of the measurement period (fig. 1D). Seasonal ET was 1044 mm.

Deficit irrigation started at the beginning of the second growth period. Little difference in cumulative ET was found between the fully and deficit irrigated treatments during the deficit irrigation period (DOY176-280), with values of 594 mm and 589 mm for the respective treatments. The ET difference was 5 mm.

Similar yields occurred for both treatments (data not shown). Yield differences for all harvests were not significant at a level of significance = 0.05 (P = 0.094 to 0.530).

At this site, little treatment differences were found in both ET and yield. The contributing factor for this behavior is shallow ground water with a water table depth of about 1 m, based on historical experience. This, coupled with the fine texture soil, resulted in sufficient upward flow of ground water to satisfy the ET requirements of the deficit irrigated treatment.

4 Conclusions

Deficit irrigation of alfalfa during the midsummer reduced both ET and yield. The amount of reduction was very site-specific. The Davis site showed the largest reduction in ET (198 mm), while the other sites showed much smaller reductions (5 to 62 mm). The Davis reduction was similar to that of the previous two years of experiments. It was expected that larger ET differences would have occurred at the Kern County and El Centro sites than were measured this year. A reason for the Kern County response may have been the soil type, which has a very high soil moisture retention capacity. This capacity appears to have been able to maintain high ET rates during periods of deficit irrigation. For the El Centro site, the effect of heat stress on alfalfa may have played a role in the ET differences between two treatments.

The seasonal measured ET values at the Davis and Kern County sites (as of October 12) were similar to the historical seasonal ET of 1,219 to 1,244 mm. The seasonal measured ET values at the Scott Valley and Tulelake sites were greater than the historical seasonal ET of 838 mm. For the El Centro site, the seasonal ET of 1,379 mm was considerably smaller than the historical ET of 1,930 mm.

We conclude that it is neither feasible nor practical to base the amount of water available for transfer elsewhere on ET differences of alfalfa. There was no consistency in these differences between sites, reflecting the site specific characteristics. Also, basing the water transfer amount on ET differences penalizes the growers since the alfalfa growth during the period of deficit irrigation was due to soil moisture supplied by previous irrigations, for which growers were charged by the irrigation or water district. The exception was the Tulelake site, where shallow ground water contributed to the ET of the deficit irrigated treatment.

References

[1] Frate, C.A., Roberts, B.A., Marble, V.L., 1991. Imposed drought stress has no long-term effect on established alfalfa. California Agriculture 45(3), 33–36.

[2] Guitjens, J.C., 1993. Alfalfa irrigation during drought. Journal of Irrigation and Drainage Engineering 119(6), 1092–1098.

[3] Ottman, M.J., Tickes, B.R., Roth, R.L., 1996. Alfalfa yield and stand response to irrigation termination in an arid environment. Agronomy Journal 88, 44–48.

[4] Putnam, D., Takele, E., Kallenback, R., Graves, W., 2000. Irrigating alfalfa in the Low Desert: Can Summer Dry-down be Effective for Saving Water in Alfalfa? Report submitted to the Bureau of Reclamation (USDI), Yuma, Arizona.

[5] Robinson, F.E., Teuber, L.R., Gibbs, L.K., 1994. Alfalfa Water Stress Management During Summer in Imperial Valley for Water Conservation. Desert Research and Extension Center, El Centro, CA.

[6] Snyder, R.L., Spano, D., Paw U, K.T., 1996. Surface Renewal analysis for sensible and latent heat flux density. Boundary-Layer Meteorol. 77, 249–266.

[7] Tanner, B.D., Tanner, M.S., Dugas, W.A., Campbell, E.C., Bland, B.L., 1985. Evaluation of an operational eddy correlation system for evapotranspiration measurements. American Society of Agricultural Engineers National Conference on Advances in Evapotranspiration, December 16–17, 1985. Chicago, IL.

Yield and canopy response of chickpea (*Cicer arietinum* L.) to different irrigation regimes

M. Rinaldi[1], A. V. Vonella[1], P. Soldo[2], G. Debiase[1] & P. Garofalo[1]
[1]*C.R.A. – Unità di ricerca per i sistemi colturali degli ambienti caldo-aridi, Italy*
[2]*Consorzio per la Bonifica della Capitanata, Italy*

Abstract

In recent years, the interest for legume crops has been increasing in the European Community, both for the agronomic improvement of soil fertility and for human and animal food source reasons. Chickpea (*Cicer arietinum* L.) in the Mediterranean environment is not usually irrigated. Knowledge about its canopy growth, water and radiation use efficiencies can improve chickpea productivity. In this study, the chickpea has been submitted to different irrigation scheduling at specific crop phases and as a function of soil moisture. Total plant biomass was related to water availability and radiation interception. But in good conditions of available water there was a lengthening of the crop cycle, with reduction of pod growth, harvest index and nutrient toward the seeds. Consequently, the best values of water use efficiency were found in the treatment irrigated with 50 mm only at flowering or at pod filling. Protein yield was higher in the treatment refilling field capacity when soil plant available water was 25%. A canopy extinction coefficient of 0.84 and a radiation use efficiency of 1.02 g MJ^{-1}, on average, were found. In conclusion, the irrigation of the chickpea at sensitive phases (flowering and pod filling) and with a low amount of water, resulted in the best strategies.

Keywords: chickpea, water use efficiency, radiation use efficiency, water deficit.

1 Introduction

The chickpea is one of the major legume crops grown in Mediterranean regions, where rainfall is highly variable and often insufficient. As the season progresses, the crop is exposed to increasing moisture deficit and heat. This results in low

www.witpress.com, ISSN 1743-3541 (on-line)
doi:10.2495/SI080131

and variable yields and discourages farmers from investing in inputs for the crop's production. Limited supplemental irrigation can, however, plays a major role in boosting and stabilizing the productivity of winter-sown chickpea (Zhang et al. [14]). Chickpea has a strong indeterminate growth habit and when growing conditions are favourable the plant continues vegetative growth without setting pods or filling few pods (Davies et al. [2]; Liu et al. [5]). The detrimental effects of drought can be modified to some extent through management options such as irrigation (Soltani et al. [12]). However, in the literature there are differing views on the effect of irrigation timing coinciding with moisture-sensitive periods in chickpea. Some authors (Jadhav et al. [4]) suggest that chickpea is more sensitive to drought during flowering. However, others (Ravi et al. [8]; Reddy and Ahlawat [9]) suggested seed filling as the critical time for irrigation. In contrast, Ramakrishna and Reddy [9] demonstrated a seed yield reduction of more than 50% in chickpea when they were irrigated due to excess vegetative growth, which leads to lodging. The present study aimed to investigate the effect of supplemental irrigation levels on the phenology, plant growth, yield, seed quality and water and radiation use efficiencies of winter-sown chickpea in Southern Italy.

2 Material and methods

2.1 Experimental site

The field experiment was carried out in 2006–2007 at Foggia (lat. 41° 8' 7'' N; long. 15° 83' 5'' E, alt. 90 m a.s.l.) in Southern Italy. The soil is a vertisol of alluvial origin (Typic Chromoxerert, fine, termic, according to the Soil Taxonomy-USDA), silty-clay with the following characteristics: organic matter, 2.1%; total N, 0.122%; $NaHCO_3$-extractable P, 41 ppm; NH_4O Ac-extractable K_2O, 1598 ppm; pH (water) 8.3; field capacity water content 0.396 m^3 m^{-3}; permanent wilting point water content 0.195 m^3 m^{-3}, available soil water 202 mm m^{-1}. The climate is ''accentuated thermomediterranean'' (Unesco-FAO classification), with temperatures below 0 °C in the winter and above 40 °C in the summer. Annual rainfall (mean 550 mm) is mostly concentrated during the winter months and class ''A pan'' evaporation exceeds 10 mm day^{-1} in summer. Daily meteorological data – temperatures, humidity, rainfall, wind velocity and solar radiation – were collected in the local meteorological station.

2.2 Field experiment

Chickpea (*Cicer arietinum* L., cv Pascià, desi type), was sown on 4th December 2006 and different irrigation scheduling were compared:

- A: one irrigation (50 mm) at flowering;
- B: one irrigation (50 mm) at pod filling;
- C: irrigation of 40 mm of water, every time that soil moisture reached the threshold of 25% of plant available water (PAW) measured with TDR probes at 0-60 cm depths;

- D: idem at 50%;
- E: idem at 75%;
- Rainfed: a not irrigated control.

To ensure uniform water distribution, a drip irrigation system was used, with one line for each plant row and drippers of 4 L h^{-1} flow. A pre-sowing with 60 kg ha^{-1} of P_2O_5 as triple perphosphate was applied. A randomised block design with four replications was used; a sowing density of 40 seed m^{-2} was adopted, with a distance between rows of 0.5 m. Harvest was performed with plot machine harvester, on 4th July 2007, when the seed moisture content was lower than 13%.

2.3 Measurements

The main crop phenological phases were recorded and expressed as degree days (GDD), considering a base temperature of +2 °C. Soil moisture was measured with TDR probes (25 cm length), placed in the soil at 30, 60, and 90 cm depth, at 1-hour time and daily averaged. Gravimetric soil water measurements were also carried out at 20, 40 and 60 cm depth at sowing, at harvest and at growth analysis sampling dates.

Growth analysis was carried out from March to June; at seven sampling dates, dry matter, separated into stems, green and dead leaves and pods was measured by taking 0.5 linear meter sample from every plot and dried at 80 °C until weight was constant. Leaf Area Index was determined measuring green leaves area with Delta T Devices (Decagon Devices Inc., WA, USA). At harvest, the total plant dry matter, and the seed yield were determined. Seed nitrogen content was determined using the elementary analyzer Fison CHN (EA 1108). Seasonal water use (*WU*) was estimated according to the following water balance equation:

$$WU = \pm \Delta SWC + R + I \tag{1}$$

where ΔSWC is the variation, between seeding and harvest date, of the volumetric soil water content in the 0-0.6 m depth layer, R is the rainfall and I the irrigations, all expressed in mm. Despite the fact that chickpea roots can reach deep layers (> 0.6 m), the presence of a compact calcareous layer reduces the root depth and allowed us to limit at 0.6 m the depth of soil samples. Water use efficiency (*WUE*), was determined as the ratio of grain or biomass yield, to seasonal water use. Irrigation water use efficiency (*IWUE*) was evaluated as being the ratio of the difference in crop yield between irrigated and rainfed plots to the difference in *WU* for the same treatments:

$$IWUE = \frac{Yi - Yr}{WUi - WUr} \tag{2}$$

where Y is the seed or total aboveground biomass yield (g m^{-2}), WU is the seasonal water use (mm), and the subscripts r and i refer to rainfed and irrigated treatments, respectively. Water and irrigation water use efficiency are expressed in kg per cubic meter and subscript *TDM* and yield indicated the total dry matter and seed respectively. Radiation use efficiency (*RUE* g MJ^{-1} of *iPAR*) was calculated as slope of regression line between cumulative intercepted photosynthetic active radiation (*iPAR*) and total dry biomass for each sampling (Charles-Edwards [1]). The *iPAR* was estimated using the following equation:

$$iPAR = PAR\, e^{-k\, LAI} \tag{3}$$

where *PAR* is equal to global radiation multiplied by 0.48 (MJ m^{-2} day^{-1}), *k* is the light extinction coefficient, calculated as slope of fitted regression between the natural logarithm of transmitted *PAR* and *LAI*, both measured with LI-COR 2000 portable area meter. Global radiation was measured daily with a thermophile pyranometer (305–2800 nm wave-length range). Analysis of variance of the data was carried out using a "randomized block" design model, and Least Significant Difference was used to compare mean values.

3 Results and discussion

3.1 Growth and yields

Total dry plant biomass was similar among the irrigation treatments, except in the last two samplings (end of May – beginning of June, Fig. 1a) when higher values were observed in the most irrigated treatments. This is due to an extension of the crop cycle, confirmed by a delayed in flowering and pod filling in D and E treatments (Tab. 1). Table 1 and figure 1b show as the irrigation before the flowering (A treatment) delayed the pod appearance, with a reduction of the useful time to mobilize nutrients from leaves and stems towards the seeds. The leaf area index (*LAI*) was positively influenced by water availability (Fig. 1c). The irrigation at pod filling (B treatment) favoured a leaf emission and vitality, as shown by an increased *LAI* after irrigation. Specific leaf area followed the same development of *TDM* and *LAI* (fig. 1d) with increased values with water supplies.

Chickpeas irrigated according soil moisture threshold (C, D and E) gave similar values of total plant dry biomass at harvest (Tab. 2), higher than A, B and rainfed treatments; this result was similar compared with results reported from Oweis et al. (2004) in a Mediterranean environment. On the contrary, the seed yield resulted greater in the plants irrigated at pod filling and at 25% of available water, with about 234 g m^{-2} of seeds, while the rainfed and the other irrigation

Table 1: Growth degree days (*GDD* in °C) for chickpea development (T base = +2 °C). PAW = plant available water.

Treatment	Emergence	Begin flowering	Pod filling	Physiological maturity
A = flowering	101	1245	1449	2329
B = pod filling	101	1245	1449	2329
C = 25% PAW	101	1284	1487	2329
D = 50% PAW	101	1284	1634	2412
E = 75 % PAW	101	1327	1681	2490
Rainfed	101	1245	1449	2249

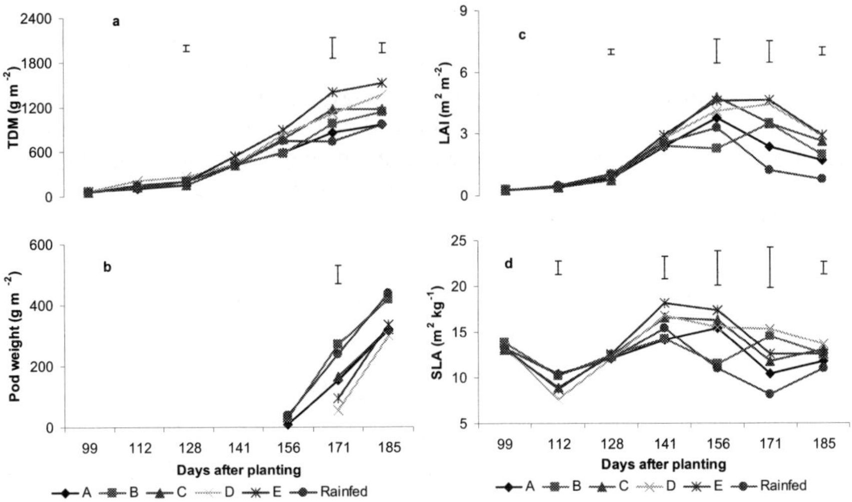

Figure 1: Growth variables for chickpea: (a) total dry matter (*TDM*), (b) pod weight, (c) Leaf Area Index (*LAI*) and (d) Specific Leaf Area (*SLA*). Bars indicate LSD at $P \geq 0.05$.

Table 2: Main yield results (different letters in each column, indicate values significantly different at $P \geq 0.05$, LSD test) of chickpea experiment.

Treatment	TDM *(g m^{-2})*	Seed yield$_{10\%}$ *(g m^{-2})*	Harvest Index *(%)*	Unit seed weight *(mg)*	Protein content *(%)*	Protein yield *(kg ha^{-1})*
A = flowering	678 C	201.4 B	0.27 A	516.0 A	22.0	444.9 ABC
B = pod filling	738 C	235.7 A	0.30 A	518.3 A	20.1	473.5 AB
C = 25% PAW	1089 B	230.9 A	0.20 B	511.1 A	21.5	495.8 A
D = 50% PAW	1474 A	204.7 B	0.13 C	464.7 C	21.2	434.0 BCD
E = 75 % PAW	1439 A	199.4 B	0.13 C	454.8C	20.3	402.7 CD
Rainfed	668 C	182.1 B	0.25 AB	486.4 B	21.6	392.7D

scheduling gave similar seed yield. The seed growth and final size were influenced by the shortening of time from ripening to maturity, with negative effects on seed unit weight, with a value of 459 mg on average for D and E the treatment, in comparison to 515 mg found in the treatments A, B and C (Tab. 2).

Also commercial seed size, influencing the final price, resulted higher in the A, B and rainfed treatments, smaller in E treatment. Seed protein content resulted not different among treatments, with an average of 21.1%, but the protein per hectare yield was bigger in C than D, E and rainfed (Tab. 2).

3.2 Evapotranspiration and water use efficiency

Under supplemental irrigation, seasonal water use increased with the amount of applied water, ranging from 354 mm in rainfed treatment, to 901 mm in E treatment. In treatments based on soil plant available water, the number of irrigation ranged from 3 (C) to 10 (E) (Tab. 3). The large number of irrigation supplies for E treatments is due to lower rainfall supply and higher *ET0* in this year than long-term values in the period ranging from 50 to 160 days after sowing (Fig. 2).

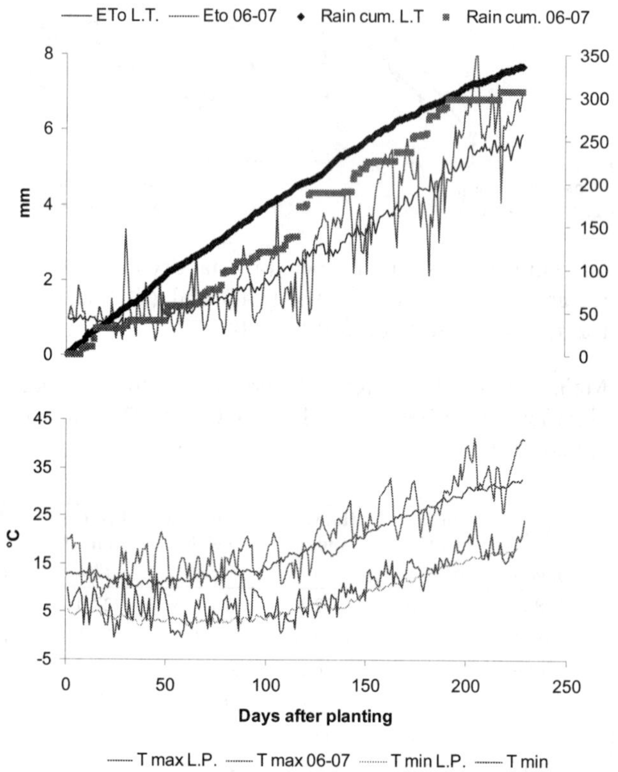

Figure 2: Comparison among some recorded climatic variables during the crop cycle compared with long term (50 years) averages.

Figure 3 shows volumetric soil content at 0-0.3 m soil depth; in A and C treatments the soil moisture fell close to the wilting point, while in D treatment it was higher and more regular. In the final phase of chickpea growth the elevated evaporative demand quickly leaded the soil water content at values lower than the wilting point, also in the well irrigated treatment.

Table 3, shows, except for the treatment D, the values of WUE_{Tdm} were similar, with values ranging between 1.6 for E and 1.89 kg m^{-3} for B treatment, close to those reported by Oweis et al. [6] and Siddique et al. [10]. The irrigation at pod filling resulted the most efficient irrigation scheduling, with 0.55 kg of grain per cubic meter of water, followed by rainfed and A treatments with 0.50 and 0.46 kg m^{-3}, respectively. The lowest values were observed in the treatment with low harvest index. Irrigation at pod filling allowed obtaining best results in term of WUE_{Tdm} and WUE_{yield} similar to those reported by Oweis et al. [6]. The efficiency of irrigation water was on average low for the good performance of rainfed chickpea; it was statistically higher in the B treatment, both for *TDM* and yield, than the other treatments. Globally, the irrigation at a specific phase resulted more efficient than soil moisture based scheduling.

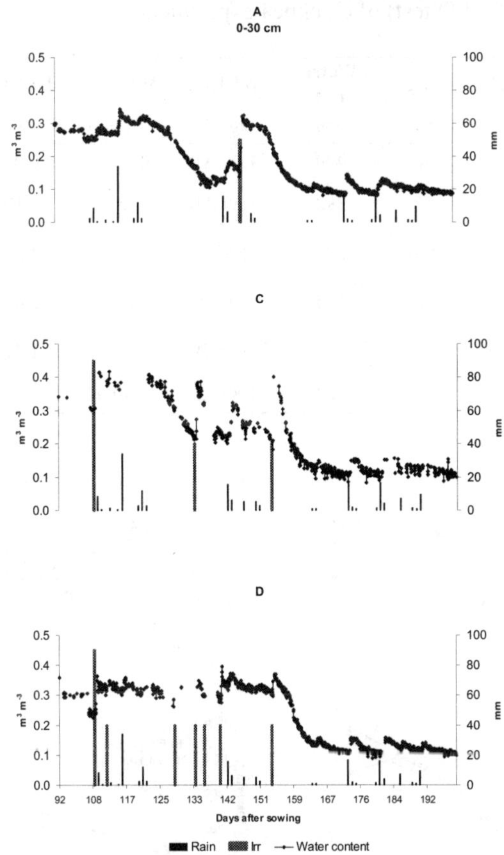

Figure 3: Volumetric soil content at 0 - 0.3 m (rhombus) measured with TDR probe, rain (thick column) and irrigation (fine column), for A (irrigation at flowering, at the top), C (irrigation at 25% of PAW; middle) and D (irrigation at 50% of PAW; below) treatments.

3.3 PAR interception and radiation use efficiency

The crop cycle duration and canopy growth, increased for the irrigation supply, influenced the leaf area index dynamic and, consequently, the fraction of intercepted *PAR*. In fact, as shown by figures 4a and 4b, intercepted *PAR* declined at the last sampling only in A and rainfed treatments. The fitted regression between the natural logarithm of transmitted *PAR* and *LAI* for all treatments, forced through the origin (*PAR* above and under canopy are equal

Table 3: Water balance components and water use efficiency (different letters in each column, indicate values significantly different at P ≥ 0.05 LSD test) of chickpea experiment.

Treatment	Irrig.	Irrig.	Water Use	WUE_{Tdm}	WUE_{yield}	$IWUE_{Tdm}$	$IWUE_{yield}$
	n.	*mm*	*mm*	*(kg m^{-3})*	*(kg m^{-3})*	*(kg m^{-3})*	*(kg m^{-3})*
A = flowering	1	50	400	1.69 AB	0.46 C	0.01 C	0.40 B
B = pod filling	1	50	392	1.89 AB	0.55 A	2.10 A	1.42 A
C = 25% PAW	3	170	520	1.55 B	0.30 D	1.21 B	0.14 BC
D = 50% PAW	7	330	685	2.15 A	0.27 D	2.44 A	0.07 C
E = 75 % PAW	10	550	901	1.60 B	0.20 E	1.41 B	0.03 C
Rainfed	0	0	354	1.89 AB	0.50 B	-	-

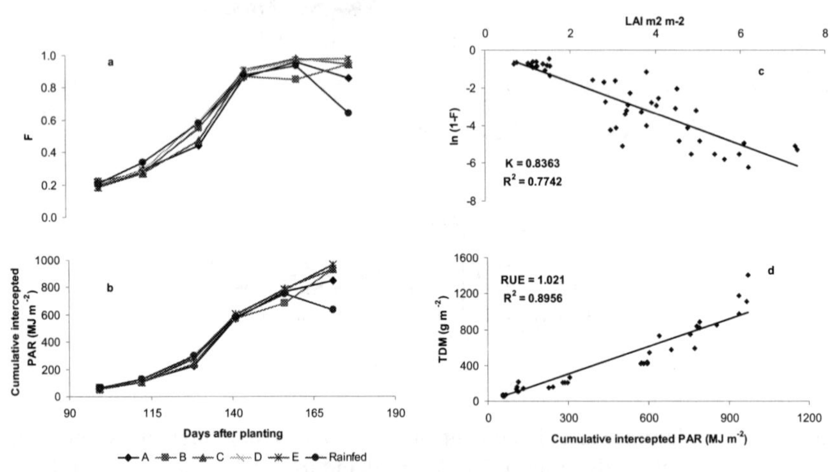

Figure 4: Fraction of intercepted *PAR* (a), seasonal cumulative intercepted *PAR* (b), canopy extinction coefficient (c) and radiation use efficiency (g MJ^{-1} of *iPAR*) (d).

when *LAI* = 0), allowed us to obtain canopy extinction coefficient (*k*) as shown in figure 4c. This coefficient equal to 0.84 ($R^2 = 0.77$) is similar to that obtained by Tesfaye et al. [13] in a semi-arid environment, for chickpea submitted to an "irrigation-late stress" treatment. Average chickpea radiation use efficiency (Fig. 4d) resulted equal to 1.02 g MJ^{-1} ($R^2 = 0.90$), a low value if compared with C4 crops, but in the range 0.30 – 1.68 g MJ^{-1} of *PAR*, reported by Hughes et al. [3], Singh and Sri Rama [11] and Tesfaye et al. [13].

4 Conclusions

In the farming systems of Mediterranean region, winter-sown chickpea, allows exploiting winter rainfall, saving on water supplying, but maintaining good productive results, as shown by the performances in terms of yield and water use efficiency on rainfed chickpea. However, in experimental season the rainfall was well distributed during flowering and this certainly favoured the yield of rainfed treatment. The results show as large irrigation supply increased seasonal evapotranspiration, but without benefit for seed yield. In fact, soil moisture was similar in the compared treatments (A, C and D); this means that a large developed crop canopy, consumes more water, but with a scarce efficiency in seed yield. The crop plasticity, important for the best adaptation to water deficit conditions, influences also the foliage orientation; in fact, at beginnings of May, when irrigation was suspended, the leaf angling of rainfed treatment resulted superior of 30% (not shown: measured with LI-COR 2000 area meter) in comparison with E treatment. This means a more leaf vertical orientation in the rainfed treatment, in response to water deficit, allowing a reduction of leaf exposition to solar radiation and consequently of transpiration; this explains the results in term of WUE_{Tdm} and WUE_{yield} in rainfed treatment.

In conclusion, despite the shortness of the experiment, it has been evident as irrigation of chickpea at specific phases of crop growth (flowering – pod filling) and with small amount of water, could be an agronomical practice to obtain good results in terms of yield, seed quality and water use efficiency.

References

[1] Charles-Edwards, D.A., 1982. Physiological determinants of crop growth. Academic Press, Sydney, Australia, 161 pp.

[2] Davies, S., Turner, N., Siddique, K., Plummer, J. and Leport, L., 1999. Seed growth of desi and kabuli chickpea (*Cicer arietinum* L.) in a short-season Mediterranean-type environment. Aus. J. Exp. Agric. **39**: 181–188.

[3] Hughes, G., Keatinge, J., Scott, A., Cooper, J. and Dee, N., 1987. Solar radiation interception and utilization by chickpea (*Cicer arietinum* L.) crops in northern Syria. J. Agric. Sci. Camb. **108**: 419–424.

[4] Jadhav, J., Shewalw, M., Mokashi, D., Gaikwad, C. and Patil, J., 1997. Crop coefficients and actual evapo-transpiration of gram (*Cicer arietinum*) grown in lysimeter. Indian J. of Agricultural Sciences, **67**: 33–35.

[5] Liu, P., Gan, Y., Warkentin, T. and McDonald, C., 2003. Morphological plasticity of chickpea in a semiarid environment. Crop Sci., **43**: 426–429.

[6] Oweis, T., Hachum, A. and Pala, M., 2004. Water use efficiency of winter-sown chickpea under supplemental irrigation in a Mediterranean environment. Agricultural Water Management, **66**: 163–179.

[7] Ramakrishna, A. and Reddy, S.L.N., 1993. Production potential and economic feasibility of different crop sequences in rice (*Oryza sativa*) fallows. Indian Journal of Agricultural Sciences, **63**: 611–615.

[8] Ravi, N., Sharma, H.M., Singh, R.N.P. and Nandan, R., 1998. Response of late-sown chickpea to irrigation and foliar nutrition in calcareous soil. Journal of Applied Biology, **8**: 5–8.

[9] Reddy, N.R.N. and Ahlawat, I.P.S., 1998. Response of chickpea (*Cicer arietinum*) genotypes to irrigation and fertilizers under late-sown conditions. Indian Journal of Agronomy, **43**: 95–101.

[10] Siddique, K., Regan, K., Tennant, D. and Thomson, B., 2001. Water use and water use efficiency of cool season grain legumes in low rainfall Mediterranean-type environments. Eur. J. of Agronomy, **15**: 67–280.

[11] Singh, P. and Sri Rama, Y.V., 1989. Influence of water deficit on transpiration and radiation use efficiency of chickpea (*Cicer arietinum* L.). Agric. For. Meteorology, **48**: 317–330.

[12] Soltani, A., Ghassemi, G.K., Khooie, F.R. and Moghaddam, M., 1999. A simple model for chickpea growth and yield. Field Crops Research, **62**, 213–224.

[13] Tesfaye, K., Walker, S. and Tsubo M., 2006. Radiation interception and radiation use efficiency of three grain legumes under water deficit conditions in a semi-arid environment. Europ. J. Agronomy, **25**: 60–70.

[14] Zhang, H., Pala, M., Oweis, T. and Harris, H., 2000. Water use and water use efficiency of chickpea and lentil in a Mediterranean environment. Aust. J. Agric. Res. **51**: 259–304.

Using individual farm management plans to manage land use change effects associated with new irrigation development, Canterbury, New Zealand

B. Ellwood[1], L. Mead[1], C. Mulcock[2] & S Dawson[3]
[1]*Meridian Energy Ltd, PO Box 2454, Christchurch, New Zealand*
[2]*Mulgor Consulting Ltd, PO Box 13240, Christchurch, New Zealand*
[3]*Boffa Miskell Ltd, PO Box 110, Christchurch, New Zealand*

Abstract

Managing the environmental effects arising from intensive agriculture has become a key issue in obtaining permission to abstract water for irrigation in New Zealand. Hunter Downs Irrigation (HDI) is a proposed irrigation scheme that would divert approximately 20 m^3/s of flow from the Waitaki River to 40,000 hectares of lands in Canterbury, New Zealand. In addition to carrying out extensive assessments of potential environmental effects of the proposed scheme, it also needed to demonstrate how it would be able to implement and maintain ongoing environmental management of irrigated agriculture.

Many of the effects arising from the new development are not readily quantifiable in advance, as they depend on the particular mix of land uses that will be established once the scheme is commissioned. These may also change over time as they will depend not only on the biophysical environment (soils, topography, climate etc.) but also on market drivers and other economic signals.

As the area supplied by this scheme is suitable for a range of land uses, the challenge was to provide sufficient certainty to regulatory agencies and to the public that adverse environmental effects generated by the irrigation scheme will be minimised and that poor practices can either be avoided or can be identified and remedied. At the same time farmers needed to have sufficient flexibility to develop a wide range of agricultural enterprises. To achieve this an environmental management system was developed that requires each farmer to develop and use an environmental farm management plan that is linked to a rigorous audit and compliance regime, and continuous improvement process, so that farmers maintain up to date practices and by doing this it is assumed that the environmental effects are the lowest possible for an irrigated agricultural system.

Keywords: Environmental Management System (EMS), irrigation management.

WIT Transactions on Ecology and the Environment, Vol 112, © 2008 WIT Press
www.witpress.com, ISSN 1743-3541 (on-line)
doi:10.2495/SI080141

1 Background

1.1 Irrigation in New Zealand

New Zealand has approximately 475,000 ha of irrigated agriculture and this is predicted to almost double by the year 2013 [1] under one scenario modelled. Most of the existing irrigated area (60%) is found on the east coast of the South Island of New Zealand in the Canterbury region.

Hunter Downs Irrigation scheme is a proposed pumped irrigation scheme which would have the capacity to irrigate 40,000 ha of land in south Canterbury (Figure 1). The scheme would take approximately 20 m^3/s of surface water from the Waitaki River (mean flow 368 m^3/s) immediately to the south.

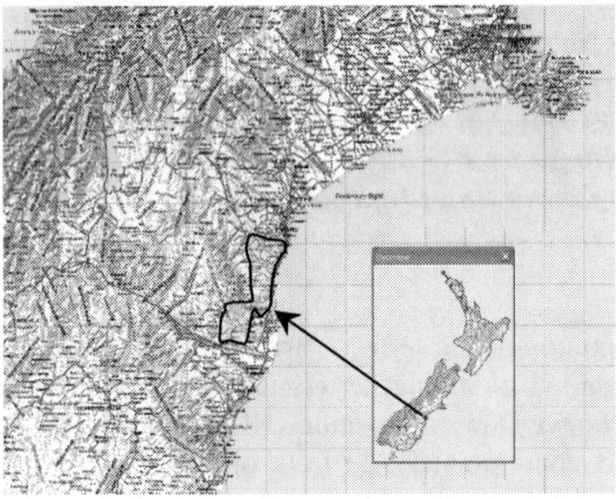

Figure 1: Hunter Downs irrigation Scheme area schematic [6].

The scheme requires authorisation under the Resource Management Act (RMA) 1991. To obtain consents the developers are required to describe their proposal and show both to authorities and the public how adverse environmental effects can be avoided, remedied or mitigated.

1.2 Environment effects management approach

To determine the effects of the proposed taking of water and using it for irrigation, the HDI scheme promoters commissioned extensive assessments of the existing environment and used models to predict the magnitude of the changes that could occur with development of the scheme, looking at the effects of both the abstraction from the Waitaki River and the use of the water for irrigated agriculture.

The models for predicting the effects of using the water for irrigated agriculture made assumptions about the land use types that would arise in this region with irrigation (Table 1). However, there is no certainty that this is the land use mix that would actually occur as the soils and climate are suitable for a

Table 1: Current and modelled land use in the HDI command area.

Land Use	Proportion of Total Area Dry Land	Proportion of Total Area Irrigated
Dairy	6%	46%
Arable	14%	19%
Sheep	58%	18%
Beef	5%	7%
Deer	12%	4%
Dairy Support	2%	7%

range of farm types and the relative profitability of different enterprises will change over time.

To be able to demonstrate that the scheme can mange the environmental effects of any land use mix that might eventuate an Environmental Management System (EMS) approach was adopted.

The proposed EMS system took the industry environmental standards that are widely agreed as best practice standards in NZ today, and incorporated the principles of continuous improvement to enable standards, and hence on-farm practices, to be revised as understanding increases over time. This enables adverse effects on the downstream receiving environment of the scheme to be minimised while maintaining flexibility of land use to optimise the benefits and opportunities.

The approach adopted with the HDI scheme takes the direction promoted by national policy discussion documents in New Zealand including the Parliamentary Commissioner for the Environment's 'Growing for Good' report [2] and "Freshwater for the Future: a supporting document for the Government's Sustainable Water Programme of Action" [3]. These reports both look towards the need to achieve irrigation performance in NZ that is clearly sustainable. "Growing for Good" signals clearly that if New Zealand wishes to remain competitive in the food and fibre industries and become more environmentally sustainable then it is not 'business as usual' for our land-based enterprises. The report focuses on the importance of two key inputs to farming productivity – nitrogen fertiliser and irrigation water. It notes that although these are fundamental to agricultural production, they also have major potential to result in adverse impacts. Therefore NZ must find innovative new directions for managing these vital inputs.

"Freshwater for the Future" points, among other matters, to the need for more efficient water use and catchment management, including more accurate information on water takes and use, and improved management of undesirable effects of land use on water quality. It identifies the need to ensure that tools and knowledge developed from science are integrated into individual rural business decisions.

2 EMS in New Zealand and Australian agriculture

Generally the EMS approaches that have been developed for agriculture in NZ and Australia have been designed as 'process' standards that can be readily integrated with other on-farm management processes and records that landholders may already maintain, such as financial accounts, food safety, occupational health and safety, and quality assurance.

In Australia use of EMS in agriculture, both on-farm and in agricultural industries is promoted and supported at government level [4] whereas in NZ use of EMS has been primarily market driven and developed by the sector concerned. In both countries the use of an EMS process is voluntary.

A number of the farm environment related codes of practice and guidelines developed in New Zealand take the EMS 'plan, act, review, revise' approach. For example: the "Code of Practice for Nutrient Management (with emphasis on fertiliser use)" [5]; the kiwifruit industry "KiwiGreen" and wine grape "Sustainable Winegrowing New Zealand" program. In New Zealand an operative irrigation company has just developed mandatory Environmental Farm Plan requirements and audit process required by their consent conditions.

With the EMS approach there are opportunities for farm businesses to be innovative in both their land use enterprises and their environmental management. The EMS approach also provides a continuous, rather than discrete adoption of new practices / technologies.

3 Key on-farm environmental management issues for HDI

It was identified through workshops and meetings between the scheme developers, farmers, scientist and environmental groups, that there are six key environmental management issues related to the on-farm effects of irrigated land use for the HDI scheme area, table 2.

Managing each of the activities and the associated issues listed in table 2 at an appropriate level allows water users to adapt the farming systems used. Two examples of the land use mix in the HDI command area are shown in table 1 a dry (unirrigated) land use mix and an irrigated system land use mix. The irrigated system example is predominately a pastoral based system. This could easily shift to a more arable system if the relative economics for either system changed. There are different potential outputs to the environment from different farming systems i.e. a pastoral system vs. an arable dominated system. As discussed earlier, the exact mix of land use types is not known, therefore a system which can adapt over time is required.

3.1 The farm management plans used for HDI

Each of the six management areas listed in table 2, covered in the HDI Farm Management Plan (FMP) has a similar template covering the management objective, and the key potential problems that the water user will avoid, remedy

Table 2: Farm Management Plan key activities, potential environmental concerns and measures to mitigate or avoid.

Activity	Key environmental concerns/ Potential impacts	Examples of Best Management Practices
Irrigation management	Wasteful use of water e.g. • irrigation during/after rainfall • ponding of irrigation water • inefficient application • drainage to other properties	Use INZ code of practice for design Use INZ evaluation code Schedule & apply water taking into account: crop type, soil type, rainfall etc Soil moisture monitoring
Soil management	• Soil compaction / pugging • Soil erosion • Soil health problems • Soil contamination	Avoid stock pugging – use stand off pads or 'sacrifice' paddock Use shelter planting & reduced tillage to avoid wind erosion Avoid irrigation during or after heavy rainfall to minimise runoff to avoid erosion & contamination of water Use only FertMark certified fertilisers to minimise soil contamination
Nutrient management	• Fertiliser getting into ground & surface waterways • Runoff and leaching of stock effluent from paddocks into water ways (including through tile & mole drains)	Follow the NZ code of Practice for Nutrient Management Use soil test results to plan fertiliser needs Use Nutrient budgeting & nutrient management Manage fertiliser applications e.g. to avoid waterways, timing re crop needs, rainfall etc.
Collected animal effluent management	Contamination of ground & surface water during disposal of collected animal effluent (e.g. dairy shed or piggery waste)	Preparing an effluent disposal plan, including spillage management Including nutrients from effluent in nutrient budget and management Include effluent irrigation in total irrigation management
Riparian management	• Damage to stream banks. • Nutrient and faecal contamination of waterways • Sediment	Stock management, including fencing to keep stock from waterways Crop management, including buffer zone around waterway Stream bank planting

Table 2: (Continued).

Biodiversity & Ecosystem management	• Loss of native plants and animals and their habitats • Loss of ecosystem diversity • Soil health problems	Protect existing habitats (e.g. wetlands) as an integral part of farm management Plantings (native & exotic) to support ecosystem diversity

or mitigate. The six templates then list the scheme 'requirements'. These are matters that are mandatory for all water users to achieve if they wish to receive water from the scheme.

The next part of each template lists the areas where water users need to implement best management practices to manage and minimise potential environmental impacts. Each user must consider each topic in relation to their specific property (e.g. soil type, slope, irrigation method, irrigated area, land uses) and determine how they will achieve best practice and what monitoring and records they will use to show their achievements. For example soil moisture measurements in a cropping system may be through use of an irrigation consultant who provides a measurement and analysis service, whereas pasture systems for dairy may have permanent soil moisture equipment that is monitored by farm staff.

Figure 2 is an example of the FMP template for Waterway and Riparian Management.

3.2 How the HDI farm management plan fits an EMS framework

The environmental management of the HDI scheme has several layers of environmental control. The overall scheme has resource consent conditions administrated by the Regional Council, and a Scheme Management Plan (SMP) which details how the conditions of consent are met and how the scheme is operated to meet those requirements. The SMP details the requirement for the FMP's and how the irrigation scheme management will provide training, FMP review, auditing and then, if necessary, compliance and enforcement of requirements. Compliance and enforcement will be undertaken by the HDI scheme management and if necessary the Regulatory Authority. The FMP is the third layer of the HDI scheme management structure. Under the structure each water user in the HDI scheme will prepare and implement an FMP tailored to that property's land use type at that time. The water users and ultimately the scheme management must demonstrate to the Regulatory Authority how the use of natural resources is being actively managed in order to minimise environmental impact while optimising productive output from irrigation.

The use of the HDI environmental plans structure provides a framework within which best practices can be implemented on farm to achieve a high level of environmental protection and enhancement. The FMP is designed so that it

Waterway & Riparian management

Our objective is to protect the waterways on our farm by maintaining healthy riparian margins.

The problems that we will avoid, remedy or mitigate include:

- *Stock damage to banks, causing sedimentation*
- *Contamination of water by stock or agrichemicals*
- *Soil loss causing sedimentation of waterways*
- *Poor water quality and stream life*

We will comply with HDI's specific requirements relating to waterway and riparian management which include:

HDI requirements	Checklist Yes	No
Riparian management to meet HDI guidelines (Appendix8) for all permanent streams		

In addition we will implement our own management policies to achieve the above objectives including:

We will incorporate the following in our waterway & riparian management & practices[1]	Checklist Yes	No
Exclude cattle, pigs and deer from waterways. Exclude other stock from waterways, if necessary.		
Leave a buffer of uncultivated vegetation beside streams to filter any runoff. This will be an appropriate distance from the stream bank depending on soil type etc. [ECAN has guidelines]		
Allow a wider buffer at low points which are more prone to potential runoff from paddocks to provide filter.		
Have field drains discharge into a riparian strip, rather than a waterway where practical.		
When applying fertiliser or other chemicals spread at a distance where it won't get into the waterway.		
Manage farm drains and races according to guidelines (Appendix 10)		
Help manage waterways with plantings of suitable trees and shrubs on waterway margins, choosing species according to guidelines (e.g. regional council) (Appendix 10)		

Self Assessment
Does my management achieve the objectives above?

Yes ☐ Objectives achieved

No ☐ Please fill out table below

List actions required	Person responsible	Timeframe for completion	Completion date

Verification
The information provided is verified as correct.

Property owner / manager

Signature ....

Date ....

[1] Note that this section refers only to waterways on or directly affected by the property that the plan applies to.

Figure 2: Waterway and Riparian management template.

can be adapted for each farm business. Many of the requirements will have both economic and environmental benefits.

The FMP of individual farms are required to be reviewed and independently audited to ensure that the best practice for the current land use mix undertaken on the irrigated property is being achieved and thus adverse effects off farm can be minimised. The audits will identify poor practice and require practices be improved for the water supply to continue.

4 Discussion

The FMP details were presented as a major supporting management tool at the HDI scheme resource consent hearings held in late 2007 to show how the HDI scheme would be able to minimise the land use intensification effects from the use of water for irrigated agriculture. While no decision has been received on the HDI hearings at the time of writing, the approach taken with the FMP's is significantly more detailed than any plans or conditions of consent offered before in New Zealand for irrigation consent.

The FMP's at the farm level are designed so different land use practices can be accommodated within the scheme. This is the advantage of using plans instead of strict conditions of consent which prescribe and limit the activities which can be undertaken.

The use of FMP's links the HDI scheme in an EMS which can adapt and adopt new technologies as they arise over time. The alternative to EMS would mean that as new practices evolved new consents or variations to existing consents would be required. In New Zealand this is not favoured because the consenting system under the RMA is very expensive and time consuming. If the scheme's resource consents were subject to continual major change this would increase the uncertainty of water availability and reduce the farmer willingness to invest either in the scheme as a whole or in new technologies.

Even if the regulatory authorities accept the approach, there will still be a number of challenges to face, including the actual development of each individual FMP and the ongoing compliance and enforcement to ensure that FMP's are implemented, and sound environmental outcomes do occur on the ground.

5 Conclusions

Using an EMS approach to manage the environmental effects of irrigated land use is a new concept for irrigation schemes in NZ. Because it is a process rather than a set of rules or prescribed practices the approach provides individual farmers with the flexibility to be innovative in their agricultural enterprises while regulatory authorities, environmental groups and others can be assured that the irrigation scheme managers have a rigorous and ongoing approach to achieving good environmental management that includes the means to address problems that may arise.

With independent audits and compliance monitoring of water users individual FMP, the HDI scheme is likely to achieve the highest benefits irrigation can bring to the farmer and the region while minimising the impact on the environment when compared with existing irrigation schemes in New Zealand.

Successful implementation will be dependent on all parties with an interest in the management of natural resources and agricultural production working together so that certainty of adopting best sustainable practices is achieved and therefore the potential adverse environmental effects arising from land use intensification are minimised.

References

[1] MAF Technical PaperNo:04/01 The Economic Value of Irrigation In New Zealand, Ministry of Agriculture and Forestry, Wellington, New Zealand, 2004.

[2] PCE, Growing for good: intensive farming, sustainability, and New Zealand's environment. Parliamentary Commissioner for the Environment. Wellington, New Zealand, 2005.

[3] MFE, Freshwater for the Future: a supporting document- a technical information paper outlining the key outcomes for the sustainable management of New Zealand's freshwater. Ministry for the Environment, Wellington, New Zealand, 2006.

[4] Australia's National Framework for Environmental Management Systems in Agriculture, 2002.

[5] Code of Practice for Nutrient Management (with emphasis on fertiliser use) NZFRMA, Fert Research, Auckland, New Zealand, 2007.

[6] Land Information New Zealand 260-J40 New Zealand Map Grid Projection, Wellington New Zealand, 2001.

Monitoring irrigation water, analysis and database maintenance for Nahr Ibrahim

J. Harb[1] & F. Comair[2]
[1]*Department of Civil and Environmental Engineering, Notre Dame University Louaizé, Zouk Mosbeh, Lebanon*
[2]*Water Energy and Environment Research Center, Notre Dame University Louaizé, Zouk Mosbeh, Lebanon*

Abstract

The present irrigation system suffers an uneven distribution of water resources. Originally, attention focused on monitoring the quantity of water allocated to irrigation. With the extensive abuse of pesticides, leaking septic systems and the spread of dumps, water quality deteriorated, rendering it unsuitable. Attention needs to be focused not only on water quantity but also on the quality that meets the minimum acceptable standards for irrigation.

In order to design a monitoring system for the country, there is a need to start implementing a pilot plant for monitoring water quality and quantity. In this paper, implementation of a pilot plant for monitoring the water of Nahr Ibrahim is proposed, due to its high relevance for electricity production, the ecosystem, irrigation, potable water, tourism and industry. This program is developed in spite of the political turmoil, economic degradations and administrative constraints the country is experiencing, shifting priorities from water management toward more basic urgent issues.

Keywords: monitoring, river, water management, quality, irrigation.

1 Introduction

Irrigation that consumes most of the water resources is best managed by an integrated water monitoring system. This system needs to be designed on the basis of impact monitoring. Traditionally, the monitoring system only covered the amount of water available. With the current practice of flood irrigation, water availability is a great concern to the water authority. The energy consumption

WIT Transactions on Ecology and the Environment, Vol 112, © 2008 WIT Press
www.witpress.com, ISSN 1743-3541 (on-line)
doi:10.2495/SI080151

associated with this practice represents a serious constraint to the agricultural industry to render its products competitive, while cheaper products from neighboring competition flood the market. Therefore, agriculture is unable to contribute favorably towards positive economic development.

The Water, Energy and Environment Research Center (WEERC) at Notre Dame University creates awareness for the need to manage water resources efficiently as a result of the upcoming forecast of the shortage of water availability and quality. Several water resources projects are accompanied by a monitoring system for the availability of water. The Technical Cooperation Project of the FAO Lebanon 3003 does incorporate special attention for water monitoring, analysis and database maintenance. This part of the project includes not only the design of a system to be implemented for the water authorities, but also capacity building through a series of seminars on the topic.

2 Macrolocation and geology of the site of Nahr Ibrahim

The selection of the area requires the gathering of information such as the extent of the area, the environmental conditions and processes that may affect water quality including human activities. The selection of Nahr Ibrahim as a case study for monitoring is based on the fact that this water course has multiple uses. Meteorological and hydrological information from Nahr Ibrahim allow the assessment of water quantities for hydroelectric production and help in the understanding of biophysical transformations of the river. Information about the water bodies in addition to actual and potential uses of water need to be established accurately, considering the variety of water uses of this river.

The geography of the water course system established shows that Nahr Ibrahim falls at the boundary of Kesrwan and Jbeil casa, and comes from the source of the Kafka, also used for potable water. The river basin extends from the coast to the summit of the Lebanese mountains to an altitude of 1800m, covering a surface area of 316.6 km^2. The hydrological network of Nahr Ibrahim contains several effluents feeding the main river. The main effluents are:

- Nahr el-Roueiss sourced from Rouaiss, which joins the main river of Nahr Ibrahim near Qartaba
- Ouadi Amsaya and Ouadi Menselit form the upstream part of Nahr Ibrahim
- Ouadi Botrayish joins Nahr Ibrahim to the right of Janneh
- Nahr El-Dib and Nahr El-Dahab
- Ouaid Ghabour

The stratigraphic model of Nahr Ibrahim extends from the Jurassic to the Miocene. The quaternary is represented by fluvial formations on the coast, landslides and mudflows (Aaqoura mudflow). Basaltic extrusions rise through sediments or are intrusive in the period from the top Jurassic (J_5) to the Albien. A complex network of faults extending from the Jurassic is accompanied by some basaltic formations that followed the faults emerging at the surface. The faults and flexures, mainly the ones of Jord Tannourine-Aaquoura play an important hydrogeologic role.

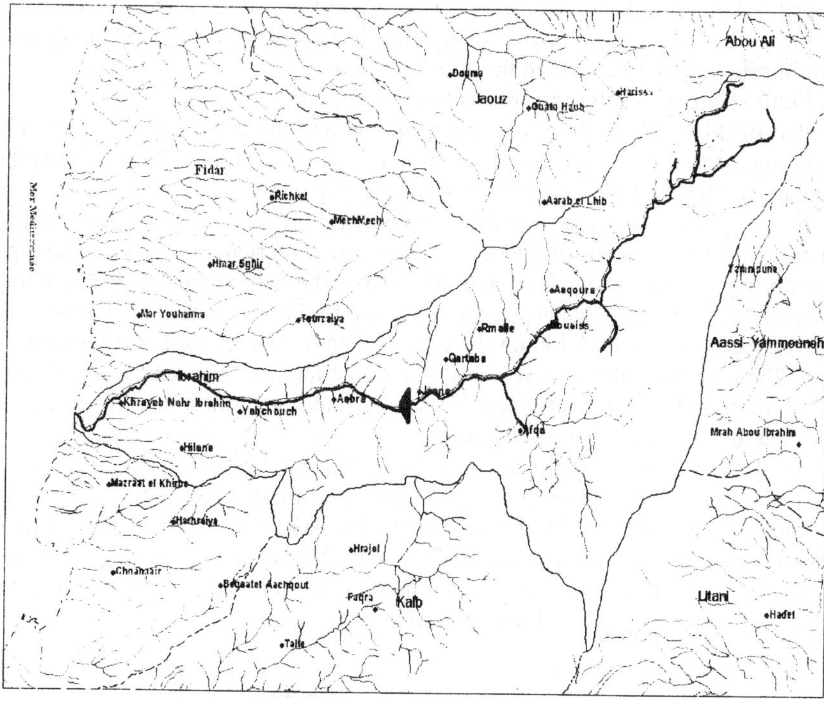

Figure 1: Nahr Ibrahim river and effluents.

Upstream from the river basin, the Cenomanien plateau occupies the lower part of the horseshoe shape of Jabal Sannine (2600m), and the Qornet el-Salouda (3093m altitude) mountains. Their side slopes are 1850m right of the Afqa source. This horseshoe layout enhances the collection of underground water: on the west part of the horseshoe, the sources of the Afqa and the Roueiss flow, and on the east part the sources of the Yammouneh.

The area is formed of 80% of karstifiable zones and the karst is dense and varied. One could find almost everywhere fields of dolines and lapiés, caves and caverns, as witness of the richness of the karstic basin of Nahr Ibrahim. The para-karstic shapes are distinct along the river.

The Nahr Ibrahim basin groups a diversity of superficial lands exploited with urban zones, agricultural fields planted with fruit trees, pasture with low vegetation and forestry. The percentage occupation of the land is distributed as follows: 75% forestry, 2% urban zone and the rest is agriculture.

Besides, the multiple usage of the water enhances the site and renders it vital for the local economy [1]. Discharge of waste is very frequent and spreads all along the river. The situation has been enhanced for the past ten years as a result of multiple efforts with the local communities, NGOs and environmentalists, to convince local industries to stop their toxic dumping. Maps and aerial photographs provide a good source of information for the monitoring system of

Nahr Ibrahim in order to decide on the location of transducers and the relevance of the collected data. A field investigation is necessary for the planning of a monitoring system to check on the above criteria.

Future water needs for the area are shown in table 1.

Part of this water will be used for irrigation and the estimated surface is 450 ha with a peak flow of 300 l/s or a total volume of 4.8 Mm^3. Downstream is the Janneh dam where a hydroelectric power plant is planned.

It is important to know that the establishment of a small scale pilot project such as Nahr Ibrahim provides a hands on experience before expanding monitoring to other sites. Such activities allow the assessment of the sampling network before any expansion is envisaged. The number of samples is established upon the decision on the type of data to be collected. For the case of Nahr Ibrahim, with the primary usage of water for irrigation and hydroelectric production and leisure activities, data on water quantities are necessary in addition to water quality as seen below.

As seen in Table 2, the water test shows that the degree of karstification is high and the contact time rock/water is minimal. During the period of snow melt, the electric conductivity of Mg^{++} and HCO_3^- decrease. The maximal and minimal rates of these elements are in correlation with the maximal and minimal water temperature. Chlorine increases during the rainy season and stabilizes during the dry season. This increase is due to rain fall.

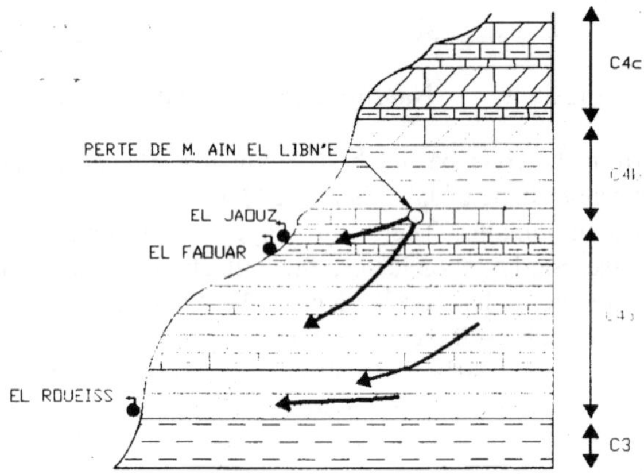

Figure 2: Roueiss spring water stratigraphy (from Ghaouche [6]).

Table 1: Daily consumption for the region of Jbeil (m^3/day).

Year	*2010*	*2020*	*2030*	*2040*	*2050*	*2060*	*2070*
Jbeil	21,400	25,600	30,600	36,600	43,700	52,300	62,500
Water pumped from Janneh	9,000	11,000	14,000	19,000	26,000	32,000	42,000

Table 2: Chemical test of the Roueiss source during a year (mg/l).

T°	Ca^{++}	Mg^{++}	Na^{+}	K^{+}	Cl^{-}	SO^{-}_{4}	HCO_3^{-}
8.83	35	8.44	1.71	0.26	12.1	5.9	122.4

3 Goals and objectives of monitoring irrigation water for Nahr Ibrahim

The deteriorated water quality raised awareness for the need to monitor and verify if it is suitable for its intended uses. Certain species under a pollution load might be in decline while others survive. In Nahr Ibrahim for example, many researchers reported the extinction of several fish species. Pollutants discharged into rivers, groundwater, lakes and the sea once they exceed the water body capacity for absorption create an imbalance in the ecosystem. Therefore data need to be collected for monitoring the evolution of the aquatic system and its response to the existing pollution load. In Lebanon, water bodies have been widely used as a means of disposal for domestic and industrial waste. In spite of the situation improvement, a large amount of pollution remains present and in certain areas, dumping practices in water still exist. Other human activities to be recalled are the illegal exploitation of shore land, unorganized recreational activities and using water for industrial activities. Several wastewater effluent pipes are directed untreated toward water bodies [2].

With this disastrous situation, there is a need for impact monitoring to provide a valuable tool to develop an environmental impact assessment, by gathering data and comparing and assessing results. An integrated monitoring system provides useful information for water management.

The objectives of establishing a monitoring system are oriented towards the global land use for the Nahr Ibrahim area. It is vital to maintain water balance calculations for any management system, and quantity monitoring is an important tool for decision makers. In Lebanon, since agriculture uses around 70% of our water resources, monitoring water availability helps in assessing the performance of irrigation systems. Therefore water reallocation could be achieved to increase the efficiency of use.

With the water quality monitoring system in place, it is easy to develop a water pollution control program. The data collected will help in the assessment of the extent of pollution and the design of a remediation scheme [3].

The pollution load is estimated from monitoring water quality over a long period of time. The location of point source pollution could be identified from plotting pollution load with distance. Contaminated areas and deterioration of water quality are then identified. Regulations covering quality and quantity of waste discharges are obtained after establishing a pollution control monitoring system [4]. Groundwater data that could be affected by the presence of waste will be used to design a regulatory system for waste management.

Water quality standards and guidelines for water usage are retrieved based on an extensive long term monitoring system. Quality data correlated with health information are the basis of establishing water quality standards. Guidelines for water use are the result of a water balance calculation after monitoring water quantity.

One of the important elements is the identification of the area concerned for monitoring. A sampling site description is needed for monitoring water. In our case, it is proposed to start the program by monitoring the area of Nahr Ibrahim, due to its ecological importance and biodiversity, the numerous industries along the river, in addition to tourist activities, archeological attractions and a large source of water supply for irrigation and hydroelectric resources.

In addition to the general geographical features information of Nahr Ibrahim, a listing of important water quality variables is needed for planning the monitoring system. In this case, knowing the kind of industries available will facilitate the identification of the pollution load present there. Then, a detailed plan on the frequency and timing of sampling needs to be established. It is useful to identify the required resources and a plan for Quality Control and Quality Assurance [5].

4 Sources of information on irrigation from Nahr Ibrahim

The collection of data on irrigation is based on farmer estimates, crop models, irrigation surveys, irrigation research, water withdrawal permits, field monitoring studies based on benchmarks and water pumping, comprehensive water metering and available water use reports. As for the quantity of water for irrigation based on area, it is collected from data and imagery, irrigation surveys, water withdrawal permits, field paper based mapping and GIS mapping from aerial photography, meter mapping and permit mapping.

Once the extent of the area is identified, monitoring of the irrigation area is feasible. In addition, the environmental conditions and processes that may affect water quality such as human activities is also an important source of information. Data for water balance and quantities is gathered based on meteorological and hydrological information. Besides, the information on the water bodies and actual and potential uses of water are also needed.

Nahr Ibrahim represents one of the most important water bodies in Lebanon. It is characterized by an abundant and variable water stream during the different periods of the year. It runs on the west side of Mount Lebanon over a length of 40 km to end up on the Mediterranean Sea 21 km from the capital Beirut. The Nahr Ibrahim valley is characterized by its depth, numerous aprons and narrow canyons. The permanent river extends 35 km from Afqa and Roueiss to the source. The remaining hydrographic system is made of seasonal water streams in spite of the highly permeable limestone formation. The flow is estimated at 3000 l/s (2700 l/s on the Mediterranean side and 300 l/s toward the east). This is equivalent to an annual volume of 95 Mm^3 considered high for a geologic formation made of karstic limestone. The average annual precipitation is estimated at 1600mm during an average period of 75 days. The mountainous area covered is not large and the slopes are steep. Snow melt starting in March is accelerated by the warm desert wind (Khamsin). A large amount of water quickly fills the karsts [6], and in the valleys torrential flows are typical. The high flow period extends for six month a year from December to May. Table 3 shows the water flow recorded over an average year during the period 2003–2004 at Afqa and Roueiss at the source. It allows the estimation of water lost expected along the Nahr Ibrahim river trajectory. As seen in the second column of Table 3, the sources of Afqa and Roueiss record flows at 38 m^3/s and 42 m^3/s respectively. This is the period of snow melt that increases the flow of the river.

Table 3: Flow recorded over an average year 2003–2004.

Station/ # of days	0	50	100	150	200	250	300
Afqa (m^3/s)	38	2.4	2.5	2.2	2.25	1.6	1.5
Roueiss (m^3/s)	42	3.6	3.5	3.8	3.75	2.4	1
Total (m^3/s)	80	6	6	6	6	4	2.5
At the Source (m^3/s)	88	19	12	4	2	1	1

5 Monitoring data

In order to store and manage collected data, a computerized data system is adopted and conforms to a well recognized and commonly used system for compatibility and exchange of information. USEPA recommend the STORET system (STOrage and RETrieval) [7]. The ministry of Water and Energy has developed its own system conforming to the standards. The system allows data to be accessed, analyzed and summarized. Then it will be easier to compare data to existing standards for water quality and produce a quality assessment report. An example of a data plot is shown below.

Monitoring media variables proposed for Nahr Ibrahim are as follows:

- Particulate matter and living organisms
- Physiology
- Morphology
- Behavior of specific organisms
- Toxicity or stress
- Phytoplankton chlorophyll pigments (algal biomass)
- Chlorophyll measurements (eutrophication)

As for agricultural and irrigation activities the following parameters are to be gathered:

- Nitrate and phosphates from fertilizers, pesticides and herbicides (dieldrin, aldrin, sum of DDTs, atrazine, lindane, aldicarb, organophosphate pesticides and 2,4-dichlorophenoxyacetic acid (2, 4-D)

- Soil permeability: plants adversely affected and livestock poisoned
- Agricultural practices influence erosion
- TDS, TSS, boron, selenium, sodium, calcium, magnesium and fecal coliform
- Sodium Adsorption Ratio (SAR) [8]

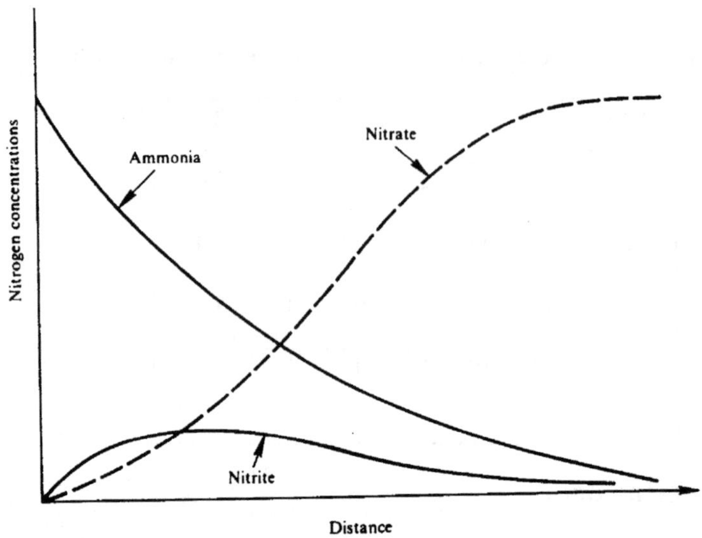

Figure 3: The nitrogen concentration plot versus distance for the case of monitoring river quality data for irrigation uses.

Sampling sites and the number of samples are identified according to the many features characterizing the river. The different subsidiaries of Nahr Ibrahim will also be monitored [9].

The system helps in the process of calibration of the model and verifying the accuracy of the information collected. Also for the case of water quantity measurements, data will help in assessing the compatibility with water accounting. Hence land data will be recompiled for adjustment of water divides.

6 Conclusion

The conclusions are summarized as follows:

- Lebanon extensively lacks water data for irrigation and there is an urgent need for data on integrated land and water resources.
- There is a need to design a real-time data monitoring system for efficient modeling.
- It is necessary to start with a small scale pilot project such as Nahr Ibrahim river for its variety of water uses and applications.

- There is a need to establish a model calibration with the collection of data.
- The bacteriological tests at the sources of Afqa and Roueiss need to be carried out to monitor a continuously safe water quality at the source.
- In the future one could expand the project and centralize data for the efficiency of water management and dissemination of information.It is important to maintain a strong data base and continuity of measurements for:
    - Rural economic activities
    - Irrigation
    - Water quality.
- Political turmoil and economic slowdown are constraints with a strong adverse impact on the execution of the project.
- The FAO support for monitoring is vital for the sustainability of the project.
- The activities of the WEERC to establish a real partnership between the public sector, water users associations and NGO's as well as research coordination with universities, awareness and training will help enhance the monitoring strategy for the country.

References

[1] American Public Health Association (APHA), American Water Works Association (AWWA) and Water Pollution Control Federation (WPCF), *Standard Methods for the Examination of Water and Wastewater* (Washington, DC, 18th ed, 1992).

[2] J.B. Bartram, Drinking Water Supply Surveillance. Robens Institute, University of Surrey, Guildford, 1990.

[3] D. Chapman, *Water Quality Assessments. A Guide to the Use of Biota, Sediments and Water in Environmental Monitoring* (London: Chapman & Hall, 2nd Ed, 1996).

[4] Environment Canada, Water Quality Branch, Inland Waters Directorate, Analytical *Methods Manual*. (Ottawa: Environment Canada, 1981).

[5] ISO, Water Quality Sampling. Part 1: Guidelines on the Design of Sampling Programmes. *International Standard ISO 5667-1, International Organization for Standardization,* Geneva, 1990.

[6] A. Ghaouche, *Etude hydrogéomorphologique et hydrochimiques du bassin amont du Nahr Ibrahim (Mont Liban)* Mémoire, (U.L., Beyrouth, 1989)

[7] B. Lloyd, and R. Helmer, 1991 *Surveillance of Drinking-Water Quality in Rural Areas*. (Area: Longman, Harlow, 1991).

[8] WHO, *Guidelines for Drinking-Water Quality. Volume 3, Drinking-Water Quality Control in Small-Community Supplies* (Geneva: World Health Organization, 1985).

[9] Khatib & Alami, *Etude Hydrogéologique de la vallée de Janneh sur Nahr Ibrahim* (Ministère de l'Energie et de l'Eau, République Libanaise, 2002)

Groundwater history and trends in Kuwait

E. H. AlAli
Kuwait Institute for Scientific Research, Kuwait

Abstract

In Kuwait, one of the GCC countries, natural resources of fresh water are very limited. Kuwait is situated in an arid coastal region characterized by high temperatures, low humidity, sparse precipitation rates, and high evaporation and evapotranspiration rates with no rivers or lakes. Therefore, Kuwait has always relied on other sources to secure freshwater to meet its growing demands. Groundwater quality deterioration in Kuwait is caused by two factors: one factor is the water quality in groundwater fields, where water is extracted for urban use could be affected by lateral flow of saline water. The other factor is that declining groundwater levels would accelerate the process of water quality deterioration. An alternative solution was suggested by the author to overcome the groundwater quality and quantity. A conceptual design system was introduced, consisting primarily of utilizing brackish groundwater in conjunction with treated wastewater augmentation and a reverse osmosis unit. The economical feasibility of the general conception design system was analyzed.
Keywords: Kuwait, groundwater, water level, water quality deterioration, conceptual model.

1 Introduction

Kuwait is an arid country with very few natural water resources (Figure 1). The only water resource in Kuwait with limited natural replenishment is groundwater. Except for some isolated fresh water lenses in northern Kuwait at Raudhatain and Umm Al-Aish, the rest of the usable groundwater is mostly saline, with some brackish zones existing in the southwestern regions. Most of the brackish groundwater fields are located in the center and western regions of Kuwait. Abdally in the north and Wafra in the south are farming areas that depend on groundwater for irrigation. The remaining groundwater in Kuwait is highly saline with a maximum of total dissolved solids (TDS) of about 200,000

WIT Transactions on Ecology and the Environment, Vol 112, © 2008 WIT Press
www.witpress.com, ISSN 1743-3541 (on-line)
doi:10.2495/SI080161

TDS. Most of the potable water comes from desalination plants (multi-stage flash distillation units) with a production cost of about \$1.5- \$3 per m^3. Fresh water is supplied to the domestic consumer at the rate of about \$0.67 per m^3 and for industrial consumers at the rate of \$0.21 per m^3 (Al-Rashed et al. [2]). Supplying water at a highly subsidized price without any limitations encourages waste. Farmers are allowed to pump groundwater without any limitations, for political reasons, which has resulted in several problems, such as a large decline in water table levels and deterioration of groundwater quality, ultimately resulting in the abandonment of several farming areas.

Average annual rainfall and evapotranspiration is about 105 mm and 2270 mm respectively, so recharging of aquifers by rainfall is negligible. Low average rainfall within the region does not result in a significant direct recharge of aquifers. Extremely high evaporation rates owing to high ambient temperatures, low humidity and persistent wind action further reduce available moisture. Evaporation and transpiration produce a soil moisture deficiency that generally does not permit significant percolation or runoff to occur, except under exceptional conditions. Where surface runoff is concentrated in wadis (valley) and playa areas, there is sufficient infiltration and percolation to produce accumulations of fresh to brackish groundwater lenses. Beneath the wadis and basins where recharge occurs, the percolating water develops slight mounds of fresh to brackish water

Groundwater quality deterioration in Kuwait is caused by two factors: one factor is the water quality in groundwater fields, where water is extracted for urban use could be affected by lateral flow of saline water. The other factor is that declining groundwater levels would accelerate the process of water quality deterioration. Water quality in the field where water is extracted for irrigation is affected by return of irrigation water with a higher TDS content, and aquifer mining will allow more lateral flow of saline water.

Figure 1: Aerial photo for Kuwait and neighboring countries.

1.1 Sources of water in Kuwait

For fresh water in earlier days Kuwaitis depended on a few artesian wells. Dhows manned by Kuwaiti seamen also brought fresh water from the Shatt Al-Arab in Iraq. With the rapid growth of population, however, the government of Kuwait built the first modern desalination plant in Kuwait city in 1953, followed by others. The Doha region had two desalination plants, whose capacity reached 138 million imperial gallons per day. A third plant was set up nearby for desalination through reverse osmosis. There are three main sources of water for agricultural and urban uses; groundwater, treated wastewater and desalinated water (Akbar and Puskas [1]).

Table 1: Stratigraphic column section of Kuwait.

Age	Group	Formation	Graphic	Lithology	Ground Water Conditions
Recent			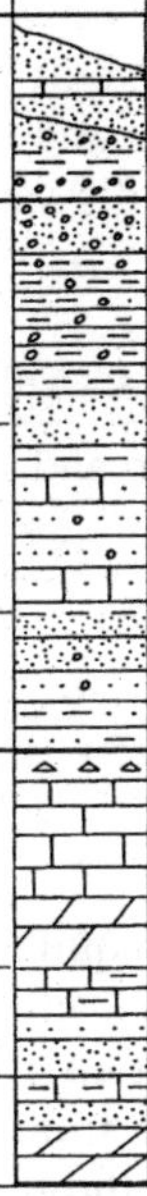	Beach sands; sand; gravel; playa silts and wadi alluvium	Above ground water saturation or locally contain brackish to saline water
Pleistocene	KUWAIT GROUP	Dibdibba		Coarse upland fluvial gravels Gravel and sand, mainly conglomeratic sandstone; siltstone shale; up to 120 m	Ground water locally fresh beneath wadis and depressions brackish at depth
Miocene		Fars		Fine to conglomeratic calcareous sandstone; sand variegated shales; fossiliferous limestone; gypsiferous; 100 m thick	Ground water generally brackish
		Undifferentiated Fars and Ghar		Quartzose sandstone; sand and conglomerate; some shale in lower parts; few meters to 250 m thick	Ground water generally brackish
Eocene	HASA GROUP	Dammam		Discontinuous chert cap; chalky and siliceous limestone; dolomite; 200 m thick	Moderately permeable; brackish water southwest of Kuwait; very brackish in east and north
		Rus		Anhydrite; limestone; marl; 70-120 m thick	Brackish/saline ground water?
		Umm Al-Radhuma		Marly limestone; dolomite anhydrite; 180-400 m thick	Brackish/saline ground water?

1.2 Topography and geology of Kuwait

1.2.1 Kuwait group

The Kuwait group underlies the unconsolidated recent and sub- recent sediments of various types, ranging from gravels and sand to fine grained coastal deposits, sandstone, clay, silts and limestone or marls covering the entire surface of Kuwait. Additionally, it extends down to the top of the underlying Dammam limestone formation. The stratigraphic sequence consists of sediments ranging in age from Miocene to Holocene. It includes conglomerates, sandstones, sandy limestones, clay stones, sand and gravel. Kuwait group aquifer was subdivided, in north Kuwait, into three formations based on the presence of intermediate evaporite deposits as follows:

1. Dibdibba Formation	Sand and Gravel
2. Fars Formation	Evaporite Sequences
3. Ghar Formation	sand and Gravels

The thickness of the Kuwait Group increases from about 150 m in the southwest to about 400m in the northeast. The Kuwait Group is completely dry in the extreme southwest, and is almost saturated with water along the coast of the Arabian Gulf. Due to the absence of the Fars formation, the Kuwait Group is undifferentiated in southern Kuwait.

1.2.2 Hydrostratigraphy

The clastic sequence of the Kuwait Group is hydrologically heterogeneous due to the uneven distribution of the clay lenses and variation in the degree of cementation both laterally and vertically. Lithological and geophysical studies show three recognizable zones of reduced hydraulic conductivity, caused by the preponderance of clayey ranges that can be correlated over large distances. Hence, these three zones act as aquitards and divide the Kuwait Group into three recognizable aquifer units. These aquitards and the aquifer sequence within the Kuwait Group, from top to bottom, are as follows:

• Dibdibba Aquifer (gravel sand).
• Aquitard (silty sand).
• Upper Kuwait Group Aquifer (sand and gravel).
• Aquitard (clay and clay sand).
• Lower Kuwait Group Aquifer (sand).
• Aquitard (basal clay and cherty limestone at the top of Dammam formation).

1.2.3 Potentiometry

All the subdivisions of the Kuwait Group, aquifers and aquitards, are connected with one another and with the underlying Dammam Formation. The regional potentiometric gradient is from southwest to northeast. Within the state of Kuwait, the Kuwait Group gains its water from the lateral inflow from Saudi Arabia and from the Dammam limestone through upward leakage (Figure 2). The potentiometric head generally increases with depth giving rise to the upward component of the flow. The mean value of the head, however, decreases toward the northeast and east direction, toward the area of discharge, toward Kuwait Bay to be discharged by evapotranspiration at the marsh lands along the shoreline and ultimately into the bay and gulf by seepage. The piezometric head of the Kuwait Group indicates that water levels were about 120 a.m.s.l. in the southwestern corner of Kuwait (Al-Ruwaih et al. [3]).

2 Problem identification

In Kuwait, one of the GCC countries, natural resources of fresh water are very limited. Kuwait is situated in an arid coastal region characterized by high temperatures, low humidity, sparse precipitation rates, and high evaporation and evapotranspiration rates with no rivers or lakes. Therefore, Kuwait has always relied on other sources to secure freshwater to meet its growing demands. The water supply in Kuwait can be obtained from three main sources: brackish groundwater, water reuse (treated wastewater), and seawater desalination. There are three classes of groundwater in Kuwait: fresh water with salinity below 1000 ppm, which is used for drinking and domestic purposes, slightly saline water with salinity ranging between 1000 and 10000 ppm, which is used for irrigation, and highly saline water with salinity exceeding 10 000 ppm. In general, groundwater quality and quantity are deteriorating due to the continuous mining of groundwater. On the Al-Wafra farms, in the south, 50% of the wells pumped water with a salinity level higher than 7,500 ppm in 1989 (Figures 3 and 4).

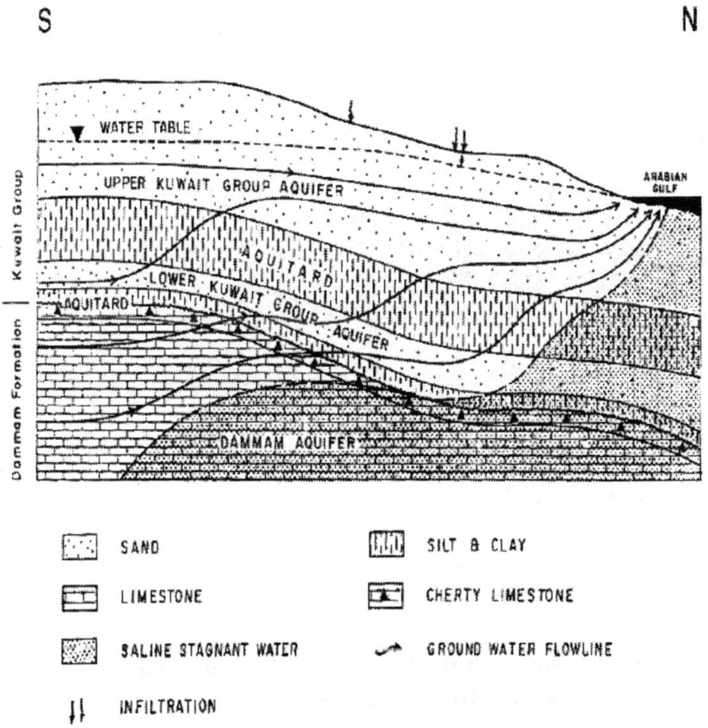

Figure 2: Schematic model of the groundwater flow system in Kuwait.

This figure reached to 75-80% and 85-90% in the years 1997 and 2003 respectively. On the Al-Abdally farms, in the north, 55% of the deep drilled

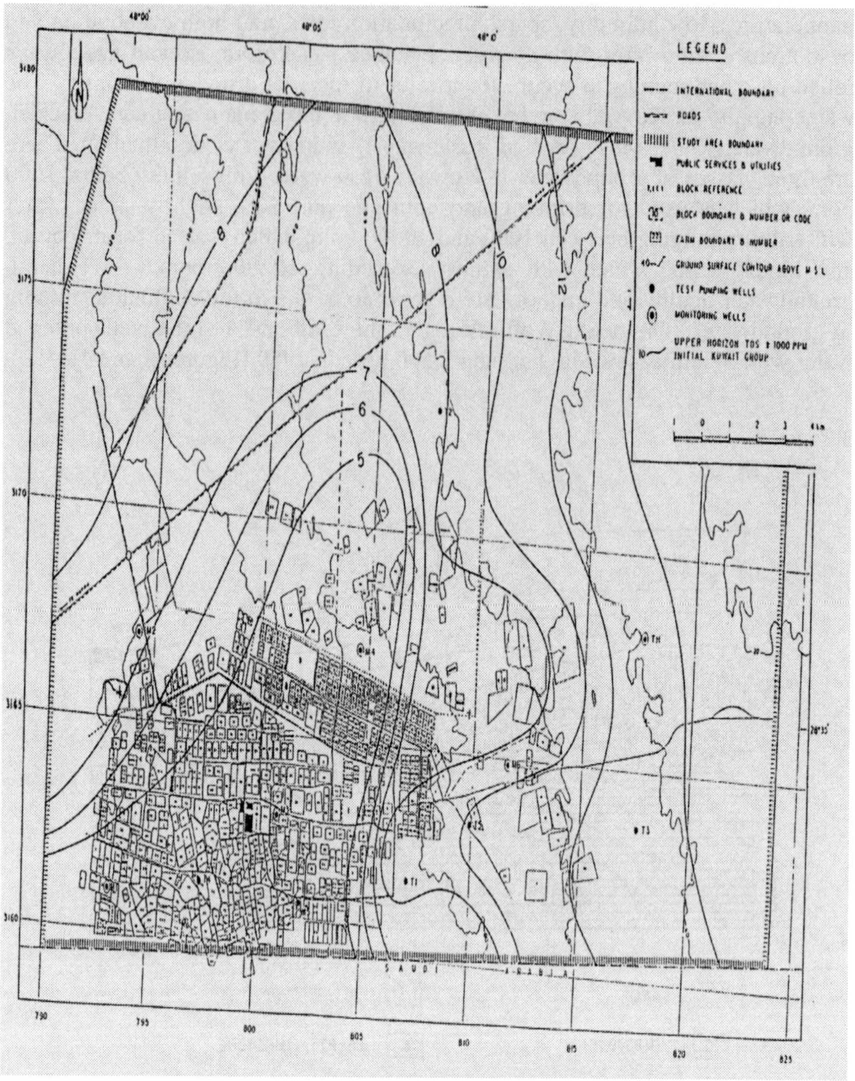

Figure 3: TDS contours of the Wafra farms – Kuwait Group 1993 – upper horizon, (ppm x100), (Al-Rashed et al. [2]).

wells pumped water with a salinity level higher than 7,500 ppm in 1989. This reached to 75% and 90% after 5 and 10 years of operation respectively.

Brackish groundwater exists in reasonable quantities. During 1999, the daily production of brackish groundwater was around 400 million imperial gallons per day (MIGD), which is almost 3 times the annual groundwater inflow (MEW, 2000 [5]). Thus, the production of brackish groundwater is exhausting the one and only vital natural water source. What is more, there are no water charges or limits on groundwater use by farmers. Additionally, brackish groundwater

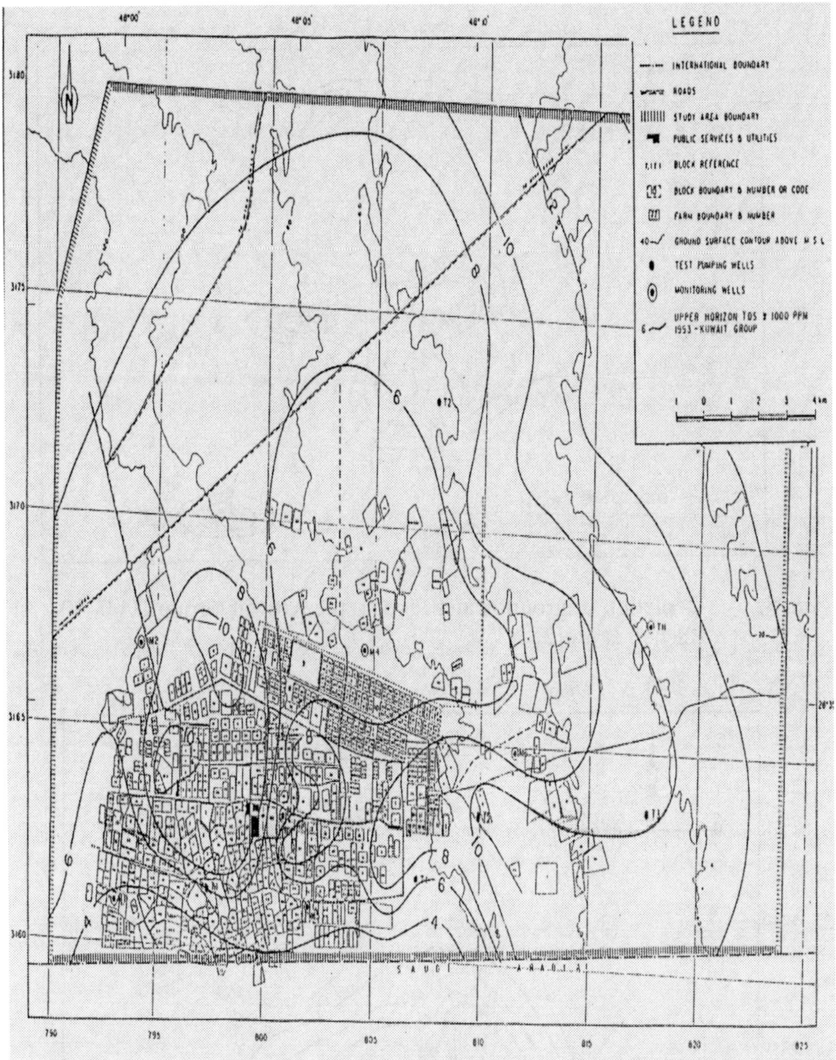

Figure 4: TDS contours of the Wafra farms – Kuwait Group 1999 – upper horizon (ppm x100), (Al-Rashed et al. [2]).

mining will have severe and serious consequences on the quality and quantity of scatters lenses of brackish water in the future due to saltwater intrusion of more saline (brine) surrounding water (Figures 5 and 6).

3 Methodology and trends

The utilization of unconventional water sources may be one of the vital solutions to over come the groundwater quality and quantity deterioration, in specific, the

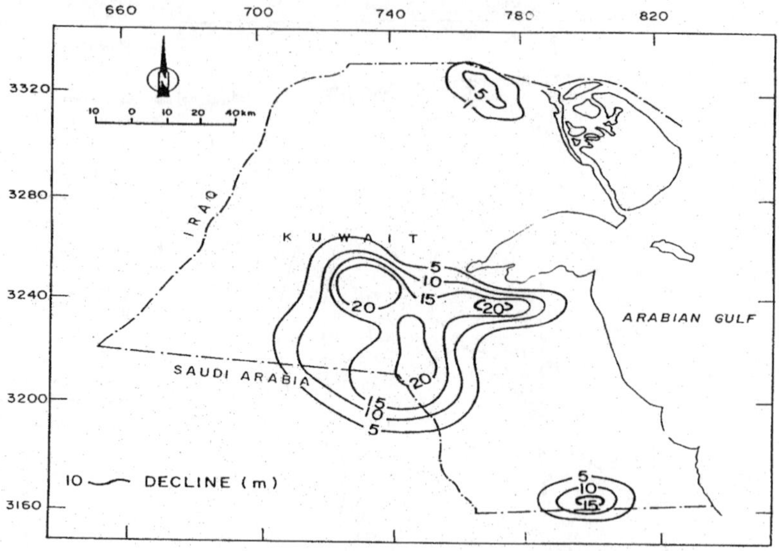

Figure 5: Depletion of groundwater levels: the Kuwait Group 1960–90, (Al-Rashed et al. [2]).

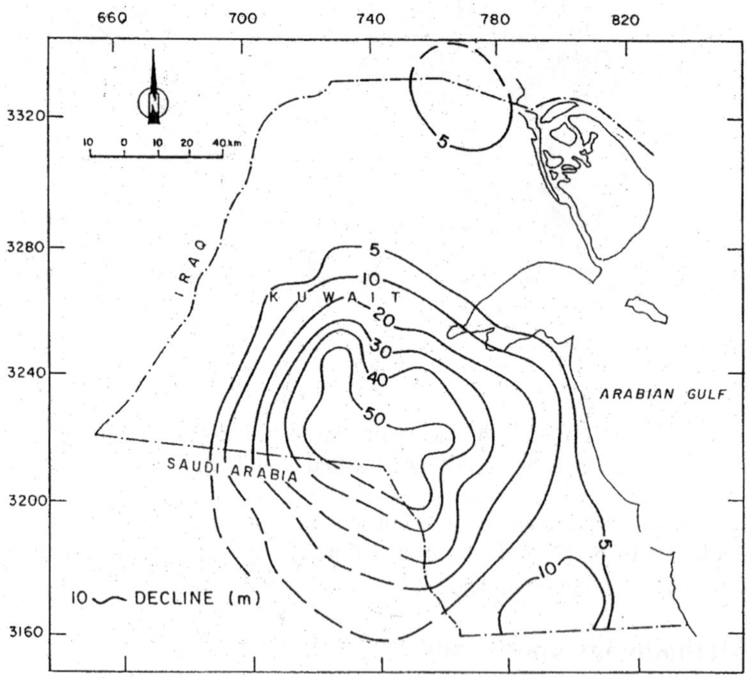

Figure 6: Predicted decline in groundwater levels: the Kuwait Group 2010, (Al-Rashed et al. [2]).

initiations of effective steps to reduce the rate of declining groundwater levels would also reduce the risk of groundwater deterioration. In Kuwait, treated wastewater is an almost unused water source. Urban wastewater is collected, treated and returned to the sea; limited quantities are utilized for landscaping purposes. In Kuwait, wastewater effluent is treated to a secondary or tertiary level. The relatively low salinity of the treated wastewater (1000 mg/l) compared with brackish groundwater of 4000-10000 mg/l makes it a potentially excellent source of good quality water. Moreover, over 100 million imperial gallons per day (MIGD) of treated wastewater production in Kuwait will have a great potential to supplement/ replace the brackish groundwater supplies, hence, by either direct application and utilization for agricultural and urban usage or by the method of augmenting the groundwater by mean of injection process, thus, reducing the burden on groundwater mining and stabilizing and enhancing the groundwater quality and quantity in the long run for the agricultural and urban sector utilization.

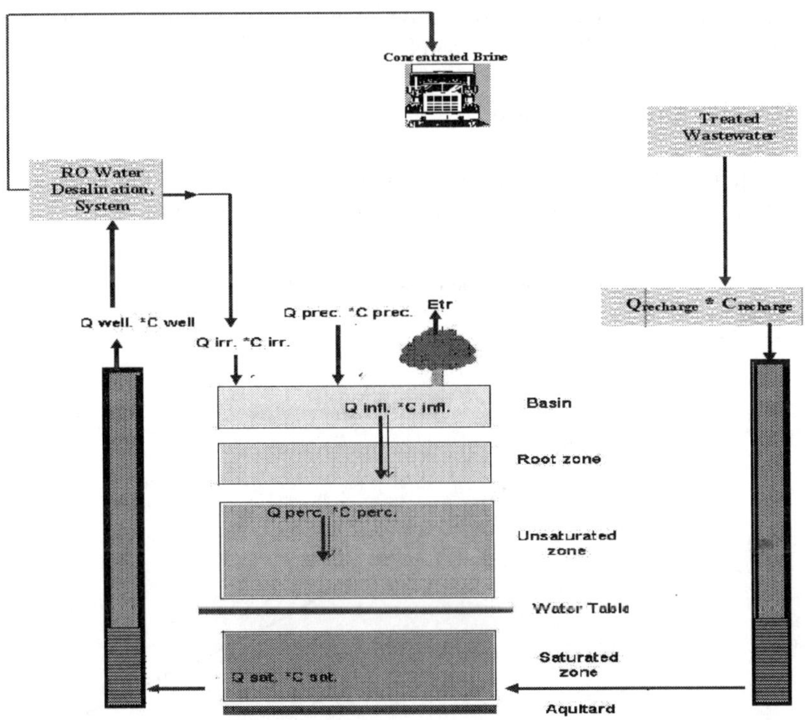

Figure 7: Conceptual design system of implementing a treated wastewater recharge technique and R.O. unit system for water sustainability.

The author conducted a study utilizing a conceptual design model system utilizing brackish groundwater in conjunction with treated wastewater augmentation and a reverse osmosis unit to achieve water sustainability (Figure

7) for the agricultural utilization . The study will result in achieving a sustainable water strategy (quantity and quality) and the relief and restoration of the aquifer after years of human over mining and uneconomical practices of brackish groundwater utilization for urban agricultural purposes.

The study employed two approaches, the lump model approach and the areal distribution model approach. The lump model approach was carried out through the construction of a simplified model approach utilizing the Visual Basic model. On the other hand, the areal distribution model approach was carried out through the utilization of the Visual MODFLOW and MT3D simulation model approach. The economical feasibility of the general conception design system was analyzed for the lump model simulation approach and for the three orientation layouts of the areal distribution model approach. The economical analysis was based on the calculations of the net present worth value of the cash flows for the capital cost, the operation and maintenance costs, and the revenues of the crop yields for the 40 years of operation. The benefit-cost ratios for the base case of the general conceptual design system using the lump model approach and the areal distribution model for the three orientation layouts were calculated for ranges of different interest rates (2%, 4%, 6% and 8%).

4 Results and conclusion

From Figures 8 and 9, the benefit-cost ratios for all the simulation approaches methods were less than 1, except for the condition at interest rate of 2%. As an

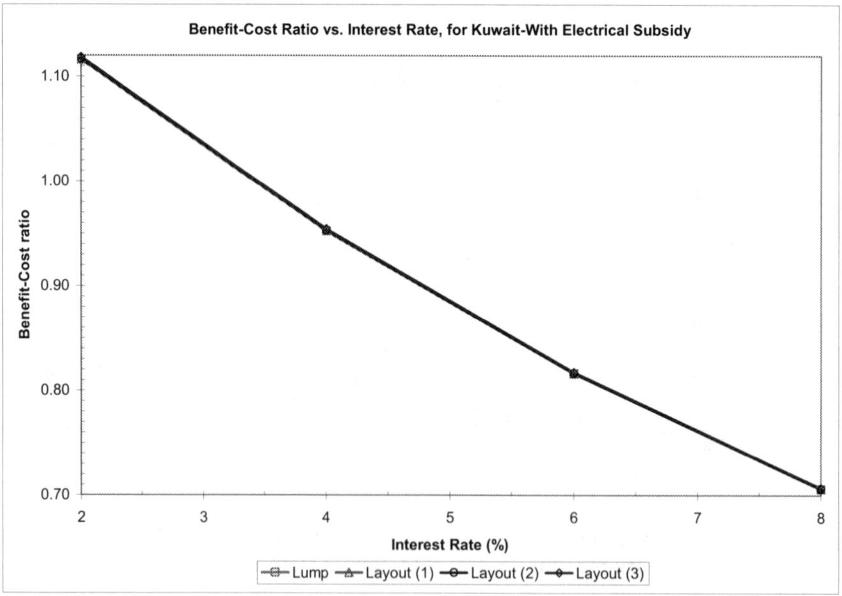

Figure 8: Benefit-Cost ratios for ranges of different interest rates for Kuwait for with electrical subsidy.

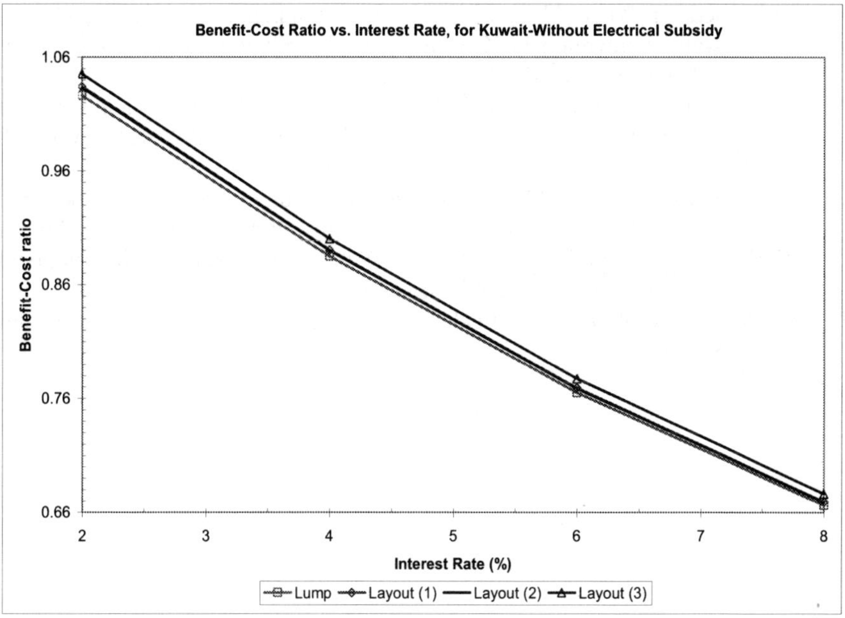

Figure 9: Benefit-Cost ratios for ranges of different interest rates for Kuwait without electrical subsidy.

initial conclusion, from the benefit-cost ratio results, the general conceptual design system is unfeasible for most of the simulation method approaches when the interest rate is higher than 2%. For Kuwait, specifically, and the GCC, in general, food-self efficiency is one of the main aspects of the government's higher political strategy. For that reason, oil wealth is used as a highly valuable commodity traded for food. Furthermore, the GCC countries, as an example, provide long period loans (up to 50 years) with 2% interest rates for small to medium size projects (Industrial Bank of Kuwait) to create more job and work activities for small investors, thus, helping to reduce the unemployment phenomena and create more job opportunities. In addition, the GCC governments adapted a strong trend in encouraging the establishment of private sectors, which in return, will help provide more job opportunities, where 90% of the GCC citizens are now employed by the government. Therefore, this will reduce the burden from the government shoulders. Thus, the general conceptual design model can be strongly economically feasible, especially with the 2% interest rate and electrical power subsidy, where the benefit-cost ratio = 1.1184.

References

[1] Akbar, A., and K. Puskas (1992). Water Availability. Sub-task 2b. In Development Study for the Agricultural Sector in Kuwait. Progress Report. Kuwait Institute for Scientific Research.

[2] Al-Rashed, M., M.N. Al-Senafy, M.N. Viswanathan, and A. Sumait (1998). Groundwater Utilization in Kuwait: Some Problems and Solutions. Water Resource Development, Vol 14, N0. 1, pp. 91–105.

[3] Al-Ruwaih, F.M., S. Sayed, and M. Al-Rashed (1998). Geological Controls on Water Quality in Arid Kuwait. Journal of Arid Environments, Vol. 38, pp. 187–204.

[4] Darwish, M.A., and A.M. Jawad (1989a). Technical comparison between large capacity MSF and RO desalting plants. Proceedings of the Fourth World Congress on Desalination and Water Reuse, Kuwait. International Desalination Association. Desalination 76 (4): 281–304.

[5] MEW (2000). Water Statistical Year Book. Ministry of Electricity and Water, Kuwait.

Section 3
Irrigation modelling

Applicability of a 4 inputs ANN model for ETo prediction in coastal and inland locations

P. Martí, A. Royuela, J. Manzano & G. Palau
Valencian Centre for Irrigation Studies CVER,
Polytechnic University of Valencia, Spain

Abstract

Artificial Neural Networks (ANNs) are simplified models of the central nervous system that can be used as effective tools to model nonlinear problems. This paper reports the study of the applicability of an ANN-based ETo predicting-model in different geographical contexts of the Valencia region. The proposed model only demands the measurement of the maximum and minimum daily air temperatures and the calculation of extraterrestrial radiation and daylight hours. The model provides acceptable approximations of Penman-Monteith ETo values, better than already existing ETo predicting tools (Hargreaves), for coastal locations, where the sea contributes to hinder drastic climatological fluctuations. Nevertheless, the mapping capability of this model in other places with higher indexes of continentality is to be questioned. Furthermore, the possibility of achieving good ETo predictions in different inland locations to those used to train the network might be looked with uncertainty, because of the local uniqueness of the complex relationships between temperature and ETo. On the other hand, models trained in coastal locations might be preferable to carry out predictions in inland locations. The proposed 4-inputs ANN can be useful and preferable to other methods when ETo models which demand a high number of variables cannot be used.

Keywords: ETo prediction, artificial neural networks, maximum and minimum temperatures, index of continentality.

1 Introduction

The precise quantification of cropping evapotranspiration has become a very important task, due to the current water shortage and the subsequent rise of water price. On the other hand, the already existing models that provide precise enough

WIT Transactions on Ecology and the Environment, Vol 112,
www.witpress.com, ISSN 1743-3541 (on-line)
doi:10.2495/SI080171

ETo-predictions demand a high number of climatic inputs for their performance. For both reasons, the development of more efficient ETo predicting tools for those cases where only scan climatic data are available turns into a task of great relevance. Between the most common applications of ANNs we distinguish: constraint satisfaction, control, data compression, diagnostics, forecasting, general mapping, multisensory data fusion, optimization, pattern recognition and risk assessment (Patterson [1]). Several studies have been carried out taking advantage of the input-output mapping capability of neural networks for ETo modelling (Sudheer *et al* [2], Kumar *et al* [3], Trajkovic *et al* [4], Zanetti *et al* [5]).

ANNs are massively parallel distributed processors made up of simple processing units, which have a natural propensity for storing experimental knowledge and making it available for use (Haykin [6]). An ANN is configured for a specific application, such as pattern recognition or data classification, through a learning process. Learning in biological systems involves adjustments to the synaptic connections that exist between the neurons. Because of their nonlinear structure, ANNs are able to capture more complex properties of the studied data than the traditional statistical techniques (Galvão *et al* [7]). Moreover, the main advantage of neural networks in comparison to conventional methods is that they do not require detailed information on the physical processes of the system (Sudheer *et al* [2]).

In the last year, Zanetti *et al* [5] used artificial neural networks for estimating reference evapotranspiration (ETo) as a function of the maximum and minimum air temperatures, extraterrestrial radiation and daylight hours in a coastal county of Rio de Janeiro state, Campos dos Goytacazes. While the first two variables were to be measured, the other two ones were calculated as a function of the local latitude and Julian data. Thus, the proposed model only depends on the measurement of two climatic variables, what can be carried out with a simple thermometer. On the other hand, the authors trained their model with Campos dos Goytacazes climatic series and tested the resulting ANN with Campos dos Goytacazes and a near location, Vicoça, data sets, achieving performance indexes of 0.875 and 0.838, respectively, which correspond to 'excellent' ($c>0.85$) and 'very good' ($0.76<c<0.85$) performances, according to the classification proposed by Camargo *et al* [8].

The high interest of the idea proposed in that paper and the local presence of an extensive weather stations network, with availability of a wide range of climatic variables, encouraged the authors to carry out an analogous study in the autonomous Valencia region, in the Mediterranean coast of Spain. Moreover, the authors have focussed their study on the applicability of this ANN-model in different geographical contexts, analysing the possible effect of the sea on the performance errors of the ANNs.

2 Materials and methods

2.1 Climatic data

The historical series of the climatic variables used in this study were obtained from the weather stations of the Valencian Institute for Agricultural

Investigations (IVIA) in the municipalities of Sagunt (latitude 39º 38' 57'' N, longitude 0º 17' 33'' W, altitude 33m), Vila Joiosa (latitude 38º 31' 46'' N, longitude 0º 15' 19'' W, altitude 138m), Camp de Mirra (latitude 38º 40' 49'' N, longitude 0º 46'18'' W, altitude 627m) and Novelda (latitude 38º 22' 42'' N, longitude 0º 44' 49'' W , altitude 244m) - figure 1 - from January 2001 to December 2006 for Sagunt and Vila Joiosa and from January 2000 to December 2005 for the others. The daily values of maximum, minimum and average temperature, average and maximum wind speed, relative air humidity, solar radiation, sunshine duration were collected by an automatic meteorological station. In Sagunt and Vila Joiosa ANNs the daily values from 2005 were used for network testing and the data series of the other five years were used for training. In Novelda and Camp de Mirra ANNs the daily values from 2004 were used for testing and the data series of the other five years were used for training. The choice of 2004 as test year was due to the breakdowns that took place during several days of 2005 in Novelda and Camp de Mirra stations. Nevertheless it seems that this fact can't interfere in the sense of the comparison results carried out, since all the data series involved in the study can be considered as normal from a climatic point of view, without sharp or noticeable changes during these years.

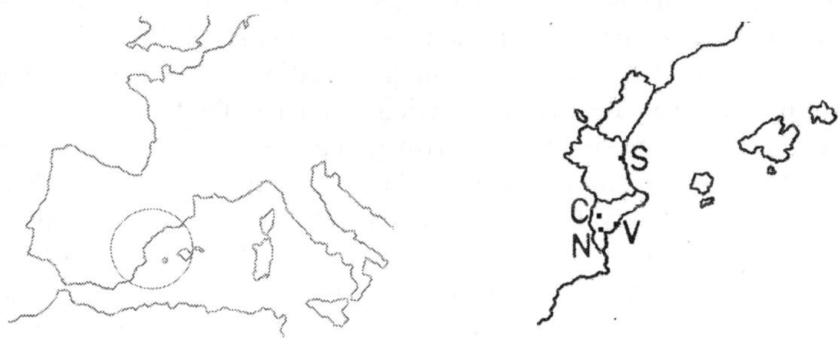

Figure 1: Map of the meteorological station locations: Sagunt (S), Vila Joiosa (V), Novelda (N) and Camp de Mirra (C).

2.2 Methods

A network is divided into layers. The input layer consists of just the inputs to the network. Then follow the hidden layers, which are middle units placed in parallel and consist of any number of neurons. Each neuron performs a weighted summation of the corresponding inputs, which then passes an activation function, also called neuron function. After the learning process the knowledge represented in the model remains stored in the weights. All used ANN neurons were configured, based on the model by Haykin [6]. The first subscript of the synaptic weight w_{kj} refers to the neuron in question and the second one refers to the input sign of the synapse to which the weight refers.

For each neuron, the activation or transfer function is applied to the corresponding summing junction, sum of all input signals weighted by the

www.witpress.com, ISSN 1743-3541 (on-line)

corresponding synaptic weights (same first subscript), in order to limit its amplitude to values between two asymptotes and to add nonlinearity to the model. On the other hand, the bias has the effect of increasing or reducing the net entrance value of the activation function. The sigmoidal function is the most common form of transfer function, it is a continuous function that varies between two asymptotic values, usually 0 and 1 or -1 and 1. Thus, the hyperbolic tangent sigmoid and the lineal functions were adopted for the intermediate and the output layers, respectively.

The used architectures correspond to multilayer feed-forward networks with back-propagation. Thus, they are feed-forward (the signal spreads layer by layer forward), fully-connected hierarchical networks that use differentiable activation functions and supervised training which involves an iterative procedure for minimization of the error function (performance function).

During the learning phase, input patterns are presented to the network in some sequence. Each training pattern is propagated forward until an output pattern is computed. The computed output is then compared to a desired or target output and an error value is determined. The errors are used as inputs to feedback connections from which adjustments are made to the synaptic weights layer by layer in a backward direction. The hidden layer weights are adjusted using the errors from the subsequent layer. The process is repeated a number of times for each pattern in the training set until the fixed criterion is fulfilled or until some limit is reached in the number of training iterations completed. The early stopping procedure was used as the criterion to finalize the training. That way, training data series were divided into two groups: one for learning/parameter estimation and one for cross-validation. According to this method, when the chosen error, of the cross validation set was lower than its value in the previous iteration, the training of the network proceeded; if not, the training was finished. Training was carried out under supervision with the Levenberg-Marquardt algorithm. Neural network minimization problems are often very ill-conditioned, that is, the Hessian is often ill-conditioned. This makes the minimization problem harder to solve, and for such problems, the Levenberg-Marquardt algorithm is a good choice.

Instead of following the methodology applied by Zanetti et al [5], where several architectures with a fixed number of neurons per layer were defined and tested, the authors have developed a general procedure which allows each time to choose the optimum architecture from a set that considers up to three hidden layers with one up to n neurons each, where the different hidden layers present always the same number of neurons. Moreover, each architecture is calculated r times and the corresponding average ANN parameters are considered. For this purpose, Matlab 2007a was used. When dealing with the creation of predicting tools based on ANNs, the corresponding statistical performance parameters should stem from an average of several training-test processes in order to avoid the effects derived from the random assignment of the weights when the training algorithm is initialized. In our case, maximum number of neurons per layer and number of repetitions were fixed in 20 each. The developed program selects the architecture that provides the best average performance and simulates the test data series of every weather station, in order to check the validity of the ANN model outside the location that has been used to train the model.

The selected training parameters are summed up in the following table.

Table 1: Parameters used in the training process.

peformance function	MSE
maximum number of epochs to train	100
performance goal	0
maximum validation failures	5
minimum performance gradient	1E-10
initial mu	0.001
mu decrease factor	0.1
mu increase factor	10
maximum mu	1E+10
maximum time to train	inf

On the other hand, the ANN estimations were compared with Hargreaves ETo predictions and several associated statistical performance indicators were computed. Thus,

$$MSE = \frac{\sum_{i=1}^{n} \left(y_{m_i} - y_{e_i}\right)^2}{n} \tag{1}$$

Where MSE is the mean squared error, y_e is the desired or target output and y_m the computed output.

$$d = 1 - \frac{\sum_{i=1}^{n} \left(y_{m_i} - y_{e_i}\right)^2}{\sum_{i=1}^{n} \left(\left|y_{m_i} - \overline{y}_m\right| + \left|y_{e_i} - \overline{y}_e\right|\right)^2} \tag{2}$$

where d is the adjustment coefficient, $\overline{y}$ is the average of the corresponding y values, m and e refer to model and experimental values, respectively.

$$c = d \cdot r \tag{3}$$

where c is the performance index and r is the correlation coefficient.

3 Results and discussion

The accuracy of each ANN prediction was measured by means of c and MSE, calculated for the ANN/Hargreaves predictions and the corresponding Penman Montheith values. The statistical outputs of the created matlab program are gathered in tables 2 and 3. Vila Joiosa and Sagunt present better performance indicators than Novelda and Camp de Mirra. While Novelda shows unacceptable associated indicators, the two coastal weather stations achieve even better performances than the ones presented by Zanetti *et al* [5] in Campos dos Goytacazes. As can be seen in table 2, the most suitable architecture varies with the studied location, although the analysis of the results that provide these average parameters reveals that one single hidden layer might be enough to represent the nonlinear relationship between the climatic inputs and the ETo, as was concluded by Kumar *et al* [3] and Zanetti *et al* [5]. In addition, the ANN model seems to generate more accurate predictions than the Hargreaves one. On the other hand, table 3 sums up the parameters that quantify the potential of these ANNs to forecast ETo outside the place where the model has been trained.

Table 2: Average indicators of optimum architecture and Hargreaves model indicators [MSE in $(mm/day)^2$].

optimum average architectures					**Hargreaves model**	
training location	*MSE*	*c*	*number of layers*	*neurons per layer*	*MSE*	*c*
C. Mirra	0.7232	0,8401	1	8	1.0446	0.8631
Sagunt	0.3299	0,9105	2	19	0.4693	0.8761
Novelda	2.2913	0,5314	1	17	1.6894	0.5437
Vila Joiosa	0.2290	0,9289	2	7	0.2686	0.9254

Table 3: MSE and performance index of the tested data sets [MSE in $(mm/day)^2$].

	test location							
	C. Mirra		Sagunt		Novelda		Vila Joiosa	
training location	*MSE*	*c*	*MSE*	*c*	*MSE*	*c*	*MSE*	*c*
C. Mirra	1,418	0,843	0,438	0,895	2,496	0,520	0,373	0,914
Sagunt	1,990	0,72	0,787	0,806	3,331	0,389	1,403	0,719
Novelda	0,899	0,863	0,810	0,822	2,324	0,525	0,512	0,874
V. Joiosa	0,819	0,852	0,479	0,870	1,675	0,536	0,289	0,913

Thus, the results corresponding to Novelda test series show very poor predicting potential of the model for this location. Nevertheless, attending to the performances achieved with the rest of test locations, especially in the similar inland weather station, Camp de Mirra, and the results obtained when the model is trained in Novelda and tested outside (for example in Vila Joiosa c=0.87 and MSE=0.51 $(mm/day)^2$), it might be concluded that Novelda test data set could contain errors or lowly reliable data. Penman-Monteith ETo estimations and the corresponding ANN predictions for the case where the model is trained and tested in the same location are presented in figure 2. It can be seen that Novelda PM-ETo doesn't show a clear annual trend like the other weather stations. In this way, as the measurement of a considerable number of climatic variables is

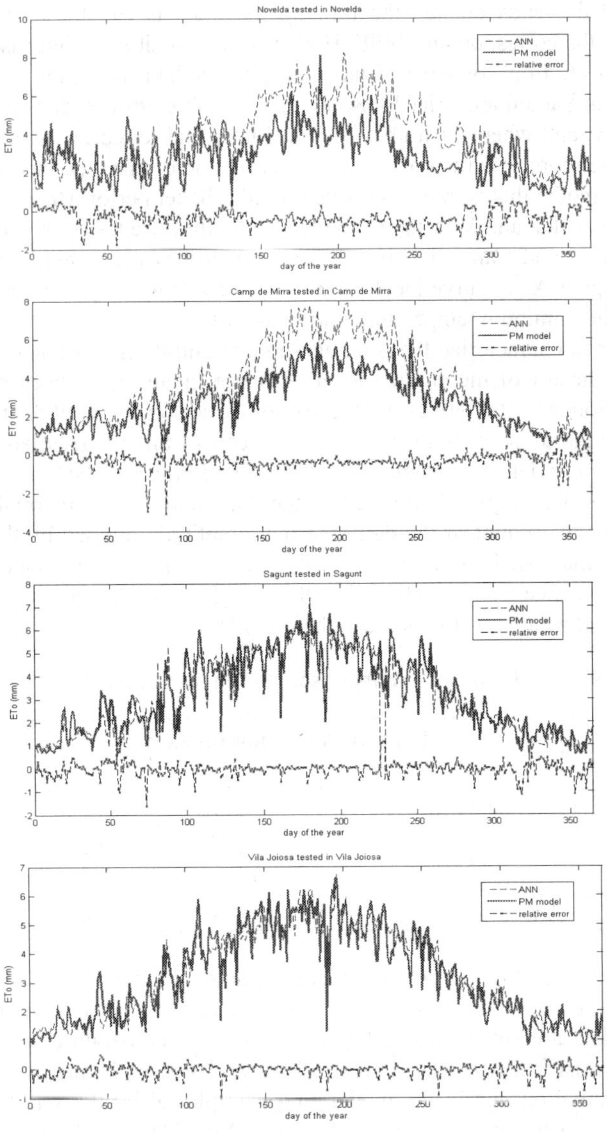

Figure 2: Comparison between the Penman Monteith ETo estimation and the 4 inputs ANN prediction in Novelda, Camp de Mirra, Sagunt and Vila Joiosa.

necessary for the application of the Penman-Monteith model, some errors might have been taking place in the climatic data acquisition during the test year. In contrast, neglecting the parameters that derive from the use of Novelda test set, the performance indexes that are reached exceed 0.8-0.85, with the exception of the model that is trained in Sagunt and tested in Sagunt and Vila Joiosa. Nevertheless, though the fact that this parameter might be acceptable, the mapping potential of this model descends outside the training location, as can be seen through the increase of the corresponding MSE. However, it's possible to find cases where the model seems to fit in some way. Thus, when the model is trained in Camp de Mirra and tested in Sagunt and Vila Joiosa the mean square error reaches 0.43 and 0.37, respectively, and when it's trained in Vila Joiosa and tested in Sagunt it comes to 0.47. Furthermore, it turns out to be peculiar in some sense the fact that for some locations a model that is trained outside provides better performance indicators (for example, Novelda and Vila Joiosa models when they are tested in Camp de Mirra or Vila Joiosa and Camp de Mirra models when they are tested in Sagunt). The peak of Sagunt ANN curve has turned out to be due to the presence of four false values of the minimum temperature during august.

In order to characterize the 4 inputs model validity in relationship with some climatic indicator of the studied location, three factors have been proposed for each location as alternatives to the existing indexes of continentality. These factors pretend to quantify in some way the scale of thermal oscillation that takes place for a year data set. Therefore, α has been defined as the annual average of the daily thermal amplitude divided by the maximum annual thermal oscillation, β is the annual amount of the daily thermal amplitude divided by the maximum annual thermal oscillation and γ is obtained by multiplying factor α by the minimum distance from the studied location to the sea. The values corresponding to the studied weather stations are gathered in table 4.

Table 4: Proposed thermal oscillation indexes for the test data sets.

Thermal oscillation indexes			
location	α	β	γ
Camp de Mirra	0,266	97	11,3
Sagunt	0,225	82	1,13
Novelda	0,283	103	6,72
Vila Joiosa	0,249	90	0,31

Consequently, it might be convenient to relate a higher predicting potential of the 4 inputs model with a lower index of thermal oscillation. Although Novelda ETo test data are suspicious to contain errors, its temperature series might be correctly measured. As it could be foreshadowed, coastal locations present lower thermal oscillation due to the moderating role played by the sea. Parameters α and β assign a higher thermal oscillation to Novelda towards Camp de Mirra, although this one is further away from the sea. The weather station where the model shows the best performance is Vila Joiosa, the location with the lowest thermal oscillation indexes.

4 Conclusions

In this paper, the applicability of the 4 inputs ANN model proposed by Zanetti *et al* [5], which considers maximum and minimum daily temperatures, extraterrestrial radiation and daylight hours, has been tested for ETo prediction in four weather stations of the Valencia region. Moreover, a general procedure to select the most suitable ANN configuration for any study case has been developed with Matlab.

The results achieved by Zanetti *et al* [5] for their 4-inputs ANNs are due to the fact that the climatic series data used for training and testing the model came from a coastal location, where the sea plays a decisive role. The mapping capability of this model in other places with more marked climatological fluctuations is to be questioned. Therefore, the authors suggest a restriction in the applicability of this model for climatic zones with low indexes of continentality and propose the creation of geographical maps for those places where this model is valid taking into consideration the corresponding indexes of continentality, which should quantify the scale of thermal amplitude that takes place for an average year data set. For the current study cases *Pérez Cueva* [9] calculated the following Gorczinsky (G.i.) and Conrad (C.i.) indexes: Novelda and Camp de Mirra (G.i. between 22 and 26, C.i. between 20 and 24) and Sagunt and Vila Joiosa (G.i. and C.i. between 16 and 17). As these ones only take into account the latitude and the average annual thermal amplitude, the authors have seen convenient to propose other parameters which considered the distance to the sea and the relationship between daily and annual thermal oscillations.

Furthermore, the authors look with uncertainty the possibility of achieving good ANN ETo predictions in different places to those used to train the network, because of the local uniqueness of the complex relationships between temperature and ETo. Therefore, further studies should be carried out in order to establish more drastic trends of the model generalization. Taking into account the air relative humidity and/or the daily average wind speed as input variables would considerably rise the potential of the ANN to estimate ETo in other places. The current 4-inputs model can be useful and preferable to other methods (e.g. Hargreaves) when ETo models which demand high number of variables can't be used, for instance, when a breakdown of the weather station takes places.

Finally, the authors suggest the interest of defining general procedures to obtain optimized architectures for each case. To achieve this objective the application of genetic algorithms can result very useful, as it was seen by Ritchie *et al* [11].

References

[1] Patterson, D.W., (ed). *Artificial Neural Networks. Theory and Applications*, Prentice Hall: Singapore, 1996.

[2] Sudheer K.P., Gosain A.K. & Ramasastri K.S., Estimating actual evapotranspiration from limited climatic data using neural computing technique. *Journal of Irrigation and Drainage Engineering*, **129 (3)**, pp. 214–218, 2003.

[3] Kumar M., Raghuwanshi, N.S., Singh R., Wallender W.W. & Pruitt W.O., Estimating evapotranspiration using artificial neural network. *Journal of Irrigation and Drainage Engineering*, **128 (4)**, pp. 224–233, 2002.

[4] Trajkovic S., Todorovic B. & Stankovic M., Forecasting of reference evapotranspiration by artificial neural networks. *Journal of Irrigation and Drainage Engineering*, **129 (6)**, pp. 454–457, 2003.

[5] Zanetti S.S., Sousa E.F., Oliveira V.P.S., Almeida F.T. & Bernardo S., Estimating evapotranspiration using artificial neural network and minimum climatological data. *Journal of Irrigation and Drainage Engineering*, **133 (2)**, pp. 83–89, 2007.

[6] Haykin, S., (ed). *Neural Networks. A comprehensive foundation,* Prentice Hall International Inc: New Jersey, 1999.

[7] Galvão, C.O., Valença, M.J.S., Vieira, V.P.P.B., Diniz, L.S., Lacerda, E.G.M., Carvalho, A.C.P.L.F. & Ludemir, T.B. (eds). *Sistemas inteligentes: Aplicações a recursos hídricos e ciencias ambientais*, Universidade Federal do Rio Grande do Sul: Porto Alegre, 1999.

[8] Camargo, A.P. & Sentelhas, P.C., Avaliação do desempenho de diferentes métodos de estimativa da evapotranspiração potencial no Estado de São Paulo, Brazil. *Revista Brasileira de agrometeorologia,* **5(1)**, pp. 89–97, 1997.

[9] Pérez Cueva A.J., (ed). *Atlas Climàtic de la Comunitat Valenciana (1961-1990)*, Conselleria d'Obres Públiques, Urbanisme i Transports. Sèrie Publicacions de Divulgació Tècnica. Col.lecció Territori: Valencia, 1994.

[10] Martin Vide J., (eds) *Fundamentos de climatología analítica,* Editorial Síntesis: Madrid, 1991.

[11] Ritchie, M.D., White, B.C., Parker, J.S., Hahn, L.W. & Moore, J.H., Optimization of neural network architecture using genetic programming improves detection and modelling of gene-gene interactions in studies of human diseases. *BMC Bioinformatics,* **4(28)**, pp. 1–14, 2003.

Modelling irrigation strategies to minimize deep drainage for two different climatic regions of Canada

G. Parkin & S. Wang
University of Guelph, Guelph, Ontario, Canada

Abstract

Irrigation is a vital part of agriculture in certain regions of Canada including the interior of British Columbia. In this study we examined the use of a soil water budget model for efficient irrigation management in two contrasting climatic regions of British Columbia: Abbotsford (AD) and Osoyoos (OS). The average annual precipitation at AD and OS are 1573 and 318 mm, respectively. The soil types (AD – silt loam and OS – sand) and major crops (AD – raspberry and OS – apple) are also quite different between the two regions. We used the Simultaneous Heat and Water (SHAW) model to estimate the amount of deep drainage and soil water content under different irrigation management strategies. The SHAW model integrates detailed physics of vegetative cover, snow, residue and soil into one simultaneous solution. The model was run on a daily basis for 28 and 32 years for AD and OS regions, respectively. Different combinations of crop and irrigation conditions were run for each region. Based on this study, the "best" irrigation management strategy involves triggering every irrigation event when the soil water content (estimated by SHAW) in crop's rooting zone reaches a prescribed amount below field capacity. At that time, 40 mm of irrigation is added as rainfall. Other strategies involved adding more irrigation and a constant weekly irrigation regardless of rainfall and soil water content. In conclusion, while most of deep drainage in the dormant seasons (no irrigation) cannot be controlled, it can be well controlled to a minimum level in the growing seasons by "best" irrigation management practice.

Keywords: irrigation modelling, minimize drainage, Canadian conditions.

1 Introduction

Irrigation of field crops is required in certain regions of Canada to maintain consistent crop yield and quality. In the province of British Columbia (BC), approximately 190,000 hectares were under irrigation in 1995 [1]. Forage crops were grown on about 85% of the irrigated land with tree fruits, vegetables, and berries together comprising about 11%. The focus of this study is on a tree fruit (apple) and a berry (raspberry) as they are high-value, intensively grown crops.

Water-use efficiency (WUE) is an important consideration in terms of designing an irrigation management system. In fact, improving agricultural WUE is a key element in coping with future water demands [2]. Excessive irrigation has been cited as a possible contributing factor to elevated levels of nitrate in some domestic wells in the two regions of BC selected for this study (Abbotsford, AD and Osoyoos, OS) [3]. The goal of this study was to investigate WUE for these two regions of BC using a model to estimate the relative amount of drainage under various irrigation management strategies. The "best" irrigation scheduling strategy is assumed to minimize the amount of drainage yet still provide sufficient available soil water for plant growth.

The Simultaneous Heat and Water (SHAW) model is a one-dimensional physical-process model, which simulates detailed heat and water movement through the vegetative cover, snow, residue cover and soil [4]. The model enables detailed simulation of water and energy flux at the atmospheric-soil interface and within the soil profile, and includes the effects of vegetation, snow, crop residue cover and soil freezing. It has been used for many applications including estimating soil water budgets and temperatures, snowmelt dynamics, components of net all wave radiation, and timing manure application [5–8]. To our knowledge, it has never been used to assess WUE of different irrigation management strategies. The SHAW model was chosen for this study as it is physically-based and includes the effects of freezing and thawing processes on water movement, which is an important selection criterion for non-growing season conditions in most of Canada.

The main objective of this study was to use the SHAW model to estimate the amount of drainage under efficient and less-efficient irrigation management systems for two vastly different climatic regions of British Columbia, Canada.

1.1 Background information

The location and climate normals for the two study sites (AD: 49°02' N, 122°22' W, 59 m a.s.l.; OS: 49°02' N, 119°26' W, 297 m a.s.l.) are given in Table 1. As discussed previously these two sites were chosen because of their vastly different climatic regimes yet irrigation of intensively-grown crops is a common practice in both regions. The average annual precipitation at AD is nearly five times that at OS; however, due to relatively low amount of summer rainfall, irrigation is still generally required at AD. The average annual temperature at the two sites is essentially the same even though summer and winter season temperatures are normally higher and lower, respectively at OS. These differences in precipitation, and to a lesser extent temperature, should lead to significant differences in the amounts of drainage even without irrigation for the two regions; the SHAW model results will confirm\deny this hypothesis.

Table 1: A comparison of some monthly climate normals for 1971–2000 for the two study sites.

Month	Precipitation (mm)		Average Daily Temperature (°C)	
	Abbotsford	Osoyoos	Abbotsford	Osoyoos
January	198	28	2.6	-2.1
February	160	26	4.7	1.1
March	146	23	6.8	6
April	120	24	9.5	10.8
May	99	37	12.5	15.1
June	79	36	15.1	18.7
July	50	24	17.5	21.7
August	49	21	17.7	21.3
September	76	16	15	16.2
October	145	17	10.2	9.8
November	241	32	5.7	3.5
December	209	34	2.8	-1.2
Average Annual	1573	318	10.0	10.1

2 Methodology

Estimates of deep drainage under different irrigation management strategies and a control were determined using the SHAW model. The control model runs were completed under local soil and climatic conditions with no crop planted and zero irrigation applied. The first irrigation management strategy (I40) was designed to minimize the amount of drainage yet maintain enough available soil water to sustain the crop grown in each region. For both regions, 40 mm (roughly equal to the amount of available water in the top 100 cm of each soil profile assuming a reasonable deficit coefficient) of irrigation was applied as additional daily rainfall when the soil water content decreased to a prescribed level as estimated by the SHAW model under a raspberry crop at AD and apple trees at OS. Soil volumetric water contents of 0.15 and 0.10 triggered irrigation at AD and OS, respectively. Irrigation increased the soil water content to about 0.19 and 0.15 in the respective regions, which are slightly below field capacity of the two local soils providing good growing conditions while presumably minimizing drainage losses. As a test of the efficiency of I40 a second strategy (I60) was tested for the AD region only by adding 60 mm to the rainfall input file of SHAW instead of 40 mm when irrigation was required according to the model estimate of soil water content as discussed previously. A third irrigation strategy was tested only in the semi-arid OS region; this strategy (IW40) applied 40 mm of irrigation to the apple trees each week regardless of the soil water content.

The main inputs to the SHAW model include: initial soil temperature and water content profiles, daily weather conditions, and parameters describing the vegetative cover, snow, plant residue and soil. General site information includes slope, aspect, latitude, and surface roughness parameters. Plant residue or litter properties include residue loading, thickness of the residue layer, percent cover and albedo. Input soil parameters are dry bulk density, saturated hydraulic conductivity, coefficients for the matric potential-water content relation, and the albedo-water content relation. Some of the physical properties of the soils selected to represent the two study regions are given in Table 2.

Table 2: Select soil physical properties used in the SHAW model to represent local soils at the two study regions.

Depth (cm)	Texture		% Clay		% Silt		% Sand		% OM		WP		FC	
	AD	OS	AD	OS	AD	OS	AD	OS	AD	OS	AD	OS	AD	OS
0-20	SiL	Lsa	15	10	65	6	20	84	1.0	1.0	0.11	0.09	0.30	0.18
20-50	SiL	Sa	15	10	65	6	20	84	0.5	0.5	0.11	0.09	0.28	0.16
50-B	Sa	Lsa	6	6	4	4	90	90	0	0	0.06	0.06	0.15	0.12

Depth (cm)	ASWC		Porosity		BD		AE (cm)		PSDI		Ksat	
	AD	OS	AD	OS	AD	OS	AD	OS	AD	OS	AD	OS
0-20	19	8.8	0.53	0.41	1.21	1.50	-47	-18	3.60	4.08	16	3.9
20-50	17	7.6	0.46	0.39	1.37	1.56	-48	-14	3.86	4.11	1.6	2.6
50-B	9.4	5.9	0.41	0.40	1.52	1.52	-18	-18	2.97	2.97	10.9	10.9

AD = Abbotsford; OS = Osoyoos; Lsa = loamy sand; Sa = sand; SiL = silt loam; OM = organic matter; WP = wilting point; FC = field capacity; ASWC = available soil water content (cmm^{-1}); BD = bulk density (gcm^{-3}); AE = air-entry pressure head (cm); PSDI = pore-size distribution index; Ksat = saturated hydraulic conductivity ($cmhr^{-1}$). B = 150 cm for Abbotsford and 200 cm for Osoyoos.

The SHAW model does not have an independent crop-growth module; it requires crop growth information as part of the input data set. The relevant crop-growth information for raspberry and apple crops grown at AD and OS, respectively as used in the SHAW model for this study are listed in Table 3. The daily climate data including maximum and minimum air temperatures, dew-point temperature, total wind run, precipitation, and sunshine hours (converted to solar radiation using the method given in [9]) were obtained from Environment Canada for the two regions. Climate data were available from 1971–1998 for AD and 1968–1999 for OS.

3 Results and discussion

3.1 Abbotsford (AD) Region

Table 4 gives average annual estimates of the water balance from SHAW for the AD region under different irrigation management strategies. Of the 351 mm of irrigation that is applied under the I40 system, it appears that only about 15% was lost as drainage increased by an average of 55 mm over the Control system.

Table 3: Crop-growth information used in this study as input for SHAW model for Osoyoos (OS) and Abbotsford (AD).

OS.

Stage of Development	Time Period	Dry biomass (kg/m^2)	LAI	Rooting Depth (m)
Initial	Apr 15 – May 5	1.0	0 - 1.0	1.0
Crop development	May 6 – Jun 24	1.0-1.5	1.0 - 2.0	1.0
Mid-season	Jun 25 – Sep 22	2.0	2.0	1.0
Late-season 1	Sep 22 – Oct 12	2.0-1.5	2.0 - 1.5	1.0
Late-season 2	Oct 13 - Nov 15	1.5-1.0	1.5 - 0	1.0

AD.

Day of year	Height (m)	Leaf width (cm)	Dry Biomass (kg/m^2)	LAI	Rooting depth (m)
91	1	0	1.3	0	1.0
120	1.5	6	1.9	3	1.0
165	2	6	2.7	4	1.0
193	2	6	3.7	4	1.0
212	2	6	3.4	4	1.0
227	2	6	3.2	4	1.0
262	2	6	2.6	2.5	1.0
273	2	6	2.4	2	1.0
319	1.5	0	1.9	0	1.0

LAI = leaf-area index.

On the other hand, on average, all of the increase in irrigation applied under system I60 was lost to drainage as estimated by SHAW. The average annual amount of evapotranspiration remained essentially the same for I40 and I60 suggesting that the additional irrigation applied under I60 did not increase plant growth.

The year-to-year variability is greatest for runoff (coefficient of variation > 100%); in fact the annual estimate of runoff ranges from 0 to 209 mm. The coefficient of variation for drainage is about 23% only slightly higher than precipitation at 16%, however, there is enough year-to-year variability to warrant caution when using long-term estimates of average drainage to guide water resource policy development for example (Figure 1). For every modelled year the SHAW-estimated drainage for I60 exceeds I40. As well, data shown in Figure 1 suggests that there is a reasonably strong correlation between annual precipitation and drainage. Since drainage is difficult to measure in the field it may be useful to be able to estimate it based on the annual amount of precipitation. Figure 2 shows that over 80% of the variability in annual drainage estimated by SHAW is explained by variability in annual precipitation, which suggests this simple approach to estimating annual drainage may be viable for the AD region.

Table 4: Summary of average (standard deviation in brackets) annual water balance components (all in mm) estimated by SHAW model for three irrigation systems at Abbotsford site.

Irrigation System	P	I	E+T	Drainage	Runoff
Control	1586 (253)	0	690 (62)	844 (222)	52 (60)
I40	1586 (253)	351 (97)	985 (48)	899 (217)	54 (68)
I60	1586 (253)	527 (146)	984 (48)	1076 (210)	54 (67)

P = precipitation; I = irrigation.; E+T = Evaporation + Transpiration.

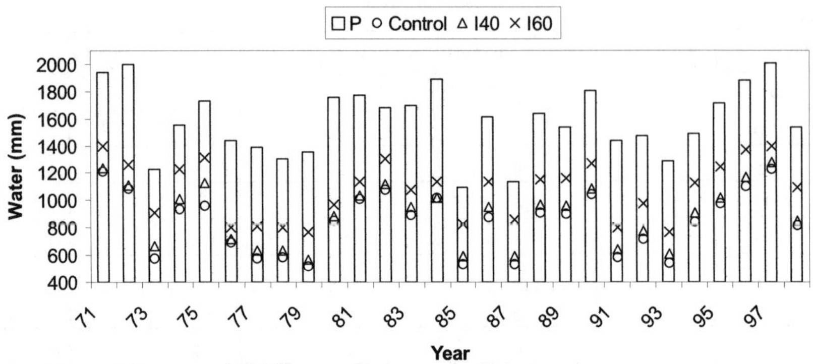

Figure 1: A comparison of annual precipitation and SHAW-estimated drainage for control, I40 and I60 irrigation systems in Abbotsford region. Note P = precipitation.

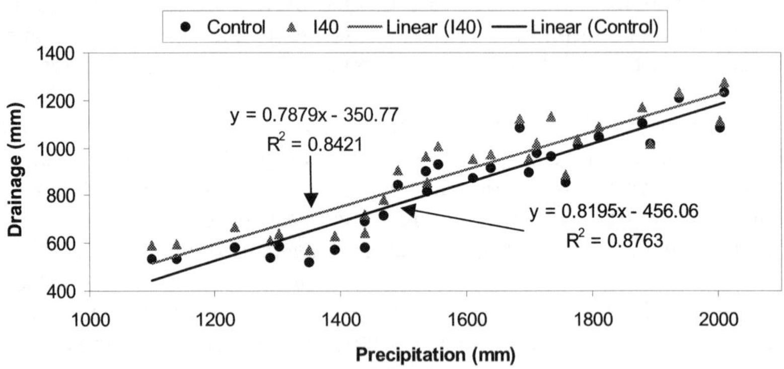

Figure 2: Relationship between annual drainage and precipitation for control and I40 irrigation system in Abbotsford region.

The SHAW model estimates that most deep drainage occurs in the winter and fall seasons; on average, the winter and fall account for 68% and 20% of total annual deep drainage, respectively (data not shown). Runoff usually occurs in the late winter\early spring seasons when the snow pack melts and a thin soil layer near surface is still frozen and thus the melted snow cannot infiltrate.

3.2 Osoyoos (OS) Region

The climate at OS is much drier than at AD; therefore, it is anticipated that the average annual drainage would be much less at OS – the SHAW model estimates bear this out (compare Tables 4 and 5). The SHAW-estimated amount of evapotranspiration at OS increased nearly 4-fold under the I40 system, which applied on average over 750 mm of irrigation per year, yet the estimated amount of drainage actually decreased in comparison to Control. On the other hand, about 84% of the additional amount of irrigation applied under IW40 (319 mm) was lost to drainage as SHAW-estimated average annual evapotranspiration only increased by 50 mm.

Table 5: Summary of average (standard deviation in brackets) annual water balance components (all in mm) estimated by SHAW model for three irrigation systems at Osoyoos.

Irrigation System	P	I	E +T	Drainage	Runoff
Control	318 (79)	0	290 (57)	25 (26)	2 (7)
I40	318 (79)	761 (107)	1062 (78)	12 (16)	4 (11)
IW40	318 (79)	1080 (0)	1112 (73)	279 (109)	5 (12)

P = precipitation; I = irrigation.; E+T = Evaporation + Transpiration.

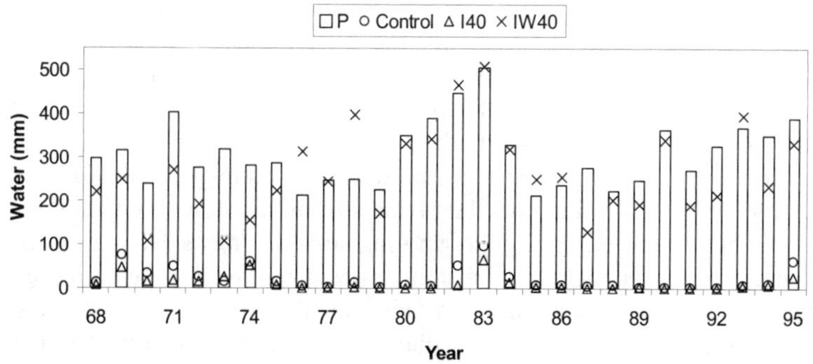

Figure 3: A comparison of annual precipitation and SHAW-estimated drainage for control, I40 and IW40 irrigation systems for Osoyoos. Note P = precipitation.

As discussed previously, over 80% of the variability in SHAW-estimated drainage for AD can be explained by the corresponding annual precipitation (Figure 2). However, at OS the linear relationship between SHAW-estimated

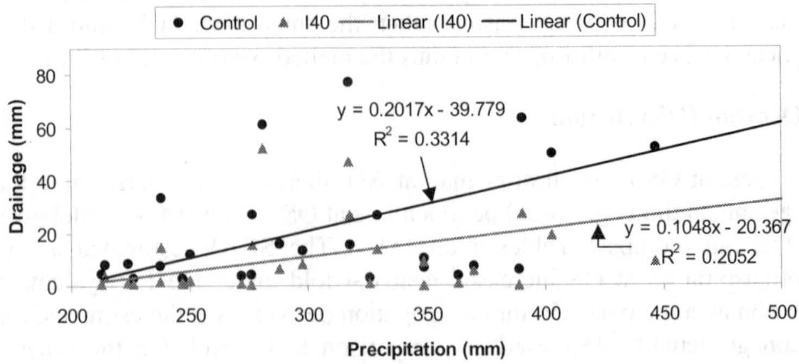

Figure 4: Relationship between annual drainage and precipitation for control and I40 irrigation system in Osoyoos region.

drainage and precipitation is much weaker (Figures 3 and 4). The weaker relationship at OS is probably related to the drier conditions for which much of the precipitation would remain in the soil profile as an increase in soil water content and not contribute to drainage. On the other hand, at AD, especially in the winter season, wetter soil conditions would result in a greater chance of drainage occurring during any precipitation (especially rainfall) event.

4 Conclusions

The SHAW model was run for about 30 years using daily climate data and local crop and soil conditions from two vastly different climatic regions in British Columbia, Canada. The model was run for bare soil\no irrigation, and efficient and less-efficient irrigation systems to compare water losses due to drainage under each scenario. Under the efficient irrigation system SHAW estimates less drainage loss than under bare soil for the drier region. For the wetter region, drainage loss was increased slightly under efficient irrigation and a raspberry crop over bare soil. On the other hand, the SHAW-estimated amount of drainage in both regions increased substantially using the less-efficient irrigation systems. In both regions the drainage that occurs during the non-growing season cannot be controlled; however, the modelling results from this study imply that drainage losses can be minimized during the growing season using SHAW or other water and energy balance models or by installing a water content or potential sensor in the plant root zone to determine when to irrigate the crops. Note that the issue of increasing salinity in the root zone due to minimizing irrigation has not been addressed in this research. Future research will examine the potential for build-up of salts in the soil profile under the efficient irrigation system using the solute transport module of SHAW.

Acknowledgements

Funding for this research was contributed by the Canadian Network of Centres of Excellence program through the Canadian Water Network. The authors wish to thank Ravindra Ramnarine for excellent assistance in checking and filling gaps in the climate data and David Fallow for translating raw climate data.

References

[1] Irrigated area in British Columbia, Irrigation Factsheet #550.100-1, BC Ministry of Agriculture and Food, 2000.

[2] Letey, J., Cardon, G.E. & Kan, I. Irrigation efficiency and uniformity (Chapter 4). *Irrigation of Agricultural Crops*, ed. R.J. Lascano and R.E. Sojka, Agronomy Monograph No. 30, 2nd Edition, ASA-CSSA-SSSA: Madison, WI, pp. 119–132, 2007.

[3] Groundwater Issues in British Columbia. BC Ministry of Environment, Victoria, BC. http://www.env.gov.bc.ca/wsd/plan_protect_sustain/groundwater/library/issues_bc.html

[4] Flerchinger, G.N., and K.E. Saxton. Simultaneous heat and water model of a freezing snow-residue-soil system I. Theory and development. Trans. of ASAE 32(2):565–571, 1989.

[5] Hayhoe, H.N. Field testing of simulated soil freezing and thawing by the SHAW model. Can. Agric. Eng. 36:279–285, 1994.

[6] Parkin,G.W., C. Wagner-Riddle, D.J. Fallow, and D.M. Brown. Estimated Annual Water Surplus for Four Regions in Ontario. Canadian Water Resources Journal, 24: 277–292, 1999.

[7] Xiao, W., Flerchinger, G.N., Yu, Q., & Zheng, Y.F. Evaluation of the SHAW model in simulating the components of net all-wave radiation. Transactions of the ASABE Vol. 49(5): 1351−1360, 2006.

[8] Fallow, D.J., D.M. Brown, J.D. Lauzon, and G. Parkin. 2007. Variation in risk of environmental contamination of manure/nutrient application during winter conditions across Ontario. Journal of Environmental Quality 36: 31–43.

[9] Selirio, I.S., D.M. Brown, and K.M. King. Estimation of net and solar radiation. Can. J. Plant Sci., 51: 35–39, 1971.

Adaptive scheduling in deficit irrigation – a model-data fusion approach

M. M. Holloway-Phillips[1], W. Peng[2], D. Smith[2] & A. Terhorst[2]
[1]*Tasmanian Institute of Agricultural Research, Australia*
[2]*Tasmanian ICT Centre, Commonwealth Scientific and Industrial Research Organisation, Australia*

Abstract

The technological demands required to successfully practice either targeted irrigation control and/or deficit irrigation strategies are currently reliant on numerical models which are often underutilised due to their complexity and low operational focus. A simple and practical real-time control system is proposed using a model-data fusion approach, which integrates information from soil water representation models and heterogeneous sensor data sources. The system uses real-time soil moisture measurements provided by an *in situ* sensor network to generate site-specific soil water retention curves. This information is then used to predict the rate of soil drying. The decision to irrigate is made when soil water content drops below a pre-defined threshold and when the probability of rainfall is low. A deficit strategy can be incorporated by lowering the irrigation "refill" point and setting the fill amount to a proportion of field capacity. Computer simulations show how significant water savings can be achieved through improved utilisation of rainfall water by plants, spatially targeted irrigation application, and precision timing through adaptive control
Keywords: deficit irrigation, wireless sensor network, adaptive irrigation scheduling, model-data fusion, irrigation decision tree.

1 Introduction

Australia is facing a severe water shortage due to below-average rainfall received over the past decade. The agricultural sector is the hardest hit by this as irrigation accounts for almost 65% of total water use nationally [1]. Long-term climate forecasts suggest that this situation is unlikely to improve [2]. Therefore

www.witpress.com, ISSN 1743-3541 (on-line)
doi:10.2495/SI080191

the agricultural industry must become increasingly innovatively in its efforts to use water more effectively. Two techniques which have been successfully used to improve water use efficiency include improved irrigation scheduling practices [3] and the reduction of water application rates using deficit irrigation strategies [4].

Optimising the timing of irrigation events involves being able to sense when the soil water reaches a threshold level estimated either directly using moisture sensors, or indirectly through meteorological data (for example evapotranspiration – ET). Incorporation of short-term weather forecasts can improve the projected need for irrigation and save water by increasing the amount of rainfall utilised by plants [5]. Solving spatial and temporal variation in soil water dynamics, however, becomes more demanding requiring real-time monitoring capabilities (e.g. soil moisture sensors, hydrologic models and remote sensing), and/or a high degree of empirical soil physical data, both of which can be very expensive and labour intensive.

Where water is particularly limiting, further reductions in irrigation water will be required. This has been addressed through the concept of deficit irrigation. Deficit irrigation strategies deliberately allow crops to sustain some degree of water deficit and sometimes an associated yield reduction through a significant reduction of irrigation water. The classic deficit irrigation strategy involves supplying water at levels below full ET throughout the season. In practice this has commonly been achieved by either irrigating at the same frequency but applying less water during each irrigation event, or maintaining the amount of water per irrigation but increasing the interval between irrigations [6]. Where water accounting has been used, irrigation decisions have generally been based on the 'trigger level' concept of available soil water.

The technological demands required to successfully practice either targeted irrigation control and/or deficit irrigation strategies are currently reliant on numerical models which are often underutilised due to their complexity and low operational focus [7]. A simple and practical real-time control system is proposed using a model-data fusion approach, which integrates information from soil water representation models and heterogeneous data sources to improve output resolution and irrigation decision making.

In comparison to other numerical models which act more in a simulation capacity for testing various water-management/allocation strategies at the basin level, the model presented here is to be applied as a functional monitoring tool for automated control at the paddock level. It has the advantage that it can adapt the irrigation plans in real-time to meet the desired soil water conditions defined by the irrigation rules. Model accuracy increases with observation as the model learns soil water flow relationships. Whilst this initial training period may result in a degree of inaccuracy, this is outweighed by the fact that the model can learn soil physical attributes *in situ*. Assigning sensors to similar soil spatial zones through an initial training phase removes some of the high labour and expense demands associated with sensor technology. The low manual inputs needed to use this technology and the limited knowledge required to interpret the output are expected to improve uptake by the broader irrigation community.

This paper presents the model-data fusion approach to irrigation control and how the technology may be applied in practising deficit irrigation. Whilst field validation has not yet been undertaken, preliminary simulation results suggest that water savings can be achieved through consideration of spatial variation and short-term weather forecasts into the irrigation decision-making process.

2 Model data fusion in deficit irrigation

Sensor network technologies are proving to be powerful tools for monitoring real-time changes in the environment. There is a need to aggregate complex data from heterogeneous sensors, providing a rich, multidimensional picture of the system. Much work has been done in the field of data fusion [8–11], which deals with the functional transformation of data into human apprehensible information. Studies looking at "situation awareness" focus on developing models based on how humans perceive and comprehend their environment and subsequently anticipate potential change [12, 13]. Situation is defined as a set of environmental conditions and system states with which the participant is interacting, and can be characterised by a set of information, knowledge, and response options [14]. Situation Awareness consists of three levels of mental models [12]:

- Perception is the process by which a participant identifies the status, attributes, and dynamics of relevant elements in the environment.
- Comprehension is concerned with prioritising and evaluating the information obtained from perception according to their relevance to current goals;
- Projection deals with forecasting future states of elements in the environment based on the awareness achieved in Perception and Comprehension.

To provide decision support in the real-time environment, a situation-aware information fusion system must handle problems in association with inflexible knowledge representation. With respect to the soil hydrologic environment, such information includes the spatial-temporal distribution and evolution of soil moisture which are non-deterministic due to complex interactions among environmental factors like soil profile, precipitation, ET, etc. This requires a system with the capability to learn new knowledge and adapt to the environment. What follows is a description of the model-data fusion approach used to develop a simple real-time irrigation control system.

2.1 A framework for model-data fusion

Model-data fusion is the approach via which an information system learns environmental behavioural models and utilises the learned models to better represent and predict environmental phenomena. Figure 1 shows the framework of a model-data fusion system. The framework consists of Sensors, Effectors, a Classifier, a Model Learner, and a Situation Projector; an Environment Representation Model which represents soil characteristics, and a Decision

Model which captures a set of irrigation decisions in relation to environmental variables.

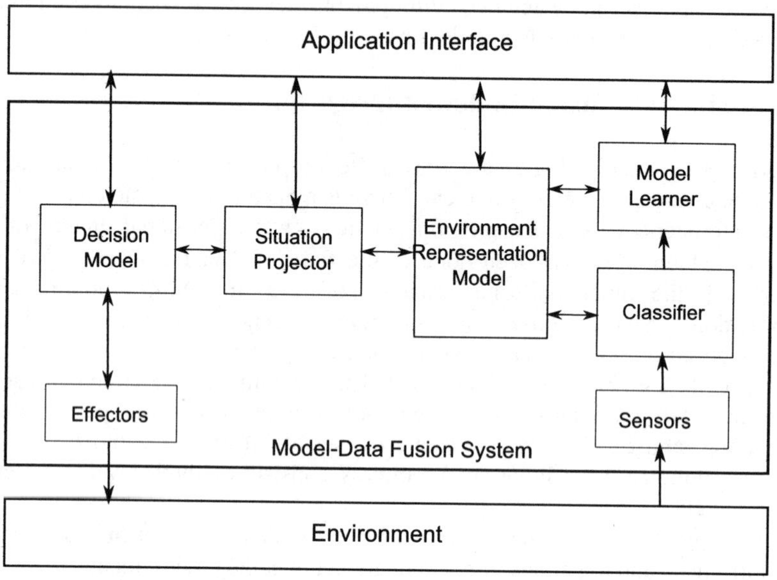

Figure 1: The framework for a situation-aware model-data fusion system.

Sensors are data collectors, which continuously observe the environment. During the learning stage, the Model Learner learns a Representation Model of the environment. The Representation Model captures empirical relationships between data inputs and outputs. The learnt model is then used by the Classifier to identify environmental properties in a perception-related process. The Situation Projector "comprehends" a situation by prioritising information according to how it impacts on the current goal. It also generates a projection (or anticipation) based on the Environment Representation Model. A Decision Model contains strategies to fulfil the goal under various circumstances. Effectors perform tasks which are specified in a decision. The application interface enables a user to view information generated at different components so as to make decisions (or choose a recommendation). Interactions between model and sensed data ground the decision making process in a dynamic environment.

2.2 Learning the environment representation model via *in situ* sensors

Soil moisture, rainfall, surface run-off, wind, humidity, temperature are some of the important types of data that can improve our understanding of the environment in the context of irrigation. The sensors adopted in this research are soil volumetric moisture sensors, soil water potential sensors (gypsum blocks), soil temperature sensors and an automatic weather station.

The Environment Representation Model, is a description of environmental variables and their interrelationships, and is used for predicting future conditions. The environment is represented as three dimensional soil cubes with each cube representing a finite element of soil behaviour. One aspect of soil behaviour of significant importance is the relationship between volumetric water content (θ) and water potential (ψ). This relationship forms what is known as the soil moisture retention curve, and is commonly used to estimate the water-holding capacity of soil horizons within the root-zone.

Within each cube, a gypsum block measures soil water potential. Correlated water potential measurements between spatially neighbouring cubes enables localised clusters to be formed. Each cluster of cubes with similar water potential behaviour can be assigned a volumetric soil moisture sensor. This approach reduces the number of expensive volumetric sensors that need to be employed within the field. For each cube, the relationship between water potential and volumetric water content is then learnt by the Model Learner during a training phase where sensor data is collected from the field. This scenario is shown in Figure 2.

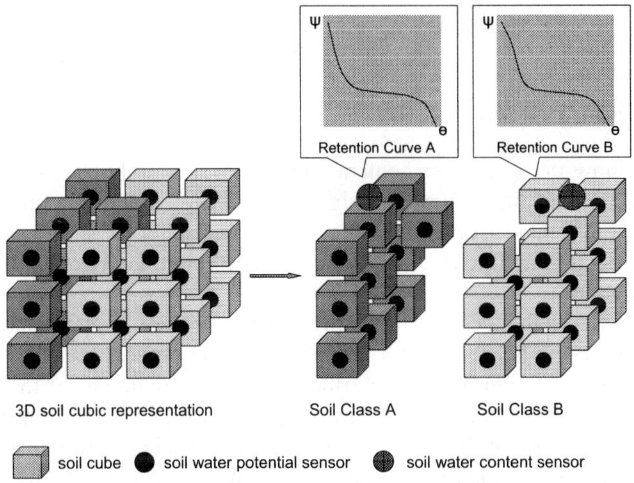

Figure 2: A 3D cubic representation of soil and classification based on water retention characteristics.

Direct measurement of retention curves in the field, particularly at such fine scale, should provide considerable improvement to the understanding of soil behaviour compared with indirect, coarser approximations, such as the use of reference soil types, which fail to take into account soil heterogeneity within the field. Reference data may also be inaccurate due to its measurement in a different environment (i.e. the laboratory) and under different conditions. There is considerable discrepancy between the retention curve of a particular soil type, measured in both the laboratory and the field [15].

Another common approach to obtain the retention curve is to use Pedotransfer (PTF) functions. PTF functions predict the water retention curve from using basic, easily measured properties of the soil. Consequently, PTF can be very crude approximations and are often only optimised for particular soil types and conditions.

Ideally, the training of retention curves within the field should cover measurements over a significant proportion of the retention curve (i.e. from saturation to the permanent wilting point), including the dynamic hysteresis within wetting and drying periods. However, it is unlikely to be possible to measure the complete range of potential and volumetric observations *in situ* during the training phrase. For unmeasured regions, the curve can be predicted by using non-linear regression or neural network methods such as the Mualem-van Genutchen model [16], to fit the measured data to retention curve models until the relevant conditions have been observed.

Retention curves are unique for different soil types as they are dependent upon the physical properties of the soil, such as porosity and particle size. Therefore deriving retention curves *in situ* also has the advantage of providing a coarse taxonomy of the soil composition of each cube, thus considering spatial variability in irrigation decisions. This can be achieved by fitting a model to reference and field retention curves and then comparing their shape parameters.

2.3 Predicting soil water content

Predicting how the soil water content changes over time becomes an important parameter for deciding when to irrigate. The high frequency of data being collected from field sensors offers the potential to predict the short term future of soil water content and enables irrigation plans to continually adapt to meet the target objectives.

One of the challenges in predicting changes in soil water content is that wetting and drying behaviour is a non-deterministic process which is dependent upon a number of environmental attributes. Hence, prediction of the soil water content through modelling is complex, as demonstrated by the Hydrus program [17]. Hydrus has not yet been proven to be an applicable solution to support real time decision making due to the complexity in the physical modelling.

We assume that the rate of soil water change can be learnt from previous soil water behaviour in the field. A relationship between the rate of soil water content change and the other dependent variables such as precipitation and ET can be learnt through observation. The benefit of this approach is that the system is not attempting to learn the physical model itself, but instead the relationship between soil water change rate and other dependent system attributes. In a sense, it is equivalent to the real time learning of PTF functions from the field.

With the learned characteristics of each soil class, the Classifier can locate environmental variables in the Representation Model (retention curves) and the Situation Projector can subsequently infer the needs for irrigation by calculating the distance to the refill point ($\theta_{threshold}$) for each soil class. Figure 3 illustrates this concept.

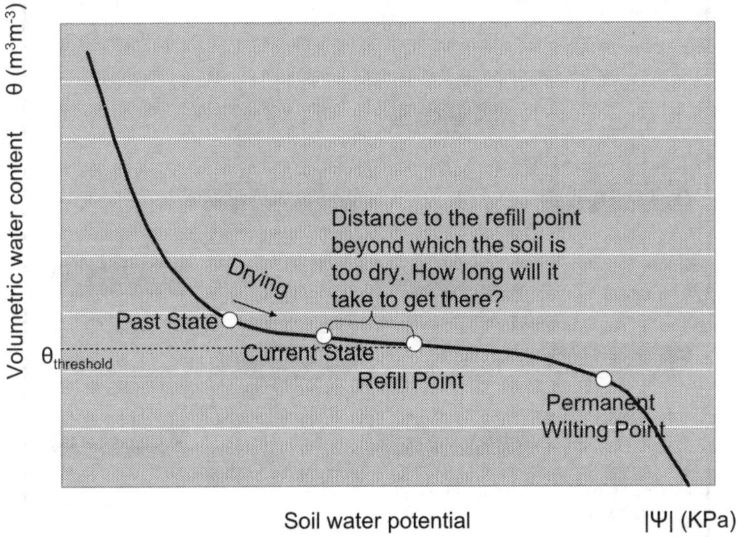

Figure 3: Predicting the rate of soil drying based on field-derived soil moisture retention curves, for irrigation scheduling decisions.

2.4 Technical-driven deficit irrigation decision model

Besides the soil drying status, the decision of whether to irrigate and the amount of water to apply at the current time, is dependent upon other information including rain forecasts, ET demand, current water allocation and other plant related factors. Decision tree learning using inductive inference has been the method adopted for deciding when to irrigate. Inductive learning methods identify features that empirically distinguish positive from negative observed training examples. Figure 4 presents the results obtained from a decision tree learning algorithm. Each non-leaf node stands for a test on an attribute. Edges of the decision tree coming from the nodes are values of attributes for that node. Leaf nodes are used to represent design decisions for selecting a deficit ratio. Numbers in parentheses illustrate an observation for the class defined in the leaf node. For example, "10.0" indicates that there are ten positive observations and no negative observations for that class.

The decision rules encoded within the model consider historical irrigation treatments, and generalises empirical knowledge that can be applied to select an appropriate irrigation treatment for a specific combination of field conditions. Three environment variables form the decision tree. These variables describe whether a soil cube is at the refill threshold point (node "reachRefillPoint"), if it is currently raining (node "currentRain"), and the probability of expected rain in 12 hours (node "forecastRain"). Leaf nodes illustrate irrigation treatments. Irrigation strategies are obtained by traversing from root node to a leaf node. For example, some interesting rules can be found:

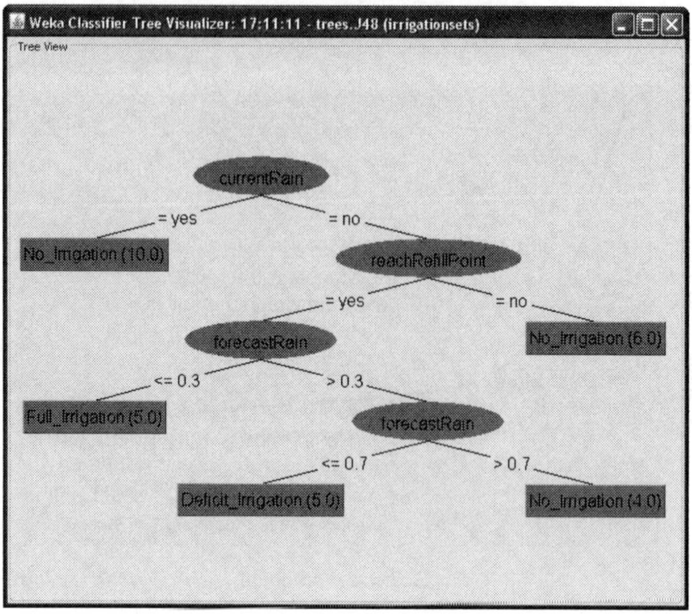

Figure 4: A decision tree structure learnt from running a C4.5 [18] algorithm from WEKA [19] on 46 instances of deficit irrigation treatments.

- if the soil cube has reached refill point whilst it is not raining, and the probability of the predicted rainfall is greater than 0.7, irrigation water should not be applied. Irrigation is postponed until the next decision time, for example, in 4 hours;
- if the soil cube has reached refill point whilst it is not raining, and the probability of the predicted rainfall is between 0.3–0.7, deficit irrigation should be applied to mitigate the risks of water stress on plant growth. The pre-defined deficit ratio is applied to the accumulated ET since the last irrigation or rainfall event to calculate the irrigation requirements.

Information fused from various sensors can be used to construct a recommendation. The system can also learn and evolve the decision model in real time. For example, by monitoring the plant response (e.g. remotely sensed canopy temperature), if the pre-defined plant-based thresholds have not been met under the applied irrigation deficit ratio, the system can re-adjust to a lower ratio when similar environmental conditions prevail. If required, the irrigation decision model can also act autonomously to schedule irrigation events.

3 Model simulation

An irrigation simulator was developed in NetLogo [20] to demonstrate how the model works and contributes to water use efficiency when taking account of rainfall information. Scenarios of a simulated deficit irrigation model and a point sensor-triggered irrigation system are discussed.

The irrigation simulator uses field data including ET and rain gathered from Elliot research farm (North West Tasmania; 41°06′S, 145°46′E) during the period of 2 November to 21 December, 2007. This data is scaled down to fit into the simulation model with essential variances unchanged. The simulator consists of two agent-based models:

1. A deficit irrigation model, which integrates local weather conditions and rainfall forecasts to decide when and how much to irrigate, and targeted irrigation control, with application to only those areas that have reached the refill threshold point;
2. A point sensor-based irrigation model, which only uses threshold soil moisture to trigger irrigation with uniformed application.

To exemplify the effect of different irrigation treatments on water usage, both models run under the same sets of environmental conditions including ET, rainfall, soil profile characteristics, and boundary conditions such as irrigation rate and drainage rate. A 3D soil water diffusion model is used to simulate soil water content changes under these conditions. The soil is represented by 3D cubes, as shown in Figure 3. The simulated soil consists of loam, clay and sand at different horizons (Figure 5). The soil diffusion model is a simplified view of the tendency of water to move from one soil cube to another, based on differences in soil water contents and soil porosities. Considering a soil cube and its neighbouring cubes, soil water diffusion can be simulated by recalculating the distribution soil water between cubes:

$$T_i = P_i \times \frac{\sum_1^n T}{\sum_1^n P} - ET - DR \qquad (1)$$

where T_i denotes soil water content unit for cube i

i is the cube for recalculation

n is the number of neighbours for cube i

P is the porosity of soil cube i

ET stands for water loss from evapotranspiration for cube i (only top level cubes have ET)

DR denotes water unit loss from drainage for cube i over boundary (only the bottom-level cubes consider DR)

Eqn. (1) describes how soil water diffusion is calculated in the simulator. Environmental changes can cause the soil water content to re-distribute. Figure 6 also shows an example of how to apply Equation 1 to calculate simulated soil water content in cube *i*. By iterating all soil cubes, the effect of environmental conditions can be represented through soil water diffusion.

Point sensor-based irrigation treatments apply water uniformly to the whole paddock. Rainfall is affected by the same environmental conditions as irrigation. Water is lost through drainage and ET with some water units retained in the soil. Figure 6 shows the accumulated irrigation water usage during 50 time units of an irrigation treatment with a model initialised to pre-defined soil water contents.

In this scenario, 6170 units of water have been consumed as a result of 21 irrigation events.

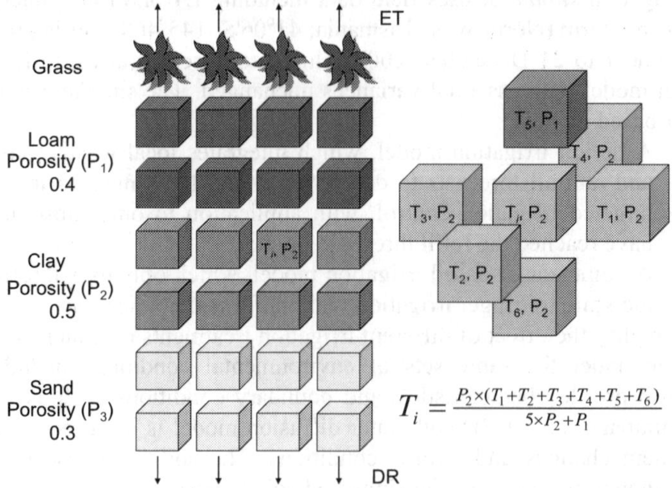

Figure 5: Soil profiles and the diffusion approach used in the simulator.

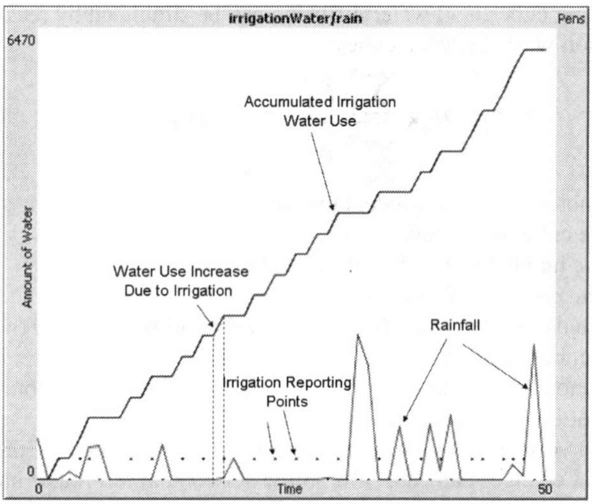

Figure 6: The accumulated irrigation water use for 50 units of time under a uniformly distributed irrigation treatment. Dots are used to report irrigation events in the previous time period.

A deficit irrigation simulation model uses heterogeneous information to make decisions on irrigation treatments (i.e. when and how much to apply). Unlike the point sensor-based uniform irrigation treatment, deficit irrigation scheduling maximises the actual and predicted short-term rainfall events. As shown in Figure 7, at the initial stage of irrigation, the sprinklers are not triggered as a result of balancing the effects of current rainfall and the future probability of rain on dry cubes in the root zone. In some circumstances, the simulation model uses a deficit irrigation treatment (0.5 of accumulated ET in this case) since the probability of rainfall is moderate. Applying a reduced amount of irrigation minimises the risk of irreversible damage to plants whilst improving rainfall utilisation. Figure 7 shows an accumulated water consumption of 4520 units over 50 units of time, which represents a water saving of roughly 27%, compared with the point sensor-based irrigation strategy. We recorded 22 irrigation events in this scenario.

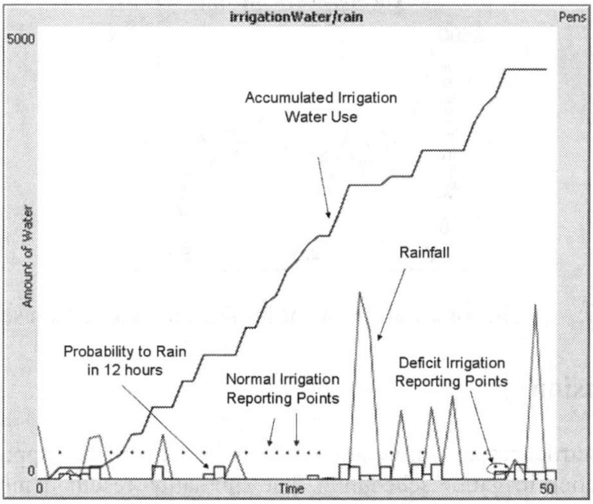

Figure 7: Accumulated irrigation water use in a deficit irrigation scenario.

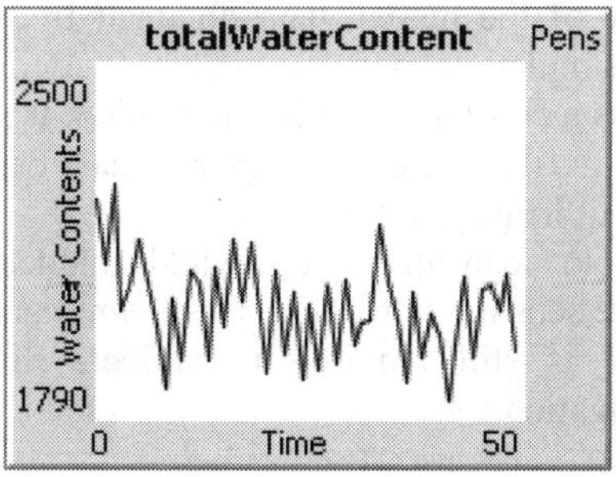

Figure 8: The total water content for point sensor-based uniform irrigation scenario.

In both scenarios, soil water contents are maintained to be below field capacity. As the deficit irrigation model integrates information about future rainfall events, it adapts irrigation frequency based on the predicted contribution of rainfall to soil water content. The frequency adaptation which is presented in Figure 7 does not occur under a point sensor-based irrigation scenario. This has a direct impact on the amounts of water consumed. Because the two scenarios are assumed to have identical environmental variables (ET, rainfall, drainage), the difference between the total soil water contents (shown in Figures 8 and 9) is an indication of over-irrigation in the point sensor-based strategy. That is, assuming production is fairly constant above the refill threshold point, which water content is maintained above this level in both scenarios, the extra water applied in the point sensor-based strategy is of no benefit to production.

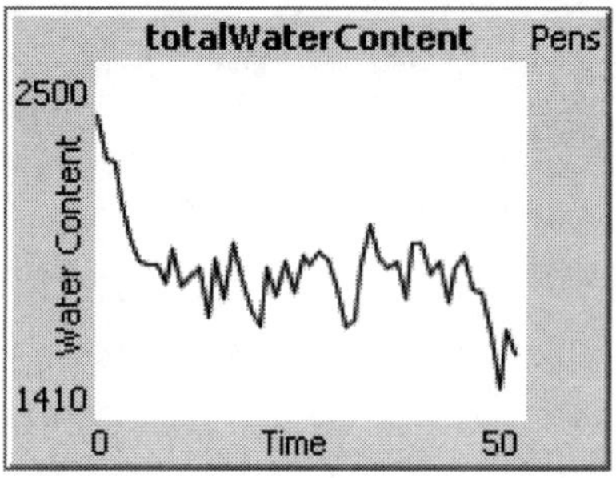

Figure 9: The total water content for the deficit irrigation scenario.

4 Conclusion

This paper introduces an information fusion framework to support decision-making in deficit irrigation scheduling. The simulation results demonstrate how real-time data collection may be used to improve water use efficiency by considering future rainfall events and practicing targeted irrigation application. The simulation model is however a reduction of a physical phenomenon. To increase the credibility of the model, data obtained from real-time sensors to enhance/calibrate the simulation model is required. At present, the simulation model treats drainage as a constant boundary condition. It is necessary to include beyond root zone soil moisture sensors to approximate deep drainage to further improve the resolution of irrigation decisions.

The intention is to consider crop yield models, multiple-objective optimisation and remote sensing in the next phase of research.

Field quantification is still required to validate this model-data fusion approach for deficit irrigation.

Acknowledgements

This work is part of a research project jointly funded by the CSIRO Water for a Healthy Country Flagship and the Australian Government through the Intelligent Island Program. The Intelligent Island Program is administered by the Tasmanian Department of Economic Development. We are grateful to all members of the CSIRO Tasmanian ICT Centre and the Tasmanian Institute of Agricultural Research for their involvement in this project. Meisha Holloway-Phillips also receives project support from Dairy Australia.

References

[1] ABS, *Water Account for Australia 2004 to 2005*, cat. no. 4610.0, ABS, Canberra, 2006.

[2] Hennessy, K.B., Fitzharris, B.C., Bates, N., Harvey, S.M., Howden, L., Hughes, J., Salinger, J. & Warrick, R., Australia and New Zealand (Chapter 11). *Climate Change 2007: Impacts, Adaptation and Vulnerability. Contribution of Working Group II to the Fourth Assessment Report of the Intergovernmental Panel on Climate Change*, eds. M.L. Parry, O.F. Canziani, J.P. Palutikof, P.J. van der Linden & C.E. Hanson, Cambridge University Press: Cambridge, pp. 507–540, 2007.

[3] Scheierling, S., Cardon, G. & Young, R., Impact of irrigation timing on simulated water-crop production functions. *Irrigation Science*, **18(1)**, pp. 23–31, 1997.

[4] Costa, M.J., Ortuno, M.F. & Chaves, M.M., Deficit irrigation as a strategy to save water: physiology and potential application to horticulture. *Journal of Integrative Plant Biology*, **49(10)**, pp. 1421–1434, 2007.

[5] Gowing, J.W. & Ejieji, C.J., Real-time scheduling of supplemental irrigation for potatoes using a decision model and short-term weather forecasts. *Agricultural Water Management*, **47(2)**, pp. 137–153, 2001.

[6] Ganji, A., Ponnambalam, K., Khalili, D. & Karamouz, M., A new stochastic optimisation model for deficit irrigation. *Irrigation Science*, **25(1)**, pp. 63–73, 2006.

[7] Bastiaanssen, W.G.M., Allen, R.G., Droogers, P., D'Urso, G. & Steduto, P. Twenty-five years modelling irrigated and drained soils: State of the Art. *Agricultural Water Management*, **92(3)**, pp. 111–125, 2007.

[8] Hall, D.L., McMullen, S.A.H., *Mathematical Techniques in Multisensor Data Fusion*, Artech House Inc.: Norwood, 2004.

[9] Steinberg, A.N., Bowman, C.L. & White, F.E., Revisions to the JDL data fusion model. *Proc. of the SPIE on Sensor Fusion: Architectures, Algorithms, and Applications*, ed. B.V. Dasarathy, pp. 430–441, 1999.

[10] Waltz, E. & Llinas, J., *Multisensor Data Fusion*, Artech House: Norwood, 1990.

[11] White, F.E., A Model for Data Fusion. *Proc. of the1st National Symposium on Sensor Fusion*, 1988.

[12] Endsley, M.R., Toward a theory of situation awareness in dynamic systems. *Human Factors* **37(1)**, pp. 32–64, 1995.

[13] Endsley, M.R., Theoretical underpinnings of situation awareness: A critical review. *Situation Awareness Analysis and Measurement*, eds. M.R. Endsley & D.J. Garland, LEA: Mahwah, pp. 3–32, 2000.

[14] Pew, R.W., The State of Situation Awareness Measurement: Heading Toward the Next Century (Chapter 2). *Situation Awareness Analysis and*

Measurement, eds. M.R. Endsley, D.J. Garland, Lawrence Erlbaum: Mahwah, pp. 33–47, 2000.

[15] Morgan, K.T., Parsons, L.R. & Wheaton, T.A., Comparison of laboratory- and field-derived soil water retention curves for a fine sand soil using tensiometric, resistance and capacitance methods. *Plant and Soil*, **234(2)**, pp. 53–157, 2001.

[16] van Genuchten, M. Th., A closed-form equation for predicting the hydraulic conductivity of unsaturated soils. *Soil Science Society of America Journal*, **44 (5)**, pp. 892–898, 1980.

[17] Šimůnek, J., and M. Th. van Genuchten, Using the HYDRUS-1D and HYDRUS-2D codes for estimating unsaturated soil hydraulic and solute transport parameters. *Characterization and Measurement of the Hydraulic Properties of Unsaturated Porous Media*, eds. M.Th van Genuchten, F.J. Leij & L. Wu, University of California: Riverside, pp. 1523–1536, 1999.

[18] Quinlan, J.R., Improved use of continuous attributes in c4.5. *Journal of Artificial Intelligence Research*, **4**, pp. 77–90, 1996.

[19] WEKA, machine learning software package, the University of Waikato, http://www.cs.waikato.ac.nz/ml/weka/.

[20] NetLogo, http://ccl.northwestern.edu/netlogo/

Hydrological water balance modelling for assessing productivity and irrigation planning

M. B. Nagdeve, G. K. Ramteke & P. A. Kamble
Dr. Panjabrao Deshmukh Agricultural University, Akola, India

Abstract

The physically based distributed modelling system, MIKE SHE, is used to simulate the hydrological water balance of *Belura* watershed with the objective of developing the irrigation plan and assessing productivity. The existing cropping system was replaced with the cropping system planned using qualitative land evaluation i.e. land capability and suitability classification for soils of the watershed. The hydrological water balance was simulated using calibrated model for *Belura* watershed. The actual yield of the sorghum, cotton and pigeonpea from watershed area was estimated by Food and Agriculture Organisation (FAO) model, using simulated hydrological water balance. Yield of sorghum, cotton and pigeonpea is estimated to be 32, 48 and 55% of the potential yield at downstream side and 28, 41 and 46% at upstream side, respectively. Considering the water requirement to obtain potential yields, these crops suffered soil moisture deficit of 61, 68 and 151mm at upstream; and 57, 65 and 157mm at downstream sides of the watershed, respectively. If irrigation is provided as and when water content in the root zone goes below allowable limit at different crop growth stages using hydrological water balance, it is possible to obtain sustainable crop yields. The overall results illustrate the applicability of MIKE SHE comprehensive hydrological modelling system for the management of water resources for sustainable agricultural productivity in a watershed.

Keywords: hydrological modelling, hydrological water balance, land capability, land suitability, storie index rating, crop planning, and irrigation planning.

1 Introduction

Land, water and vegetation are the most important natural resources for survival of mankind, to satisfy the need for food, fibre and fuel and other material. In

WIT Transactions on Ecology and the Environment, Vol 112, © 2008 WIT Press
www.witpress.com, ISSN 1743-3541 (on-line)
doi:10.2495/SI080201

recent year, the land productivity is continuously declining because of overexploitation and degradation of natural resources. It resulted in static and some times even a significantly lower agricultural production. Watershed management has become the cornerstone of planning and development of land and water resource. Hydrological simulation models are often used to provide information as a basis for decisions regarding the development and management of water and land resources.

MIKE SHE (Refsgaard and Storm [11]) is a comprehensive deterministic, distributed and physically based modelling system for the simulation of all the major hydrological process occurring in the land phase of the hydrological cycle. MIKE SHE is a further development based on the SHE modelling concept. It has been widely adopted for catchment studies (Barthurst [3, 4]; Refsgard *et al.* [12]; Jain *et al.* [8]). It simulates water flow, water quality and sediment transport.

Land evaluation is a process of assessment of land performance, which is essential step in crop planning. It is the process of grouping the soils in to various groups in relation to their crop productivity. The productivity of soil is defined with the realized yields of economic crops. The evaluation of soils may be qualitative and quantitative. The modified FAO relationship can estimate the actual yield of different crops on the basis transpiration. With this background a present investigation was formulated to develop crop plan and irrigation plan by assessing productivity of *Belura* watershed.

2 Methodology

2.1 Study area and data

This study concentrates on *Belura* watershed, which is located in the Patur Tahsil of Akola district. The watershed lies between 76° 53′00″ to 76° 54′55″ E longitudes and 20°32′20″ to 20° 33′54″ N latitudes with an altitude ranging from 300 to 343m above MSL covering an area of about 577ha. The location map of the *Belura* watershed is presented in fig.1. The climate of the area is semi-arid, characterized by three distinct seasons viz., hot and dry *summer,* warm and rainy *monsoon,* and dry mild *winter.* Average maximum and minimum temperature is 42°C and 10°C in the month of May and December, respectively. The total rainfall received during the year 2000 was 573.4mm as against normal rainfall is about 824.7mm during the last 25 years (1971 to 1995). Most of the rainfall in the area occurs during June to September.

2.2 Land use

Land use/ land cover classification and mapping of the study area was carried out with the help of Geocoded (1:12,500) IRS 1C AND 1D LISS III and PAN images. The PAN sharpened image of the study area was visually interpreted. Based on the photographic elements like tone, texture, shape, size and pattern etc., the various land unit boundaries were delineated. These land units were then confirmed by ground verification and finally the study area was grouped in to seven land use/land cover classes and presented in fig. 2.

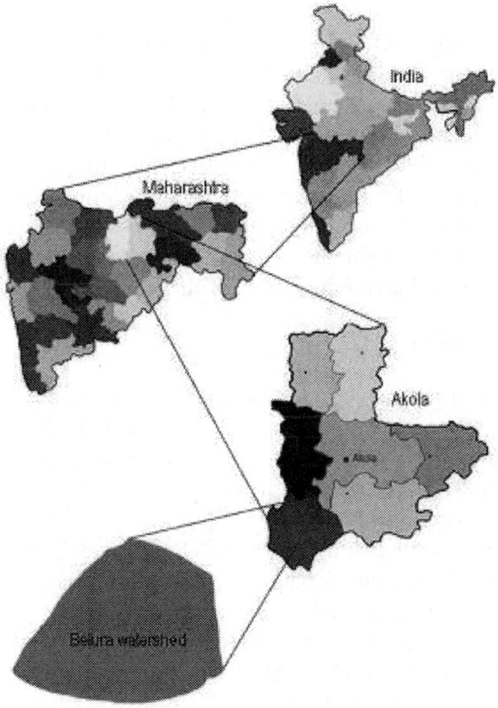

Figure 1: Location map of Belura watershed.

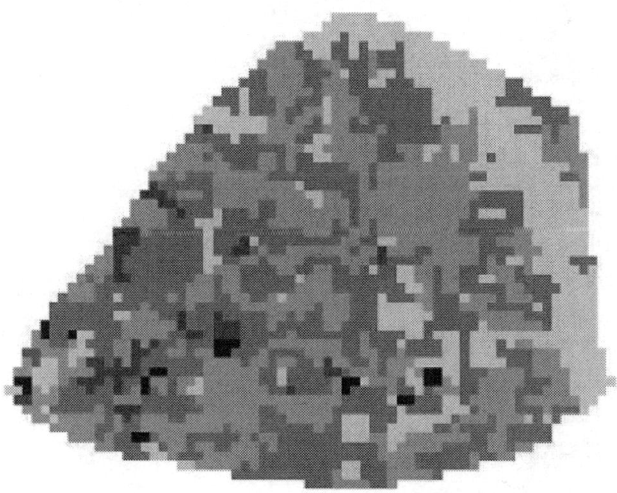

Figure 2: Landuse/land cover map of the watershed.

2.3 Land evaluation

The productivity of soil is defined with the realized yields of economic crops. As the crop yields are dependent on climate as well as soil-site factors, integration of these factors are necessary to obtain yields. The evaluation of soil may be qualitative or quantitative. In the present study both types of evaluation of seven soils have been done as suggested by USDA (Klingebiel and Montgomery [9] and FAO [5]. The levels of limitations were used as defined by FAO [6, 7]. The crop planning has been done using qualitative land evaluation i.e. land capability and suitability classification for the soils of the watershed. Storie index rating (Storie [14]) was used for quantitative land evaluation.

2.4 Description of MIKE SHE

MIKE SHE is a comprehensive deterministic, distributed and physically based modelling system for the simulation of all major hydrological processes occurring in the land phase of the hydrological cycle. Two analogous horizontal-grid square networks discretize the model area for surface and subsurface groundwater flow components. A vertical column of nodes at each grid representing the unsaturated zone of MIKE SHE model. A finite difference solution of the partial differential equation, describing the process of overland and channel flow, unsaturated and saturated flow, interception and evapotranspiration, is used for water balance modelling. The overall model structure is illustrated in fig. 3. Description of its components is given in Abbott *et al.* [1, 2]; Refsgaard and Storm [11].

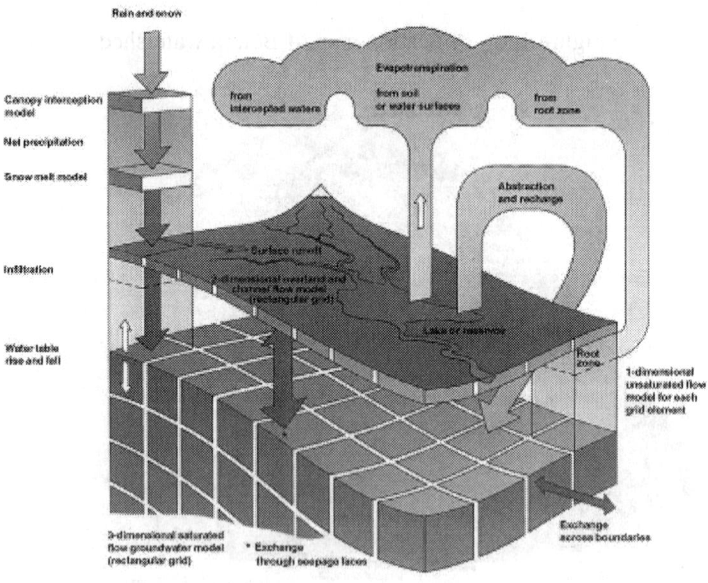

Figure 3: Schematic representation of MIKE SHE.

2.5 Model simulation

MIKE SHE model calibrated for the *Belura* watershed was used to simulate the hydrological water balance using suggested crop planning for the season 2001. The daily hydrological water balance is essential for assessing the yield of the different crops, which is suggested by the qualitative evaluation.

2.6 Evaluation of actual yield

The actual yields of different crops from the area was evaluated by using the output of MIKE SHE model of the *Belura* watershed and the modified FAO relationship (Singh *et al.* [13]), with replacing evopotranspiration term with transpiration, which is given as

$$\left(1-\frac{Y_a}{Y_m}\right)=\sum_{i=1}^{n}K_y^i\left(1-\frac{E_{at}}{E_{mt}}\right) \qquad (1)$$

where, Y_a = the actual yield, Y_m = maximum attainable yield, K_y = yield response factor, E_{at} = actual transpiration, E_{mt} = maximum transpiration, i = crop growth stages. The value of K_y for different crops is based on the evaluation of numerous research results, which cover a wide range of growing condition.

3 Results and discussion

3.1 Land evaluation

The evaluation of soils may be qualitative and quantitative. In present investigation both types of evaluation of soils were used.

3.1.1 Qualitative evaluation

The qualitative evaluation of the soils for crop production has been done by two ways viz. land capability classification and land suitability classification and presented below:

3.1.1.1 Land capability classification Soils of the watershed are described and according to their properties the land capability classes are assigned and presented in table 1.

3.1.1.2 Land suitability classification Important climatic and soil-site characteristics were evaluated to determine the suitability of the soils for sorghum, cotton and pigeon pea, as these are major crops of the region. The results are presented in table 2a and 2b for these crops. The crop requirement table compiled by NBSS and LUP [10] with some modification of crop requirement as given by Sys [15] were also used. The result indicated that the soils Bk5CB(A)1 and Bh5CB(A)1 are very suitable for cotton, sorghum and pigeon pea. Bh4CB(A)1is moderately suitable for cotton, sorghum and pigeon pea. Similarly, Bh3CB(A)1 is also moderately suitable for sorghum. Tn2CB2, Tn2C1B2 and Tn2C1C3 are unsuitable for cultivation.

Table 1: Capability classification of the soils.

Soil series	Description	LCC
Bk5CB(A)1	Clay soils, deep, moderately well drained with 1-3 slope and non-to slightly eroded.	IIe
Bb3CB(A)1	Clay soils, shallow, well drained with 1-3 percent slope and moderately eroded.	IVes
Bh4CB(A)1	Clay soils, moderately deep, well drained with 1-3 percent slope and moderately eroded.	IIIes
Bh5CB(A)1	Clay soils, deep, moderately well drained with 1-3 percent slope and non-to slightly eroded.	IIe
Tn2CB2	Sandy clay loam soils, shallow, well drained, 1-3 percent slope and moderately eroded.	VIes
Tn2ClB2	Sandy clay loam soils, very shallow, well drained, 1-3 percent slope and moderately eroded.	VIes
Tn2ClC3	Sandy clay loam soils, very shallow, well drained with 3-5 percent slope and severely eroded.	VIes

Table 2: (a) Climatic characteristics of soils and climatic suitability for Sorghum, Cotton and Pigeon pea. (b) Soil site characteristics selected for evaluation.

(a)

Climatic Characteristics	Sorghum	LL*	Cotton	LL*	Pigeon pea	LL*
Annual rainfall mm (1971-95)	824.7	0	824.7	1	824.7	1
Rainfall in growing season, mm	537.4	1	537.4	2	537.4	2
Length of growing period, days	138	0	138	1	138	2
Mean temp in growing season, °C	26.55	1	26.55	0	26.55	0
Mean maximum temp. in growing season, °C	34.0	2	34.0	-	34.0	-
Mean minimum temp. in growing season ,°C	19.1	1	19.1	-	19.1	-
Mean relative humidity in growing season, °C	62.0	0	62.0	0	62.0	-
Suitability class	S2		S2		S2	

*Level of limitations.

(b)

Soil-site characteristics	Soil types						
	Bk5CB(A)1	Bb3CB(A)1	Bh4CB(A)1	Bh5CB(A)1	Tn2CB2	Tn2ClB2	Tn2ClC3
Site characteristics							
Slope, %	1-3	1-3	1-3	1-3	1-3	1-3	3-5
Erosion	e1	e2	e2	e1	e2	e2	e3
Drainage	Mod. well	Well	Well	Mod. well	Well	Well	Well
Soil characteristics							
Texture	C	c	C	c	Scl	Scl	scl
Depth, cm	118	30	76	112	20	15	15

3.2 Quantitative evaluation

The productivity index of upstream and downstream of soils of watershed has been calculated and presented in table 3a and 3b. The result showed that the Storie Index Rating (SIR) of the downstream and upstream sides of the watershed falls between 27.45 to 54.15 and 6.12 to 48.45. The SIR of the downstream side is more than the upstream side of the watershed. The productivity class was found poor to fair class at downstream side and very poor to fair at upstream side of the watershed, as downstream side soils are always better than upper reaches. The higher SIR of soils produce more yields than the lower SIR of the soils

Table 3: (a) Storie index rating of different soils of downstream side watershed. (b) Storie index rating of different soils of upstream side watershed.

(a)

Soil types	Factor A Depth	Factor B Textural	Factor C Slope	Factor X Drainage	SIR	Productivity Class
Bk5CB(A)1	100	60	95	95	54.15	Fair
Bb3CB(A)1	60	60	95	85	29.07	Poor
Bh4CB(A)1	85	60	95	85	41.18	Fair
Bh5CB(A)1	100	60	95	85	48.45	Fair
Tn2CB2	40	85	95	85	27.45	Poor
Tn2ClB2	40	85	95	85	27.45	Poor

(b)

Soil types	Factor A Depth	Factor B Textural	Factor C Slope	Factor X Drainage	SIR	Productivity Class
Bk5CB(A)1	95	60	85	100	48.45	Fair
Bb3CB(A)1	40	60	85	95	19.38	Poor
Bh4CB(A)1	60	60	85	95	29.07	Poor
Bh5CB(A)1	95	60	85	95	46.02	Fair
Tn2CB2	20	85	85	95	13.72	Poor
Tn2ClB2	20	85	85	95	13.72	Poor
Tn2ClC3	20	85	60	60	6.12	Very poor

Table 4: Suggested cropping plan.

Present land use	Slope %	Soil series	Area, ha	Suggested crop
Double crops (Green gram & Black gram)	1-3	Bb3CB(A)1	219.49	Sorghum
Single crop (Short duration, Sorghum)	1-3	Bh4CB(A)1	120.46	Cotton
Single crop (Long duration, Cotton)	1-3	Bh5CB(A)1	51.57	Pigeon pea

3.3 Crop planning

The crop planning has been done using qualitative land evaluation i.e. land capability and suitability classification for the soils of watershed. The suggested crop planning of the study area has been briefly summarised in the table 4.

3.4 Water balance of the watershed

MIKE SHE model calibrated for *Belura* watershed, to simulate hydrological water balance of the watershed with suggested crop scenario. The simulated values were used to evaluate the actual yield of the suggested crop.

3.5 Evaluation of yield

The actual yield of sorghum, cotton and pigeon pea crops from the area was evaluated by using the output of MIKE SHE model of the *Belura* watershed and the modified FAO relationship. The actual transpiration and maximum transpiration was estimated by deducting the soil evaporation from the actual and maximum evaptranspiration values, where the soil evaporation and actual evaptranspiration was estimated by using Kristensen and Jensen model. The actual yield of the crops *viz.* sorghum, cotton and pigeon pea at the upstream and downstream of watershed were calculated and presented in the table 5.

The result indicates that the actual yields of the sorghum, cotton and pigeon pea at downstream side of watershed were estimated to be 16.28, 5.87, and 7.72q ha^{-1} and 14.09, 4.91 and 6.48q ha^{-1} at the upstream side, respectively. The actual yields calculated by modified FAO relationship of sorghum, cotton and pigeon pea are estimated to be is 32, 48 and 55% of the maximum yield of 50, 12 and 14 q ha^{-1}, respectively

Table 5: Actual yields of Sorghum, Cotton and Pigeon pea estimated using modified FAO relationship and irrigation requirement.

Crops	Maximum yield, q ha^{-1}		Actual yield, q ha^{-1}		Irrigation required, mm	
	Down	Up	Down	Up	Down	Up
Sorghum	50.00	48.00	16.28	14.09	61.18	64.70
Cotton	12.00	10.00	5.87	4.91	67.92	57.21
Pigeon pea	14.00	12.00	7.72	6.48	110.10	119.20

3.6 Irrigation plan

The amount of water required to obtain maximum yield of the crops, the maximum evapotranspiration at different crop growth stages were taken from the model output and total irrigation requirement were calculated (table 5). The irrigation requirement at downstream side was found to be 61.18, 67.92 and 110.10 for sorghum, cotton and pigeon pea, respectively. Similarly, at upstream side these values are found to be 64.70, 57.21 and 119.20, respectively.

4 Conclusions

The actual yields, calculated using modified relationship developed by Food and Agriculture Organisation, of sorghum, cotton and pigeon pea are estimated to be 32, 48 and 55% of the maximum yield of 50, 12 and 14 q ha^{-1}, respectively. The estimated yield of the crops coincides with respective Storie Index Rating (SIR), which confirm the utility of SIR for crop planning of the watershed. The overall results illustrate applicability of the comprehensive hydrological modelling system, MIKE SHE, for management of water resources for sustainable agricultural productivity in a watershed.

References

[1] Abbott, M. B., Barthurst, J. C., Cunge, J. A., O'Connell, P. E. O. & Rasmussen, J., An introduction to European Hydrological System – Systeme Hydrologique Europeen, "SHE" 1: History and Philosophy of a physically based distributed modelling system. *J. Hydrol.*, **87**, pp. 45–59, 1986a.

[2] Abbott, M. B., Clarke, R. & Preissmann, A., An introduction to European Hydrological System – Systeme Hydrologique Europeen, "SHE" 2: Structure of a physically based distributed modelling system. *J. Hydrol.*, **87**, pp. 60–79, 1986b.

[3] Bathurst, J. C., Physically based distributed modelling of upland catchment using the Systeme Hydrologique Europeen. *J. Hydrol.*, **87**, pp. 79–102, 1986a.

[4] Barthurst, J. C., Sensitivity analysis of the Systeme Hydrologique Europeen for an upland catchment. *J. Hydrol.*, **87**, pp. 103–121, 1986b.

[5] FAO, *A frame works for land evaluation.* FAO Soil Bull. No.32, Rome, pp. 72, 1976.

[6] FAO, *Land evaluation for rainfed agriculture,* FAO Soil Bull. No.52, Rome, pp. 60–63, 1983.

[7] FAO, *Sustainable agriculture production for international agriculture research. Report.* Tech. Advisory Committee, Consulting Group on Intl. Agril. Res. (CGIAR) Pergamon, New York, 1989.

[8] Jain, S. K., Storm, B., Barthurst, J. C., Refsgaard, J. C. & Singh, R. D., Applications of the SHE to catchments in India - Part 2: Field experiments and simulation studies on Kolar subcatchment of the Narmada River. *J. Hydrol*, **140**, pp. 25–47, 1992.

[9] Klingebiel, A. A. & Montgomery, P. H., *Land capability classification*, US DA Handbook, **210**, Washington, pp. 21, 1961.

[10] NBSS and LUP, *Proceedings National meet on soil site suitability criteria for different crops*, National Bureau Soil Science and Land Use Planning, Nagpur, pp. 30, 1994.

[11] Refsgaard, A. & Storm, B., *'MIKE SHE', Computer models in watershed hydrology,* V. P. Singh Ed., Water Resources Publication, Colorado, USA, pp. 806–846, 1995.
[12] Refsgaard, J. C., Seth, S. M., Barthurst, J. C., Erlich, M. & Jorgensen, B., Application of SHE to catchment in India - Part II: General Results. *J. Hydrol.*, **140**, pp. 23–59, 1992.
[13] Singh, R., Subramanian, K. & Refsgaard, J. C., Hydrological modelling of a small watershed using MIKE SHE for irrigation planning *Agricultural Water Management*, **41(3)**, pp.149–166, 1999.
[14] Storie, R. E., *Storie index soil rating (revised)*. Spec. Publ. Agric. Sci. Univ. California, **3203**, 1978.
[15] Sys. C., *Land evaluation Part I, II and III*, State University, Ghent Publications. pp. 352, 1985.

Section 4
Irrigation systems and planning

An integrated framework for sustainable agriculture land use and production practices

T. Leitão[1], M. C. Cunha[2], I. Laranjeira[1],
J. P. Lobo-Ferreira[1] & E. Paralta[3]
[1]*LNEC, National Laboratory for Civil Engineering, Portugal*
[2]*IMAR, Department of Civil Engineering, University of Coimbra, Portugal*
[3]*CVRM, Technical University of Lisbon, Portugal*

Abstract

Diffuse pollution is among the major environmental concerns in terms of prevention of further deterioration and restoration of water to a "good status" in terms of ecological and chemical parameters. A step forward should be made for finding better land use, crop irrigation and fertilization practices to avoid damaging water bodies. This paper presents a framework for obtaining the decisions to be implemented in order to define a sustainable agriculture land use and production practices. A multidisciplinary approach is proposed for building a decision model capable of representing all the issues involved in the scope of an integrated management of land and water resources. The framework proposed is the basis of a research project recently financed by the Portuguese research foundation (Fundação para a Ciência e a Tecnologia). The main objectives of the project can be summarized as follows: 1) Improve the scientific knowledge regarding the interrelation between the agriculture land use and the production practices and the groundwater and the surface water quality protection, towards a more sustainable agriculture; 2) Contribute to support future decisions in terms of more adequate policies regarding rural land use planning (type of crops and associated fertilizers and treatment techniques), taking into consideration the protection of the environment based on vulnerability and risk concepts.

Keywords: diffusion pollution, Sustainable agriculture practices, decision models.

WIT Transactions on Ecology and the Environment, Vol 112, © 2008 WIT Press
www.witpress.com, ISSN 1743-3541 (on-line)
doi:10.2495/SI080211

1 Introduction

There is a major challenge ahead for Europe in what concerns the management of the environment through the concept of integrated river basin management. This challenge is being driven by an important piece of European legislation concerning water protection, the Water Framework Directive. The overall aim of the Directive is to establish a legal framework to protect surface water and groundwater using a common management approach and following common objectives and principles.

Diffuse pollution is among the major environmental concerns in terms of prevention of further deterioration and restoration of water to a "good status" in terms of ecological and chemical parameters. This objective is stated in the Water Framework Directive paragraph "Member States shall implement the measures necessary to prevent or limit the input of pollutants into groundwater and to prevent the deterioration of the status of all bodies of groundwater".

In fact, one of the basic principles of the Directive says that "Surface waters and groundwater are in principle renewable natural resources; in particular, the task of ensuring good status of groundwater requires early action and stable long-term planning of protective measures, owing to the natural time lag in its formation and renewal. Such time lag for improvement should be taken into account in timetables when establishing measures for the achievement of good status of groundwater and reversing any significant and sustained upward trend in the concentration of any pollutant in groundwater".

This paper presents a framework for obtaining the decisions to be implemented for accomplishing a sustainable agriculture land use and production practices. This framework is the basis of a research project recently financed by the Portuguese research foundation (Fundação para a Ciência e a Tecnologia). In this project we want to go a step further on analysing how can land planning be used to improve the surface and groundwater quality. An improved knowledge of the vadose zone role (usually an area where the phenomena are more difficult to measure and to study) would be crucial. The construction of a decision tool integrating all the knowledge gathered in the foreseen multidisciplinary approach will contribute to more rational decisions.

This project will concentrate efforts in gathering and integrating the knowledge of the processes that interfere in the migration of pollutants originated by soil fertilization, for the different media (soil, vadose zone, groundwater and surface water), encouraging the future utilization of more sustainable crops and fertilizer practices that can decrease the risk of groundwater and surface water quality degradation.

This project will give scientific proves for the implementation of decision-making strategies for environmental protection of water resources. It will support political decisions from county administrations, responsible for regional planning, taking into account the sustainable development of the region and the appropriate use of nitrogen fertilizers on the basis of several European and national directives: EU-Nitrate Directive (91/676/EEC); Drinking Water Directives (80/778/EEC); Environmental Farming Proposals (91/2078/EEC);

Delimitation of Vulnerable Zones to Nitrates (91/676/EEC) and EC Water Framework Directive (2000/60/EC).

2 The decision model

The decision tool will be built to maximize the net benefits from land use, taking into consideration the different soils present in the area (and its behaviour measured in the field and lab experiments), the fertilization practices for each crop, the groundwater and surface water availability and vulnerability, the costs of land use changes, socio-economic aspects in such a way that agricultural practices can be compatible with water consumption and respecting the ecological status of associated water bodies.

Having Figure 1 as a reference scenario to implement the decision model, and taking into account the interrelationship between practices and consequences in water bodies (Figure 2), the objective function of general decision model can written as follows:

$$\max\ BLA = \sum_{ijk}\left(p_j y_{aijk} A_{ijk}\right) - \sum_{ijk}\left(ca_{ijk} A_{ijk}\right) - \sum_{ijk}\left(ci_{ijk} I_{ijk}\right) - t_p \sum_{i} N_i$$

BLA: net benefits;

i: zone index;

j: crop index;

k: agriculture practice index;

p_j: unitary price of the crop j;

y_{aijk}: actual production of the crop j in the zone i with the agriculture practice k;

A_{ijk}: area of the zone i, concerning the crop j with the agriculture practice k;

ca_{ijk}: unitary cost for cultivating the zone i, with the crop j, with the agriculture practice k;

ci_{ijk}: unitary cost of irrigating the zone i, with the crop j, with the agriculture practice k;

t_p: unitary cost (tax) for pollution;

N_i: pollution from the zone i.

The pollution from zone i is determined, through the use of a transfer function χ_{il}, taking into account the area and the quantity of fertilizer used:

$$N_i = \sum_{jkl} \chi_{il} n_{ijk} A_{ijk}$$

χ_{il}: transfer function that represents the fraction of the pollution from zone i that reaches the l water body (vadose zone, groundwater and the surface water), depending on the soil type, the infiltration, the piezometric level, etc.;

n_{ijk}: quantity of fertilizer used in the zone i, for the crop j with the agriculture practice k.

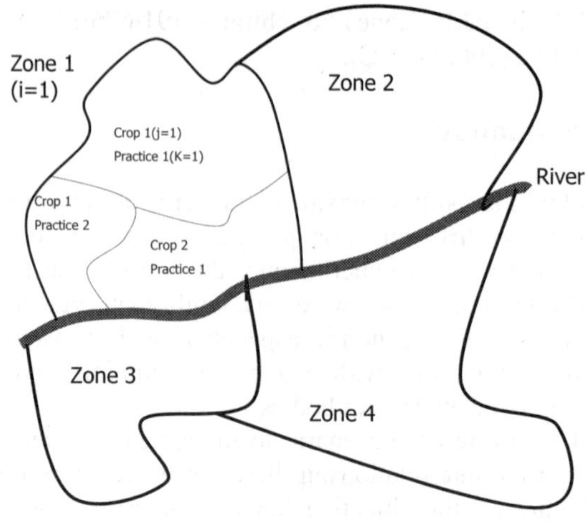

Figure 1: Reference scenario.

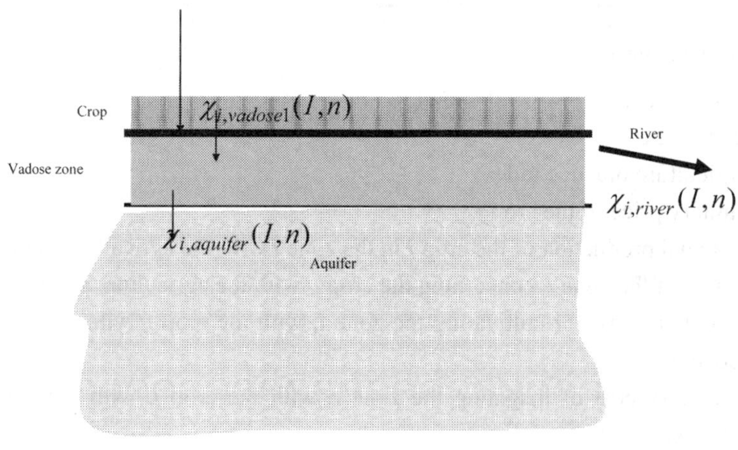

Figure 2: Reference water bodies transfer functions.

The decision model will include constraints representing water and land availability, limits on the amount the pollution allowed in each water body, physical limitation relating crop production with the production factors.

To built the decision tool the following tasks have to be accomplished:

- Development of cost and benefit functions from land use. Production functions relating the agriculture production and the production factors like water, soil and crop type, fertilizers, agriculture response factors will be considered.
- Development of the transfer functions to evaluate the consequences of land use and agriculture practices. For such purposes, the project will thus focus on assessing the impacts of agriculture on groundwater and

surface water quality through an integrated and multidisciplinary approach including:

i) Agro-Hydrosystem Characterisation - Inventory and quantification of diffuse pollution sources, inventory of crops, identification of associated fertilizers and treatment techniques, irrigation schemes for each crop, and after all the movement of pollutants in the soil, vadose zone an, groundwater, surface water associated with each crop.

ii) Soil and Vadose Zone Characterization and Modelling - Characterisation of soil and unsaturated zone hydraulic properties and mass flow and transport modelling calibration through the unsaturated zone using deterministic models for nitrate and pesticides (HYDRUS 1D or 2D).

iii) Aquifer Monitoring and Modelling - mass flow and transport modelling (MODFLOW), interactions between groundwater and surface water; water balance modelling; recharge assessment; and groundwater seasonal monitoring (nitrates, pesticides, phosphates, and water level).

iv) Surface water quality modelling - mass flow and transport modelling (QUAL2K), evaluation of the ecological status resulting from non-point sources pollution coming from the different agriculture practices.

- Development of decision models (objectives and constraints). These models will include the cost and benefit functions, the agriculture production function, the groundwater and surface water quality model. They will incorporate also the information and the models used in the previous tasks in order to link the groundwater and surface water response to the agriculture practices regarding quantity and quality issues. Socio-economic aspects will be also taken into account in the benefits and in the cost terms.
- Assessment of the more suitable optimization methods to solve the decision models developed. The decision models previously built will be stochastic non-linear mixed integer models. In fact these models will involve uncertainty, non-linear functions, numerical approximations of differential equations with partial derivatives and combinatorial decisions (yes or no decisions). Robust optimization techniques combined with new metaheuristics like simulated annealing or genetic algorithms will be used for solving such decision models.
- Resolution of the decision models through the use of the chosen methods. Some work calibration of the metaheuristics algorithms have to be accomplished and different robustness measures have to be evaluated.

3 Models calibration and validation

A pilot area has been defined to conduct field experiments and analysis that will provided the understanding and the data for the application of the decision model to a real world case study. Diffuse pollution is a problem of increasing concern in Alentejo rural region (South Portugal) and particularly in the new areas of Alqueva Irrigation Project. Several previous studies have been conducted to analyse the effects of diffuse pollution in the water resources, in Portugal and elsewhere. The main hydrogeological and agriculture features of the study area were previously identified by the project partners and published [8]. Project partners have large experience on this type of studies [2–9]. Also the study of nitrate pollution associated to agriculture and its direct and indirect effect on the rates and compositions of groundwater recharge flux and aquifer biogeochemistry have been studied by several other authors, namely [1] and [10]. The study area is located in Alentejo region (South Portugal), with semi-arid climatic conditions. That area represents the first irrigation perimeter (64 km^2) starting 2004 in the context of Alqueva Irrigation Project (the biggest artificial lake in Europe) expected to achieve 110 000 ha of irrigated land in 2025. It is an intensively agricultural site of arable land with different cultivations scenarios and techniques located above a regional aquifer used for water supply. The different tasks to be developed for models calibration and validation are described next.

3.1 Task 1

In the first task a complete characterisation of the agro-hydrosystem in the study area, and its seasonal variations, is to be carried out for a 2-year period. The results expected in this task are: field quantification of nitrogen, pesticide and phosphates inputs for 3 different crops, considering their different irrigation schemes, and fertilizers and treatment techniques; field quantification of nitrogen, pesticide and phosphates migration paths in 2 different soils and for 3 different crops: pollutants quantification in soils, vadose zone and groundwater; laboratory soil-column simulation of field experiments: pollutants quantification in soils and water.

The results are expected to be obtained through the following actions:

• Selection of a small watershed basin area within the “Infra-estrutura 12” 64 km^2 total area, with different crops along the year.

• Delimitation of two irrigation plots with different soil types (A and B). Selection and characterisation of 6 irrigation portions with different soil types (2) and crops (3).

• Inventory of the historical and existing annual land use plan at the site.

• Analysis of previous studies in the area and inventory of existing infrastructures that can be used for further groundwater monitoring.

• Analysis and interpretation of previous groundwater and surface water quantity and quality data and its connection to the land use practices. Analysis of the following relations, by using statistical modelling (trend and cluster analyses):

i) Type of land use / type and amount of fertilizers applied (including the season for its application);

ii) Type of land use / agriculture techniques / groundwater quality/ surface water.

• Installation of tensiometer to measure soil humidity, which will allow following the progression of water in the soil.

• Installation of Teflon capsules in the vadose zone for water sampling.

•Assessment of different land use scenarios effects in the migration of pollutants (nitrates, phosphates, and some applied pesticides). This action will be carried out in (a) two field crops and in (b) laboratory soil-columns:

i) The field monitoring of crop effects will be carried out for two soil types (A and B) in which three different types of crop (corn, sunflower and melon) will be tested. For both soil types, the effects of the migration of nitrates, chlorides, phosphorous and some pesticides, will be monitored and analysed in three media: soil (soil itself), vadose zone (Teflon capsules) and groundwater (piezometers).

ii) The laboratory tracer tests, to be performed in soil-columns, will be used to reproduce the field experiments in the saturated zone and possibly also in the vadose zone. In that sense, this task will be done by measurements in the same (two) soils and the effect of the three crops will be reproduced in lab conditions by the type and intensity of fertilizer applications that characterise each crop.

3.2 Task 2

This task is devoted to a complete characterisation of soil and unsaturated zone hydraulic properties and will be made on the base of field permeability tests, soil bulk density, granulometry and laboratory pF curves determination on undisturbed samples. After that, it will be developed and calibrated a mass flow and transport modelling through the unsaturated zone using deterministic models for nitrate and pesticides. These models are widely used and accepted by international scientific community (HYDRUS 1D and 2D) using Richards equation, convection-dispersion equation and van Genuchten algorithm and considering half-life decay of pesticides.

In the end of this task, an evaluation of the effects of different land use scenarios in the concentration of pollutants (nitrates, phosphates, and some applied pesticides) in the soils and the vadose zone will be done. This will be made by using the field data and by interpreting it using deterministic models for nitrate and pesticides transport.

The results are expected to be obtained through the following actions:

• Determination of seasonal variations of soil moisture over unsaturated zone profile and fertiliser leaching. This will be supported by automatic soil moisture probes and appropriate ceramic blocks during 2 years for major pollutants (nitrates, phosphates, and some applied pesticides).

• Measurement of agro-meteorological data (rainfall, ET0, soil moisture, groundwater level, etc.) for soil and unsaturated zone.

• Determination of seasonal water balance using deterministic models supported by field calibration.

• Evaluation of different land use scenarios in the concentration of pollutants (nitrates, phosphates, and some applied pesticides) in the soils and the vadose zone taking account specific conditions of the study area. This will be done by using deterministic models for nitrate and pesticides (HYDRUS 1D and 2D).

3.3 Task 3

Under this task the development and calibration of flow and transport models (MODFLOW and QUAL2K) for saturated media and river quality will be done. The input parameters results obtained in the previous task, for unsaturated zone modelling, will be used.

Groundwater quality modelling in order to ascertain the correspondence between the land use and water composition and to calculate geochemical mass balances will be made using PHREEQ2C model. Also the study of main interactions between groundwater and surface water, recharge assessment and seasonal monitoring of pollutants (nitrates, pesticides, phosphorous) and water level will be carried out.

The results are expected to be obtained through the following actions:

• Soil and water samples will be collected and analysed for major elements and common pesticides. All information will be stored in Database and user-friendly GIS.

• Groundwater and surface water quality modelling in order to ascertain the correspondence between the land use and water composition and to calculate geochemical mass balances will be made using PHREEQ2C model.

• Hydrogeological and stochastic modelling including:

i) Definition of the hydrogeological conceptual model at the site.

ii) Determination of the spatial distribution of the main hydrogeologic parameters using stochastic models for further incorporation on deterministic models.

iii) Flow and transport modelling using the appropriate deterministic models.

iv) Simulation of different pollution scenarios for different land uses and soils.

This task intends to help interpreting the field and lab results carried out for the 6 different scenarios (2 soils and 3 crops). For the field experiments, the following tasks will be done:

• Definition of the hydrogeological conceptual model at the site (partially already done in previous work at the site, although needing to be adapted to the small watershed basin area to be selected).

• Determination of the spatial distribution of the main hydrogeologic parameters using stochastic models for further incorporation on deterministic models (also using some previous data for the permeability values, dispersivity, etc.).

• Flow and transport modelling of the vadose zone using the appropriate deterministic models (possibly HYDRUS-2D).

• Flow and transport groundwater modelling using the appropriate deterministic models (possibly MT3D); some previous modelling of the site was done and the values of recharge calculated will also be used.

• Simulation of different pollution scenarios for different land uses and soils.

For the laboratory experiments, the following tasks will be done:

• Modelling flow and transport for each crop at laboratory scale for the 2 different soils (possibly using CANALT model). The relevance of the physical properties, namely the permeability and the porosity of the media, as well as the chemical properties like cation exchange capacity, adsorption capacity as well as the conditions of temperature and pH, will be analysed.

4 Conclusions

An integrated framework for a sustainable agriculture land use and production practices is presented. Diffuse pollution is among the major environmental concerns in terms of prevention of further deterioration and restoration of water to a "good status" in terms of ecological and chemical parameters. The importance of building a decision model is enhanced and the different tasks to be accomplished in the scope of a multidisciplinary approach, for such purposes, are systematized. These kind of models are very important to tackle the challenge of an integrated river basin management, thus fulfilling the objectives of the Water Framework Directive.

The results achieved until now within this project are presented in [11] and [12].

Acknowledgement

The authors gratefully acknowledge support received from the Fundação para a Ciência e a Tecnologia through grant POCTI/AGR/57719/2004.

References

[1] Bohlke, J-K., Groundwater recharge and agricultural contamination. *Hydrogeology Journal*, 10: 153–179, 2002.

[2] Leitão, T.E., Quigley, S. & Lobo Ferreira, J.P.C., Propagação de Nitratos na Zona Vadosa do Solo. Apresentação de um Ensaio de Traçador Realizado no "Aquífero Artificial" do GIAS/DH", *Comunicação apresentada ao 4.º Congresso da Água*, APRH, Lisboa, 11 pp., 1998.

[3] Leitão, T.E., Lobo Ferreira, J.P.C., Quigley, S. & Theves, T., On Flow and Transport of Miscible Tracers in the Vadose Zone, *European Science Foundation (ESF) Groundwater Pollution Conference*, Goteborg, Sweden, 8 pp., 1998.

[4] Lobo Ferreira, J.P.C. & Oliveira, M.M., On the Experience of Groundwater Vulnerability Assessment in Portugal, *1st International Workshop – Aquifer Vulnerability and Risk*, Salamanca, Gto. México, 10 pp., 2003.

[5] Novo, M.E., Ikavalko V.-M., Diamantino, C. & Lobo Ferreira, J.P.C., MANPORIVERS - Management Policies for Priority Water Pollutants and their Effects on Foods and Human Health: Generic Characterization of Mathematical Models to Simulate the Pollutant Transport in Groundwater, *Relatório 310/03-NAS*, LNEC, Lisboa, 113 pp., 2003.

[6] Nunes, L.M., Paralta, E., Cunha, M.C. & Ribeiro, L., Groundwater nitrate monitoring network optimisation with missing data, *Water Resources Research,* Vol. 40, W02406, doi:10.1029/2003WR002469, 2004.

[7] Paralta, E. & Ribeiro, L., Stochastic Modelling and Probability Risk Maps of Nitrate Pollution in the Vicinities of Beja (Alentejo, South Portugal), in Ribeiro L. (editor). *Proc. of 3rd International Conference on Future Groundwater Resources at Risk* (FRG'01), CVRM/IST, Lisbon. Portugal, 251–261 pp., 2001.

[8] Paralta, E., Francés, A. & Sarmento P., Caracterização Hidrogeológica e Avaliação da Vulnerabilidade à Poluição Agrícola do Aquífero Mio-Pliocénico da Região de Canhestros (Alentejo), *6.º Congresso da Água*, Porto, 16 pp, 2002.

[9] Ribeiro, L. & Paralta, E., Stochastic Modelling of space-time variability of nitrate pollution using indicator geostatistics and transition probability, *4th International Conference on Calibration and Reliability in Groundwater Modelling*, MODELCARE 02, IAHS, Prague, Czech Republic, 7 pp. 2002.

[10] Tindall, J.A, Petrusak, R.L. & Mcmahon, P.B., Nitrate transport and transformation processes in unsaturated porous media, *Journal of Hydrology*, 169 51–94, 1995.

[11] Leitão, T.E., Laranjeira, I., Henriques, M.J. e Paralta, E., Metodologias para um Melhor Planeamento e Gestão do Uso do Solo Agrícola atendendo à Vulnerabilidade dos Aquíferos à Poluição Difusa. 2.º Relatório 75/07 - NAS, LNEC, 2007, 65 pp.

[12] Leitão T.E., Laranjeira, I., Oliveira, L., Paralta E. e Cunha, M.C. - Metodologias para um Melhor Planeamento e Gestão do Uso do Solo Agrícola atendendo à Vulnerabilidade dos Aquíferos à Poluição Difusa. 3.º Relatório. Relatório 48/2008-NAS, 2008, 68 pp.

Sustainable irrigation in areas with an arid and semi-arid climate in the province of Alicante

C. J. Sánchez-Rubio
Institute of Water and Environmental Science,
University of Alicante, Spain

Abstract

The lack of water in sufficient quantity and quality is the main cause of the reduction of the irrigated area in the region studied in the province of Alicante. The irrigated area has incorporated the most advanced agricultural techniques (greenhouses) and irrigation systems for greater efficiency in water use (localized) obtaining economic, social and environmental sustainability. This vanguard agriculture is sustainable, profitable, exports on both a national scale and to the European Union, generates employment, enjoys a favourable income situation and has a set of objective conditions and advantages confirming its importance in the recent process of agrarian economy. This new agricultural irrigation has been the engine of economic and social development but it is not a final solution to solve the problems of regional imbalance.
Keywords: water scarcity, irrigation, vanguard agriculture, export, profitable and sustainable.

1 Objectives

This work aims to study and analyse the economic, social and environmental aspects of sustainable irrigated agriculture in areas with an arid and semi-arid climate in the province of Alicante to establish a diagnosis of the deficits, advantages and potential. Also, to address the lack of guarantee of water for agricultural use through better management and optimization of this resource: saving, reuse of reclaimed flows, desalination, and even new posts alien to recover the comparative advantages of sustainable vanguard agriculture (vegetables, citrus fruits, flowers and ornamental plants) in the domestic market and the European Union. The region studied includes both the areas with an arid

WIT Transactions on Ecology and the Environment, Vol 112, © 2008 WIT Press
www.witpress.com, ISSN 1743-3541 (on-line)
doi:10.2495/SI080221

climate (Bajo-Segura and Bajo-Vinalopó) and with a semi-arid climate (Alto-Vinalopó, Medio-Vinalopó, Campo de Alicante and La Marina Baja).

2 The traditional model of productive agricultural and the environmental problems generated

2.1 A physical environment with risks and opportunities for agricultural use of land

The biggest limiting factor of agricultural irrigation is the non-availability of water in sufficient quantity and quality in the region studied (Fig. 1). Irrigation means multiplying profitability in the area of upland between 10 and 15 times. An Alicante farmer, skilled, dynamic and favoured by their income situation is

clima arido = dry climate, clima semiarido = semi-dry climate

Figure 1: Type climate of the province of Alicante.

synonymous with exporter but influenced by endogenous processes (erratic and little rainfall) and exogenous Common Agricultural Policy (CAP), World Trade rganization (WTO), Framework Directive of Water (FDW), the non-agricultural water demands and technology). The ability to regulate water for irrigation involves converting a strangulation risk into a large productive strength that does not allow the cultivation of upland dependent on the timing and amount of precipitation.

The control the of surface water flow, retaining it through dams and storing it in reservoirs or diverting it through channels enjoys a long tradition in the region of Alicante. The existence of irrigation farming goes back to the Muslim era. Characterized by the exclusive use of the resources of the Segura river surface to transform unproductive soil in areas of considerable wealth (Bajo-Segura), is growing with the construction of reservoirs in Tibi (1579) to supply the Huerta de Alicante and in Elche the Vinalopó area in the following century. The search for new water resources led to the construction of important waterworks throughout the 20th century such as: the Canal de la Huerta of Alicante, which was opened in 1909, which supplies water from artesian wells in Alto Vinalopó (Villena) and the Compañia de Riegos de Levante (The Levante Irrigation Company) by granting flow surplus from the Segura river in 1918, whose infrastructure ranges from the Bajo Segura to the Campo de Alicante.

The processes described are always related to the use of agricultural water, attempting to promote the economic development of the area through the continued expansion of the irrigated area.

The use of surface water was instigated by the Government to regulate the use of a particular flow of public water is a paradigmatic example in the Plan for the use of the River Segura, according to the Bill of 25th April 1953, which assesses the future availability, allocation to the various irrigated areas and the land subject of enlargement. The physical and ecological potential, the high production and the income generated (6–7 six times higher, when the average in Spain is only four times higher) encouraged the exploitation of groundwater between 1960–1980 as a private initiative, as an alternative to ensure the availability of water for production of the irrigated area. This is the source of the new irrigation agriculture which incorporates new technologies (localized irrigation), to reduce water consumption per unit area, that are intensive (greenhouses) and productive and preferably located with preferential in the coastal areas. This new agriculture is based on small technified production units, flamboyant crops and preferential sales to foreign markets.The extraction of water in the region studied led to an alarming decline in the aquifers which made it necessary to transfer water from the headwaters of the River Tajo to the River Segura. The scarcity of water resources in relation to the growing demand obliges the State to take certain restrictive measures in determined areas which are also applicable in others (Bajo-Segura, Bajo Vinalopó and Campo de Alicante). Irrigated agriculture, industry, tourism and leisure with the corresponding urban consumption have increased pressure on existing water resources. The greatest increase in the need for water occurs in the region studied, making it necessary to transfer surplus flow from the headwaters of the

River Tajo, based on final production criteria and economic and social efficiency.

The use of surface water was instigated by the Government to regulate the use of a particular flow of public water is a paradigmatic example in the Plan for the use of the River Segura, according to the Bill of 25th April 1953, which assesses the future availability, allocation to the various irrigated areas and the land subject of enlargement. The physical and ecological potential, the high production and the income generated (6–7 six times higher, when the average in Spain is only four times higher) encouraged the exploitation of groundwater between 1960–1980 as a private initiative, as an alternative to ensure the availability of water for production of the irrigated area. This is the source of the new irrigation agriculture which incorporates new technologies (localized irrigation), to reduce water consumption per unit area, that are intensive (greenhouses) and productive and preferably located with preferential in the coastal areas. This new agriculture is based on small technified production units, flamboyant crops and preferential sales to foreign markets.

The extraction of water in the region studied led to an alarming decline in the aquifers which made it necessary to transfer water from the headwaters of the River Tajo to the River Segura. The scarcity of water resources in relation to the growing demand obliges the State to take certain restrictive measures in determined areas which are also applicable in others (Bajo-Segura, Bajo Vinalopó and Campo de Alicante). Irrigated agriculture, industry, tourism and leisure with the corresponding urban consumption have increased pressure on existing water resources. The greatest increase in the need for water occurs in the region studied, making it necessary to transfer surplus flow from the headwaters of the River Tajo, based on final production criteria and economic and social efficiency.

Act 2/1980 of 16th October on "Regulation of the economic system of exploitation of the Tajo-Segura aqueduct", in its first additional provision, assigns a volume of 127.8 hm^3/year for agricultural use in the Lower Segura, Lower Vinalopó and Campo de Alicante regions. These flows of water are the reason for the expansion of the irrigation area to a maximum of 123,936 ha. In 1986 (100%) which was reduced to 97,550 ha. by the year 2006 (78.7%) due to the insecurity and uncertainty of receiving the volume of water allocated because of the continuing drought periods at the head of the Tajo river in the Segura basin (Table 1). Therefore, the objective and strategies in the irrigated area is estimated at consolidation by guaranteeing supply (transfer, sewage and treated water, desalination plants, surveys and recharging of aquifers), and modernization (savings, localized irrigation, improved management efficiency). This is the best way to have sufficient allocations, high efficiency and environmental improvement.

2.2 The development model of productive agriculture: economic benefits and environmental impact

The Spanish policy in rural areas has been defined by two criteria: economic/political/social and productive. The first is aimed at increasing the returns to supply the growing food demand. The solution offered was the expansion of irrigated area to make it a vanguard, in the mid-fifties, agricultural modernization of the country. This productive agriculture is profitable, competitive and dynamic as a basic factor for the development of the rural world. One area favoured by bringing Spain into the former European Economic Community (EEC) that acquires major economic impact (for ease of marketing their products derived from citrus, vegetables, fruit etc.) and social (creates a lot more jobs than the same rain-fed surface and fixed population in rural areas). The irrigation of the new territorial unit aimed to increase production, productivity and be of a family character. Demand for these products has grown significantly as a result of demographic changes, the income level of the population and the new diet. The farmer's response has been the intensification of agricultural production through increased food and its variety. This process has been carried out by incorporating technological advances, chemicals (fertilizer) and mechanization. The farm owner decided on the idea that investing more would increase his earnings. He began dialectic mechanization (without reducing wage labour) – productivity-emigration. The traditional model of agriculture was therefore abandoned and capitalism began in the agricultural sector so the model of subsistence was done with and production began to be intensified to meet the demands of new products. This process resulted in the appearance in rural areas of new functional and socio-economic relationships which altered traditional ways. The modern agricultural sector is linked with extra-regional commercial channels as opposed to the old system of city-country relations.

The irrigated area in the region studied adapted its products (citrus, fruits and vegetables, ornamental plants, flowers etc.) to the market thus converting traditional agriculture into modern agriculture dependent on exports. Few Spanish agricultural areas have experienced a change so radical in their agriculture as in this geographical area, confirmed statistically by improving familiar agricultural structures, the introduction of new crops, intensive vegetable development, the proliferation of greenhouses, replacing traditional irrigation by localized or drip, increased productivity and income of farmers and harvesters-exporters and the employment generated which in the irrigated coastal ranges from 30 times to 50 times that of rain-fed areas. A more practical way to acknowledge the crucial role played by vanguard agriculture in economic growth and level of income, both for its own monetary value as the multiplier effect generated in other sectors of its economy. Model consolidated land use, from the 1960's sixties in the territorial unit to capture 90% of irrigated area and contribute the same percentage in the final agricultural production. The intensification of modern vanguard agriculture has significantly improved productivity through high prices and guarantee of sales in the domestic and community markets. The increased profitability comes from the utilization of land through: tomatoes, peppers, artichokes, lettuce, flowers, ornamental plants, citrus and almond in irrigation, constantly threatened by the lack of guarantee of water availability. Therefore, operation of a capitalized and intensive family nature, of greenhouse crops localized irrigation, cooperatives, marketing, etc., is being debated so as to match the size to the most favourable crops in the market, the optimum use of water (greenhouses, drip), the use of soluble fertilizers, construction of a framework for closer planting and saving in manpower. This is to face the future and the surrounding uncertainty in an intelligent and pragmatic manner to resolve the problems of more productive, competitive and dynamic agriculture in the province.

This intensification of soil irrigation and the advantages mentioned have caused problems for the rural community such as depletion of water resources (by disproportion between the irrigated surface and constant periods of drought, over-exploitation of groundwater), soil erosion and contamination of agricultural environments. Problems generated for productive agriculture created by Sustainable Agriculture. In other words, the old concept of growth based on the use of ever increasing water flow, of energy and raw materials is unsustainable and should give way to a less intensive exploitation of natural resources.

3 The territorial model of sustainable irrigation: risks and opportunities

Sustainable development was introduced as an explicit objective of the European Union in the Single European Act of 1987. The treaty of Maastricht 1992 obliges members to integrate environmental matters into all Community policies. Thus, irrigation agriculture of the territorial unity of Alicante must adapt its traditional approach to a framework where the predominant factors are the reduction of unit costs, increased product quality, preservation and maintenance of natural resources and environmental integration. Present and future irrigation must be able to manage water resources, obtaining an agricultural product competitive in quality and price in an increasingly global marketplace through compatibility with the economic, social and environmental criteria. This new concept identifies a form of irrigation capable of remaining indefinitely (durable) fulfilling the basic functions of being economically viable, socially useful and environmentally sustainable.

Under this new approach, it is necessary to abandon the symbolic nature which irrigation has traditionally had as an engine of development. Sustainable agriculture has to compete with other uses of water. Also, the Common Agricultural Policy (CAP) must involve the development of irrigation and sustainable development in force since 1992. The result of the CAP in Spain is the consolidation of a dual agriculture: intensive and hardly competitive or non-subsidized, as compared to extensive cultivation of uncompetitive products (cereals) which is almost entirely subsidized. That is to say, the most promising horticultural sector in the regions of Alicante is at a disadvantage compared with less economically viable crops. We must produce what sells, rather than trying to sell what has been produced.

Sustainable agriculture in Alicante is conditioned by the new policy of agricultural markets and limited water resources. As a result of such constraints, irrigation has to be redirected from the traditional model to the new model which is sustainable through the rational use of water and the efficiency of the utilization of the resource from the perspectives of production and the environment must be increased.

3.1 Ensuring water availability as a requirement of sustainable irrigation

Sustainable irrigated agriculture in the regions of Alicante aims to ensure the need of water for crops and achieve greater efficiency in its use. Renewable indigenous resources and those from other regions, subject to irregularities in rainfall (drought) are not able to secure agricultural production under irrigation. The need to make the use of land more competitive involves ensuring adequate flow to the existing irrigated areas, before expanding new areas of irrigation. This approach is essential in the province of Alicante where water management is a very difficult and complex problem that coincides with a growing demand for agriculture, industry and urban supply, particularly tourism (in expansion) , with increasingly scarce surface and underground water resources and a few outside sources dependent on availability of the headwaters of the River Tajo. Irrigation is the single largest user of water in the areas studied. This consumes 75% of water and produces, referring to the territory climate, between 5 and 8% of the gross domestic product in the agricultural district that falls to 2% when the value refers to the whole provincial. This agricultural sector contributed 15% of the total value of exports in 2006. The economic importance of the value of exports of agricultural products improved significantly from1999 to 2006, increasing by 35%, from an amount of 263 to 550 million Euros. An amount that should be much higher but that was affected by the impact of drought that plagued the province of Alicante and the headwaters of the River Tajo. This lack of assurance of water resources has prompted both the incorporation of technological improvements (drip irrigation) and the increased availability of flows through regeneration that produces better drinking water than simple debugging. The latter is a process that involves improvement in the environment and optimizes the management of water. These are new technologies that extend to the practice of desalination of brackish and marine water to make new resources available to the farmer because the current systems do not ensure efficiency, fairness (ensuring water quantity and quality of the less favoured farming population) and sustainability. By constructing new desalination plants to supply the Irrigators Communities (Comunidades de Regantes) (40 hm^3/year in Torrevieja and 60 hm^3/year, according to requirements in Guardamar de Segura and another 16 medium and small size plants) at a price of 0.30 € (double the current rate of the Tajo-Segura aqueduct) it is hoped that agricultural development will be enhanced. This is one way to subsidize productive and sustainable agriculture in the Mediterranean. Accordingly, it is necessary to overcome the shortage of water and its impact on economic development in vanguard agriculture by offering a guarantee of water availability. Economizing on water will permit the comparative and profitable advantages of the territorial unit of Alicante to be recovered. The rise in horticulture has more to do with the dialectic integration-production-export than with aspects relating to ownership of land or poor structure of farms. And it is that by controlling the technical processes of greenhouses and localized irrigation, that reduces costs by the induced effects arising from fertigation and saving on manpower) higher yields and quality are obtained that reflect the status of the agricultural land boom in Alicante. There is an offer of agricultural products which is concentrated in the months of winter-spring during which there is little production and a high demand exists in the countries of the European Union. There is the advantage of a favourable climate and timetable if the problems due to drought and the price received by the farmer are overcome. The benefit obtained by the most significant crops shows that vegetables, citrus and table grapes because of their economic social and environmental viability, can assume the utilization of reclaimed, desalinated water and even of the transfer from other territories (ability of payment).

3.2 Impact of the international market for trade

The irrigation of crops in greenhouses and outdoors in the region is one way of producing land compatible with the CAP and the globalization tendencies of agricultural free trade that the WTO is trying to establish. This aims at liberalizing world agricultural trade by for example eliminating aid for agricultural production. Both the high level of production technology and the quality of the products of the region studied territorial unit of analysis allow it to compete in world markets. A skilled workforce is available that can maintain and increase production with no problem in the future. Irrigated agriculture will remain profitable and competitive even when applying the real costs of water contemplated in the Water Framework Directive of the European Union. The modern exploitation of crops is primarily competitive with respect to with land use in the area to achieve annual returns that exceed the selling price of land for urban use.

Horticultural exports and their value have gradually increased from 1999 to 2006. There has been a 25% increase in value 398,400 € to 498,470. Increased exports related to the demands of the emerging markets of the developing countries, especially in the Asia-Pacific axis. Exports which are concentrated in the period from October to June and represent 85% and then decrease during the summer months when weather conditions allow the access of products from other territories to the markets.

The productive strength indicates that agricultural activity should continue improving product quality, health conditions and provide for the expansion of consumption by improving the standard of living of all EU countries. It will also improve transportation and phytosanitary control to be able to penetrate the market in the USA and Japan. This is the priority objective of the Horticultural Cooperative Pilar de la Horadada (Surinver), which manages the sale of the California pepper variety in those countries. This is a modern design of agricultural exporting that demands a guarantee and quality of water to improve on the size of farms and degree of marketing in order to compete more profitably in the increasingly globalized market.

4 Conclusions

The only agriculture that is sustainable, economical, social and environmental is that requiring the assurance of the necessary irrigation and water quality. The region of study brings together a number of objective and beneficial conditions to maintain and enhance the role of agricultural production in both the EU and the globalized markets. Agriculture in the region studied must be an engine of regional development although this is not the final solution to solve the problems of territorial imbalance.

References

[1] Juárez, C. (1991). Planificación hidrológica y desarrollo económico: El trasvase Tajo-Segura. Instituto "Juan Gil-Albert, Alicante : 167 pp.

[2] Juárez, C. (2004). Asignación de recursos de agua para uso agrario y crecimiento económico en la comarca meridional agraria de Alicante. Estudios Agrosociales y Pesqueros, nº 202, pp. 135–166.

[3] Juárez, C.; Ponce, G.; Ramón, A.; Canales, G. "Las agriculturas alicantinas" en Estructura económica de la provincia de Alicante. pp. 147–171. Diputación de Alicante 1993.

[4] Rico, A. M. (2006). Políticas agrarias, eficiencia socioeconómica y retos de futuro en los regadíos intensivos. B. de la A. G. E nº 41 pp. 113–149.

Predicting environmental sustainability for proposed irrigation schemes

N. Norton[1], S. Dawson[2], B. Ellwood[3] & L. Mead[3]
[1]*National Institute of Water and Atmospheric Research, New Zealand*
[2]*Boffa Miskell Ltd, New Zealand*
[3]*Meridian Energy Ltd, New Zealand*

Abstract

We have developed an innovative approach for predicting the environmental sustainability of proposed irrigation schemes on rivers. There are three key elements to this approach. First, the water resource potentially available to be taken for out-of-stream use must be defined by setting a river flow regime that sustains in-stream and other environmental values. Second, the capacity of the environment to support intensified land-use must be evaluated, by predicting effects on water quality and related values, and identifying ways to mitigate adverse effects. Third, the predictions must be integrated in a way that allows iterative adjustment of the proposal and assists decision-makers with the holistic evaluation of positive and negative effects. We trialled the approach for a proposed irrigation scheme that would divert 20 m^3/s from the Waitaki River, Canterbury, New Zealand, to irrigate 40,000 ha of pasture. This river has a mean annual flow of 369 m^3/s downstream of a large hydropower dam. Water is used for irrigation and environmental flows, as well as electricity generation. Under New Zealand legislation an assessment of environmental effects (AEE) is required for proposals to use natural resources. Our team of scientists, engineers and planners collaborated to prepare an AEE for the proposed irrigation scheme. Predictive models were coupled in novel ways to achieve integration between the technical disciplines. For the first time in New Zealand we were able to make quantitative predictions of key cumulative effects at the catchment scale, of both the water abstraction and the use of water for irrigation. The predictions were integrated and the proposal adjusted to achieve a balance between the needs of conflicting values. This approach should achieve better information about the true environmental costs of alternative future water use scenarios and thus lead to better decisions on the environmental sustainability of irrigation proposals.
Keywords: environmental impact assessment, irrigation management, river flow regime, water quality, land use intensification.

WIT Transactions on Ecology and the Environment, Vol 112, © 2008 WIT Press
www.witpress.com, ISSN 1743-3541 (on-line)
doi:10.2495/SI080231

1 Introduction

Achieving sustainable management of the quantity and quality of freshwater is both a high priority and a serious challenge for most countries worldwide [1]. In New Zealand, a recent review of environmental performance by the Organisation for Economic Co-operation and Development (OECD) [2] concluded that two of the country's major environmental pressures, agriculture and energy production, have expanded in the last ten years resulting in a significant increase in the use of water, fertiliser and pesticides. They concluded that, while the use of water and fertiliser in New Zealand is still on the low side for OECD countries, water quality in rivers and lakes has declined in regions dominated by pastoral farming. About half of the total length of New Zealand's rivers occurs in these areas [3], and in lowland areas high nutrient inputs and microbiological contamination have regularly exceeded national guidelines [2]. Subsequently the New Zealand Ministry for the Environment's own environment report has highlighted the decline in water quality caused by intensifying agricultural production, together with global climate change, at the top of its list of environmental challenges [3].

Water managers in New Zealand have witnessed a recent explosion in the demand for water for both irrigation and hydroelectricity generation. Both of these uses have the potential to cause large-scale changes to the environment in New Zealand. With increasing pressure on water allocation decisions, it is becoming well recognised that there is a need for robust methods of predicting the full environmental consequences of alternative future scenarios for water use.

2 A brief history of changing approaches

In the days before much thought was given to environmental sustainability, decisions about large scale irrigation schemes were based largely on the available volume of the water resource and the economics of transferring that water to irrigate land. Often all of the water in a river was considered to be available for abstraction. This approach had obvious and immediate effects on the river environment. All in-stream values, including aquatic ecosystems, recreation, visual landscape, cultural and social values, were lost or significantly altered. Little thought was given to the effects of using water to change the use of land, other than the economic benefit.

Predicting the environmental effects of the abstraction of water from a river requires determination of the amount of flow that needs to be left in the river in order to sustain in-stream values at a satisfactory level. In the last 30 years a range of methods have been developed for predicting the in-stream flow requirements of river geomorphic processes, aquatic ecosystems, recreation activities, landscape and cultural needs, while contemplating an abstraction for economic and social needs. An entire field of science and engineering continues to develop and advance these methods today, e.g., [4–7].

Predicting the environmental effects of using water for irrigation requires consideration of the potential adverse effects of land-use intensification, such as degraded water quality and ecosystems, changes to landscape, cultural and social values, as well as economic and social benefits [8]. In the last 10 years technical methods for predicting these effects have advanced rapidly, e.g., [9–12].

Until recently, decisions on the allocation of water from rivers in New Zealand were based largely on consideration of the effects of the abstraction from the river. The potentially adverse effects of water use were considered separately and generally less comprehensively. One reason may be that land-use intensification effects are cumulative, only appearing some time after numerous separate water allocation decisions have been made, and often in different locations to the source of the water. Another reason is that in New Zealand, as in many countries, water and land resources have traditionally been managed by separate authorities and under separate pieces of legislation, albeit with some recognition of the inter-relationship between the two resources.

In New Zealand the Resource Management (RM) Act (1991) has facilitated closer integration of water and land management but it has taken time for practitioners to implement effectively [13]. The result is that environmental effects of both the abstraction of water and the use of that water are now being considered together in water allocation decisions.

3 A contemporary approach to environmental sustainability

Today's approach to predicting environmental sustainability for irrigation proposals is concerned with both water quantity and water quality. It is concerned with how much water is left in the river, as well as the amount of water applied to land and the consequences of the land-use changes that result. Environmental sustainability today encompasses a holistic consideration of effects on a wide range of values, including aquatic biodiversity, recreation, landscape, cultural and social values, as well as economic and community interests, e.g., [14]. Assessing environmental sustainability requires professionals from multiple technical disciplines in planning, science and engineering, as well as participation by the community at large. The challenge is to find ways of predicting the effects of change on all elements of the environment and then to integrate this knowledge to inform decisions about environmental sustainability.

We suggest there are three important elements to the contemporary approach:

i) Predict the effects of future water abstraction scenarios;
ii) Predict the effects of future water use scenarios;
iii) Integrate the predictions to form a holistic assessment for decision-makers.

At the outset it is necessary to identify all of the technical work-streams that may be necessary to predict effects. Once the relevant technical experts are assembled, they can identify the predictive tools that are available, or that need to be developed, in order to predict effects in their area of expertise. The key to successful integration is recognising the specific role of the integrator(s) in the team. The role of the integrators is not only to assist the technical experts to identify effects and couple their predictive models between disciplines, but also to facilitate inter-disciplinary communication and relationship management. It is important that integration is not just a phase at the end of the individual technical assessments, but occurs throughout the predictive assessment. Effective integration allows the knowledge provided by each technical expert to be built upon by other experts. It also allows for knowledge feedback that can be used to alter the proposal to manage adverse effects as they are identified. Such an iterative process should tend to produce irrigation proposals that move progressively closer to being truly environmentally sustainable.

The key elements of our approach are illustrated in the following case study.

4 Case study: new best practice in New Zealand

4.1 Proposed hunter downs irrigation scheme

The proposed Hunter Downs Irrigation Scheme (HDIS) would divert 20 m^3/s from the lower Waitaki River to irrigate 40,000 ha of pasture (Fig. 1). The Waitaki River is braided with an average of seven braids across a 700m wide fairway (see [15]). It has a mean annual flow of 369 m^3/s downstream of a large hydropower dam. Water is already used for irrigation, environmental flows and electricity generation.

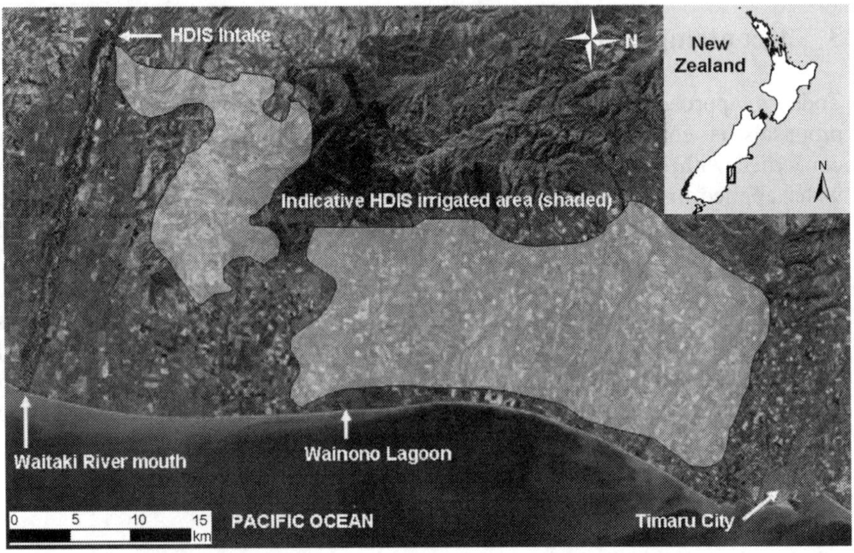

Figure 1: Satellite image showing location of the proposed HDIS (see [15]).

4.2 Legislative framework

The purpose of the RM Act is to promote the sustainable management of natural and physical resources. The Act defines the functions and powers of regional and territorial authorities in New Zealand. Regional authorities are responsible for the sustainable management of water, land and air.

The RM Act requires regional authorities to prepare regional policy statements and guides the preparation of regional plans that define policies, objectives and rules for the management of water, air and related land-use.

Preparation of these plans must follow a prescribed public participation process so that the final documents reflect the local community's view on what constitutes the sustainable management of natural and physical resources.

Anyone proposing to abstract and use water from a river, lake or groundwater must obtain resource consent from the relevant regional authority. The application for such consent must include an assessment of environmental effects (AEE) and consultation with stakeholders and the affected public. The application is considered by the regional authority under the framework provided by the RM Act and any relevant regional policy statements and plans. For decision-makers, the test of environmental sustainability is the extent to which the proposal meets the purpose of the RM Act and the regional policies and plan objectives.

4.3 Assessing environmental effects of the HDIS

A team of scientists, engineers and planners was assembled to prepare an AEE for the proposed HDIS. There were two technical work-streams that aimed to determine: i) a proposed sustainable water abstraction, by determining a sustainable river flow regime and thus the water that could be available for out-

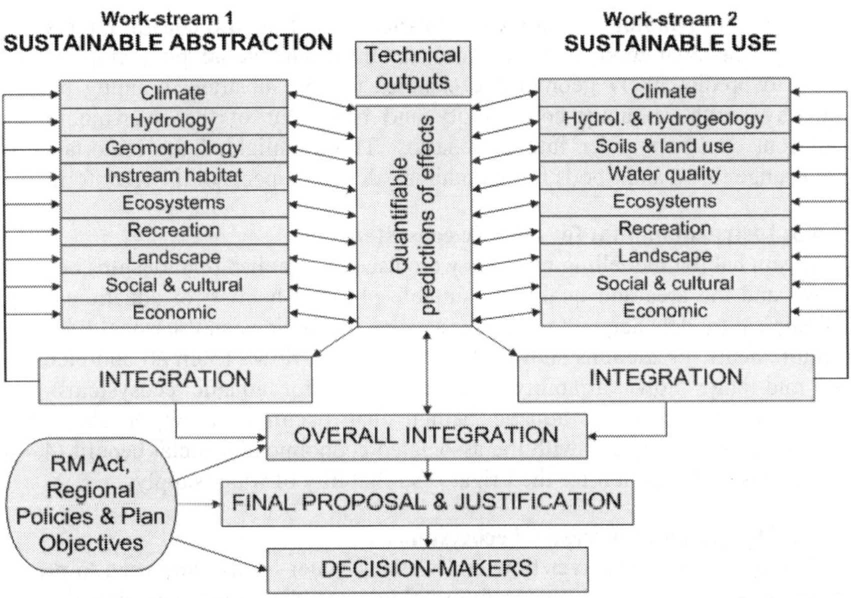

Figure 2: Concept diagram of technical work-streams and integration process.

of-stream use; and ii) a proposed sustainable use of water, by determining the capacity of the environment to support intensified land-use, as well as the economic and social value of that water use. Reports were prepared documenting the methods, results and predictions for each technical discipline [16, 17]. Our assessment approach, including the roles played by each technical expert and our integration process, is summarised in Fig. 2. The two technical work-streams and integration process are described in sections 4.4 to 4.6 using examples that illustrate key aspects of the approach.

4.4 Work-stream 1: defining a sustainable water abstraction

4.4.1 Hydrology - climate variability and change

Understanding how river flow varies over days, seasons, years and decades, is fundamental for assessing how the existing flow regime supports current values, and for determining what irrigation abstraction might be sustainable. HDIS water abstractions and river flow were modelled for a range of future scenarios. The future scenarios included various combinations of irrigation abstraction, further hydro-electric power generation, and environmental flow regimes with rules requiring minimum flows, mid-range flow variability and floods. Various options for flow regime rules were modelled based on recommendations made by the other technical assessments (outputs from 4.4.2, 4.4.3 and 4.4.4). The hydrological data outputs for each scenario were used by technical experts in other disciplines for their predictive models (Fig. 2).

4.4.2 Geomorphology – sediment transport, flooding and coastal effects

Two-dimensional models, aerial laser scanning and aerial photography were used to predict likely geomorphic changes to bed substrate, braiding pattern, fairway width, tributary connectivity and behaviour of the river-mouth and adjacent coastline under future scenarios. This highlighted the importance of mid-range flows and floods for maintaining desired aspects of the river form.

4.4.3 Instream habitat for aquatic ecosystems

Instream habitat modelling methods were used to establish relationships between flow, and the area and quality of suitable physical habitat for aquatic species. Biological models helped to consider species interactions and seasonal life-cycle requirements. Predictions showed that a minimum flow ($150m^3/s$) and elements of mid-range flow variability were important for aquatic ecosystems. The quantified reduction in aquatic habitat that occurred as water abstraction increased was compared with the associated economic and social benefit (4.4.6) of increased abstraction for the irrigator's reliability of water supply.

4.4.4 Wetland and terrestrial ecosystems

Relationships between river flow and wetland water levels were used to predict changes to wetland area under future flow scenarios, thus demonstrating the importance of a minimum river flow for maintaining hydraulically connected wetlands. Minimum flows, floods and vegetation control were shown to be important for maintaining braided river islands free from rodent bird predators and for creating bare fairway breeding habitat for endangered riverbed birds.

4.4.5 Recreation, landscape and cultural values

Effects on these values were assessed by experts using consultation techniques including surveys, interviews and public workshops, as well as knowledge of the physical (see 4.4.2) and biological (see 4.4.3) predictions of others (Fig. 2).

4.4.6 Economic, social and community effects

Economic and social analyses demonstrated the value of the HDIS to individual farmers and the local community as monetary value at the farm gate, value added to the regional economy, and increased levels of community well-being. Monetary value and social benefits depend on the reliability of water supply. Because reliability of supply is directly affected by the required minimum flow in the river, and this in turn affects the extent of river and wetland habitat area, landscape, recreation and cultural values, a link could be established between these latter environmental outcomes and the economic and social outcomes. This was crucial for the integration phase described next.

4.4.7 Integration and feedback-informed adjustment of the proposal

In addition to ensuring the coupling between technical disciplines, the integrators facilitated exercises where the whole project team weighed the identified negative and positive effects to inform adjustment of the proposal (Fig. 2). One example was the conflict between irrigators, who desired a highly reliable water supply to avoid hardship in dry summers, and the needs of aquatic ecosystems, whose habitat area is reduced by lower minimum flows. Being able to quantify the negative and positive outcomes allowed the project team to weigh these, decide where they felt the balance lay, and adjust the proposal accordingly. This decision involved value judgements, which will be discussed further in section 4.6. By way of example, this exercise resulted in a proposed minimum flow for the HDIS of 100 m^3/s instead of 150 m^3/s, but only for a limited period (7%) of the time.

4.5 Work-stream 2: defining a sustainable use of water

4.5.1 Scheme area hydrology, hydrogeology and soils

Scheme area hydrology, hydrogeology and soils were characterised to provide the basis for predictive methods of other technical experts (Fig. 2).

4.5.2 Land-use change scenarios

Experts considered soil mix, topography, capital costs and financial returns to predict future land-use mix scenarios if the HDIS scheme were to proceed. Key predictions were a significant increase in dairying and cropping, with corresponding reductions in dry-land sheep and deer farming.

4.5.3 Water quality: nutrients, sediment and micro-organisms

A nutrient budgets model was used to predict nitrate nitrogen leaching and dissolved phosphorus run-off from predicted land-use scenarios (outputs from 4.5.2 and 4.5.1). The outputs were coupled to a mass mixing model and used to predict increases to dissolved nutrient concentrations in groundwater, in rivers and in the local coastal Wainono Lagoon (see Fig. 1). Increases in sediment and micro-organisms in waterways were predicted qualitatively. Predictions showed that improved farm management practices could lessen water quality effects.

4.5.4 Aquatic ecosystems: rivers, wetlands and a coastal lagoon

Mass mixing model nutrient outputs (4.5.3) were coupled to an empirical model that predicted algae biomass would increase by 60% in rivers and 50% in Wainono Lagoon, with smaller increases for future scenarios that implemented improved farm management practices and riparian vegetation enhancement. The mass-mixing model also predicted increased flow, with associated positive ecosystem benefits, in some sections of some rivers.

4.5.5 Recreation, landscape and cultural values

The degraded water quality effects (4.5.3 and 4.5.4) had flow-on consequences for recreation, landscape and cultural values (Fig. 2). Increased algae growth would reduce visual aesthetics and be a nuisance for anglers. Reduced water quality and algae-smothered aquatic habitat degrades the *mauri* or 'life force' of waterways for maori people. In contrast, the predicted increase in flow for some river sections was a positive effect for angling, landscape and cultural values.

4.5.6 Economic, social and community effects

The identification of adverse environmental effects of intensified land-use led to a proposed mitigation package that is described in the next section. The cost of this mitigation to farmers was quantified to assist with the integration process.

4.5.7 Integration and feedback-informed adjustment of the proposal

The project team weighed the negative and positive effects of land-use intensification, just as was done previously for effects of the water abstraction on the river (Fig. 2). As a result a mitigation package was proposed that included: i) a mandatory requirement for farm management plans that defined efficient water use protocols, nutrient budgets and other on-farm actions; ii) financial incentives for riparian vegetation enhancement; iii) a contestable community environment enhancement fund; and iv) a robust environmental monitoring programme.

The benefits of mitigation measures were evaluated against the costs to farmers of implementing them. Furthermore, because some environmental degradation of waterways had already occurred due to past practices, it was recognised that some of these measures were desirable regardless of whether HDIS proceeded or not. The economic analysis showed there was doubt about whether the local community could afford to undertake, or would undertake, these measures if they did not have the additional income generated by the HDIS. All of these factors were considered by the project team when deciding on the financial quantum of these measures to be proposed to decision-makers (Fig. 2).

4.6 Overall integration: weighing the negative and positive effects

Deciding on what is truly environmentally sustainable involves value judgements about the relative merits of different values. Our iterative integration and proposal adjustment process (Fig. 2) involved value judgements by the project team that were not always agreed. The team used the relevant regional policies and plan objectives, as well as consultation with stakeholders and the affected public, to assist with this judgement, but it was the resource consent applicant (the project investor) who decided, based on advice from their expert consultant team, the nature of the final proposal put before decision-makers at a hearing.

While this process tended to produce an increasingly environmentally sustainable proposal, the ultimate test of environmental sustainability is a matter for decision-makers. To aid their evaluation at a hearing, we described a series of options that linked positive and negative environmental outcomes with social and economic outcomes. These were similar options to those the project team had used for their weighing exercises. The project investor highlighted its preferred option, but the alternative options were provided as context for the decision-makers and to assist their evaluation of the most environmentally sustainable option. Under the New Zealand legislative framework this would be the option that best met the purpose of the RM Act, regional policy statements and objectives of the relevant regional plans.

5 Discussion

At time of writing, a decision is still pending on the HDIS proposal. However, regardless of the decision outcome, the case study illustrates useful aspects of our approach. The environmental assessment process involved a large team of experts that was costly, but necessary for making informed decisions about the full range of potential effects, some of which could be very costly, or impossible, to reverse. Our approach (Fig. 2) can be scaled down for smaller proposals, with each of the technical disciplines being addressed at an appropriate level of detail for the likely scale and significance of effects. The multidisciplinary integration process is a key aspect, and can be undertaken with large or small teams. The iterative proposal adjustment process encourages the project investor to pro-actively address the broad spectrum of environmental sustainability at the project formulation stage, while reserving the final decision for independent decision-makers. This costs the project investor time and money, but their cost of not doing so is to risk environmentally precautionary decisions necessitated by inadequate information. In addition to these project benefits, each of the discussed aspects of our approach helps advance New Zealand's three national sustainability outcomes for freshwater [18]. The approach also aligns with United Nations (UN) Millennium Development Goal 7 (Ensure Environmental Sustainability) and relevant recommendations of the UN Millennium Project Task Force on Environmental Sustainability [14]. In this way our approach also addresses national and international policy goals for sustainable development.

6 Conclusions

Predicting the environmental sustainability of irrigation schemes requires a holistic approach to the assessment of effects of future water abstraction and use scenarios. Our approach has:

- Produced quantitative predictions of the positive and negative effects of the abstraction and use of water on a range of environmental values.
- Integrated the predictions and iteratively adjusted the irrigation proposal in an attempt to balance the needs of conflicting values.
- Provided decision-makers with a series of options that link environmental outcomes with social and economic outcomes, so that they may form a decision as to the most environmentally sustainable option.

This approach should achieve better information about the true environmental costs of alternative future scenarios and thus lead to better decisions on the environmental sustainability of irrigation proposals.

Acknowledgements

We thank HDIS team colleagues, reviewers (H. Rouse) and the Foundation for Research, Science & Technology (New Zealand) for Contract CO1X0308.

References

[1] UNEP, Challenges to International Waters – Regional Assessments in a Global Perspective. United Nations Environment Programme, Kenya, 2006.

[2] OECD, Environmental Performance Reviews – New Zealand, Organisation for Economic Co-operation and Development, Paris, France, 2007.

[3] MFE, Environment New Zealand 2007, Ministry for the Environment, Wellington, New Zealand, 2007.

[4] Jowett, I.G., Flow Management (Chapter 12), New Zealand stream invertebrates: ecology and implications for management, eds. K. J. Collier, M. J. Winterbourn, New Zealand Limnological Society, Christchurch, pp. 289–312, 2000.

[5] Jowett, I.G. & Biggs, B.J.F., Flow regime requirements and the biological effectiveness of habitat-based minimum flow assessments for six rivers. Journal of River Basin Management **4(3)**, pp. 179–189, 2006.

[6] Arthington, A.H., Bunn, S.E., Poff, N.L., Naiman, R.J., The challenge of providing environmental flow rules to sustain river ecosystems. Ecological Applications **16**, pp. 1311–1318, 2006.

[7] Harding, J., Mosley, P., Pearson, C., Sorrell, B., (eds.). Freshwaters of New Zealand, New Zealand Hydrological Society and New Zealand Limnological Society, Caxton Press, Christchurch, 2004.

[8] PCE, Growing for good – Intensive farming, sustainability and New Zealand's environment, Parliamentary Commissioner for the Environment Wellington, New Zealand, 2005.

[9] Biggs, B.J.F., Eutrophication of streams and rivers: Dissolved nutrient-chlorophyll relationships. Journal of the North American Benthological Society **19(1)**, pp. 17–31, 2000.

[10] McDowell, R.W., Biggs, B.J.F., Sharpley, A.N., Nguyen, L., Connecting phosphorus loss from agricultural landscapes to surface water quality. Chemistry and Ecology **20**, pp. 1–40, 2004.
[11] Wilcock, R.J., Monaghan, R.M., Thorrold, B.S., Meredith, A.S., Betteridge, K., Duncan, M.J., Land-water interactions in five contrasting dairying watersheds: Issues and solutions. Land Use and Water Resources Research **7(2)**, pp 1–10, 2007.
[12] Rosen, M.R. & White, P.A. (eds.). Groundwaters of New Zealand, New Zealand Hydrological Society, Caxton Press, Christchurch, 2001.
[13] Ford, R., Taylor, K., Managing nitrate leaching to groundwater: an emerging issue for Canterbury. Proceedings of the Fertiliser and Lime Research Centre Workshop for Implementing Sustainable Nutrient Management Strategies in Agriculture, eds. L.D. Currie & J.A. Hanly, Occasional Report No. 19, Massey University, Palmerston North, pp. 48–67, 2006.
[14] UN Millennium Project, Environment and Human Well-being: A Practical Strategy, Summary version of the report of the Task Force on Environmental Sustainability, The Earth Institute at Columbia University, New York, 2005.
[15] Google Maps New Zealand, http://maps.google.co.nz/maps
[16] MEL, Hunter Downs Irrigation Scheme Assessment of Environmental Effects – incl. 15 appended technical reports (Annexure A-O), Meridian Energy Ltd & the South Canterbury Irrigation Trust, Christchurch, 2006.
[17] MEL, Water Consents for a North Bank Tunnel Concept: Assessment of Environmental Effects – incl. 29 appended technical reports (Appendix 1–29), Meridian Energy Ltd, Christchurch, 2006.
[18] MFE, Freshwater for the future: A supporting document, New Zealand Ministry for the Environment, Wellington, 2006.

A framework for evaluating the consumption patterns and environmental impacts of irrigation methods: a case study from South-Eastern Australia

T. M. Jackson[1,3], S. Khan[1,2,3] & M. M. Hafeez[1,3]
[1]*International Centre of Water for Food Security, Charles Sturt University, Australia*
[2]*CSIRO Land and Water, Australia*
[3]*Co-operative Research Centre for Irrigation Futures, Australia*

Abstract

Irrigated agriculture is an essential tool for increasing food production to meet global demand. However, along with benefits of high yields there are environmental impacts that must be considered. The tendency in Australia to evaluate irrigation systems in terms of water use efficiency only is problematic, and the criteria used to assess the sustainability of agricultural systems should reflect broader issues. A representative broad acre farm in the south east of Australia was used as a case study to explore the different resource consumption patterns and environmental impacts of flood, centre pivot and sub-surface drip irrigation methods. The energy and water consumption of each method was determined, along with greenhouse gas emissions and groundwater impacts, and the systems then ranked in order of highest to lowest resource use efficiency and environmental impact. It was found that when an irrigated system was evaluated by more than just water use efficiency, its ranking often changed. This method also allowed for the identification of areas where improvements could be made. Assessing irrigation methods and their appropriate selection for a given situation can aid in efficient and environmentally sound production systems at the local and global scale.
Keywords: energy, water, environmental impacts, indicators.

www.witpress.com, ISSN 1743-3541 (on-line)
doi:10.2495/SI080241

1 Introduction

At both national and global levels, concerns over the state of the environment are well established throughout the wider community, including issues such as salinity, water shortages, energy use and carbon emissions. From an agricultural perspective, these are serious matters that impact greatly on the production of food and fibre. Farming has entered a new era, whereby it is not simply enough that a farm produce food and fibre as economic goods; there is now a second important function for farms, which is to produce or protect environmental services [1]. With an increasing public awareness of the environmental impact resulting from food production, it is becoming more important to consider the associated resource inputs and environmental impacts. In addition to this, population pressure means that food production must increase from current levels. Given that irrigated agriculture can be doubly as productive as rainfed agricultural land [2], it can contribute greatly to increases in food production and is therefore a vital part of world agriculture. However, poorly managed irrigation systems can have detrimental environmental impacts, and it is therefore necessary to conduct irrigated agricultural production in such a way that these environmental impacts are minimised.

It is a common practice in Australia to evaluate the effectiveness of an irrigation system in terms of its water use efficiency, using this figure (often given in ML/ha) as a means for comparing different crops and irrigation methods. However, it is imperative that the criteria used to assess the sustainability of agricultural systems reflect the issues of the time [3]. Major global concerns at present are the need to increase food supply, competition among water users and the threat of global warming due to greenhouse gas emissions. Efficiency therefore needs to be considered in a broader sense, incorporating technical and environmental aspects. Khan et al [4] identify the importance of the paddock as the basic decision-making unit, where choices impact on land and water management and salinity dynamics. A consumptive and environmental assessment is important in order that improvements to systems can be made on a local scale, leading to efficient and environmentally sound production systems at the global scale [5].

This paper uses a basic conceptual structure of an irrigated Lucerne seed production system built in Vensim™ to understand the consumption patterns and environmental impact of irrigation with regard to water and energy consumption. This allows us to explore how changing climatic patterns affect energy and water consumption and the consequences associated with these changes. Salinity impacts in soil and on the watertable are explored using Swagman Farm™.

2 Materials and methods

In order for farms to achieve the dual goals of production and protection of environmental services, it is essential to develop sound evaluation methods that can be used as decision support tools for sustainable agricultural production. As cited by van der Werf et al [1], there are generally five common stages in any given method used to assess the environmental impacts of farms.

1. *Defining the broad objective of the method.* For the purpose of this study, the broad objective is the evaluation of the environmental impacts resulting from the consumption of energy and water for irrigated crop production at the field scale.
2. *Define the environmental objectives.* Several environmental impacts linked to energy and water consumption are considered in this paper. These include water and energy consumption, carbon emissions linked to energy use, impacts on soil salinity content and local watertable responses associated with irrigation.
3. *Define the system to be analysed.* The system to be analysed is three irrigated fields, each using a different irrigation method but located on one farm. Some indirect impacts are considered in terms of indirect energy inputs.
4. *Identification of indicators.* Inputs and outputs will be considered per unit of area and per unit of input (productivity). Salinity responses will be in change of salinity levels in soil and changes to watertable depth.
5. *Calculation of results.* Values for each indicator are calculated for the system or in this case the irrigation method used, in order that comparisons can be made.

These methods have been adapted to the field level by defining the objectives as impacts that can be measured at the field scale. These impacts directly affect the farm at both an environmental and production level, and the results of these may influence the decisions made by farmers, making them extremely important at the field/farm scale.

2.1 Site description

The study sites are located on an irrigated farm producing Lucerne seed, approximately twenty kilometers south of Keith in South Australia. Three paddocks using different irrigation methods were selected for analysis, with the aim of comparing the differences between flood (F), centre pivot (CP) and sub-surface drip (SSD) irrigation systems. The entire region is dependent on groundwater for irrigation, with bores pumping from the shallow watertable. Groundwater quality is variable, ranging from 1.8 dS/m (CP) to 4.0 dS/m (F and SSD).

The region is typified by hot, dry summers and cool, wet winters, with average summer maximum temperatures of 27.5 to 29.9°C and mean annual rainfall of 466 mm. There has been a general trend of increasing temperatures and decreasing amounts of rainfall since the 1950s [6], increasing the level of dependence on irrigation for crop production. Soils in the region are generally a combination of sand and calcium carbonate, with small areas of loam and clay. The soil type associated with the study sites was sandy loam over limestone.

2.2 Data collection

On-farm data was collected via a survey and personal communication with the farmer. In order to calculate the energy used on-farm, a range of data was collected pertaining to irrigation methods, water use and energy consumed for pumping and pressurizing systems, land preparation, machinery operations and fertiliser and chemical application. These inputs were selected based on common categories for the consideration of energy inputs as used in previous studies [7, 8]. Energy categories include diesel, electricity, machinery hours, chemicals, fertiliser and seed. Energy and CO_2 equivalents were used to quantify energy inputs and emissions. The values used in this study are shown in Table 1.

Table 1: Energy and CO_2 emission co-efficients.

Input	Unit	Sequestered energy (MJ)	Emissions (gCO^2/MJ)	Reference
Diesel	litre	56.31	80.8	[5,8,9]
Electricity	kWh	11.93	43.1	[5,9]
Fertiliser				
Nitrogen	kg	65	50	[10,11]
Phosphate	kg	11.96	60	[5]
Potash	kg	11.1	60	[12]
K_2O	kg	6.7	60	[5,12]
Phosphorous	kg	12.44	60	[8]
Potassium	kg	11.15	60	[8]
Sulphur	kg	5	60	[11]
Lime	kg	0.6	720	[11]
Lubricant (Oil)	litre	47.6	43.4	[13]
Fungicide	kg	92	60	[10]
Herbicide	kg	240	60	[10]
Insecticide	kg	200	60	[10]
Seed - general	kg	14		[10]

2.3 Analysis of water and energy productivity

The Vensim™ modelling environment was used to construct a model of on-farm energy consumption and greenhouse gas emissions for Lucerne seed production using on-farm data from the study site. Vensim™ is a visual modelling tool for conceptualising, documenting, simulating, analysing and optimising models of dynamic systems such as farms and irrigated regions. Once a model is built that can be simulated, Vensim™ allows the behaviour of the model to be thoroughly explored.

2.3.1 Water application, pumping energy and emissions

In order to determine the amount of water applied for each irrigation method, the following procedure was used. The crop water requirement (CWR) for Lucerne seed production for the past six years was determined from regional ET_0 and crop factor information [14]. This then resulted in the calculation of net CWR from 2001–2007. Finally, to calculate the amount of water applied using each irrigation method, the net CWR was divided by the given efficiency of the irrigation method, resulting in a total amount of water applied by each method (ML/ha). Energy requirements for pumping were determined using standard equations [15].

2.3.2 Other inputs and emissions

Other energy input categories included diesel, fertiliser, chemicals and seed. The energy associated with diesel use was calculated by determining the total number of hours of machinery operations per hectare relating to sowing and fertiliser and chemical application and relating this to the size of the tractor (hp) to calculate litres of diesel. This quantity was then converted to an energy and CO_2 emission equivalent using the appropriate conversion factor given in Table 1.

Energy associated with fertiliser inputs was calculated by using the amount of phosphorous and trace element fertiliser applied per unit area and converting this to an energy and CO_2 emission equivalent using the appropriate conversion factor given in Table 1.

Energy associated with chemical inputs was calculated by using the amount of herbicide and pesticide applied per unit area and converting this to an energy and CO_2 emission equivalent using the appropriate conversion factor given in Table 1.

Seed input energy was averaged over six years, which is the length of certification of a Lucerne seed crop.

2.4 Calculation of impacts on watertable and soil salinity

SWAGMAN (Salt, Water and Groundwater Management) Farm™ is a lumped water balance model that predicts changes in the depth to watertable, salinity of the rootzone and gross margins. As a farm scale hydrologic economic model, it combines agronomic, climatic, irrigation, hydrogeological and economic aspects of irrigated agriculture [16] at a paddock or farm level. For the purposes of this study, the focus was on changes in depth to watertable and rootzone salinity under separate paddocks with different irrigation systems. While there are several other models available for modelling salt and water movement, SWAGMAN Farm™ is customised for situations of shallow watertables and soil salinity and as such is an appropriate model for this situation. Good agreement between SWAGMAN Farm™ predictions and field observations has been demonstrated [16], further validating its use in this study.

2.5 Energy and water productivity

Productivity refers to the benefits derived from inputs into a system, and physical productivity is a ratio of the quantity of yield produced and the quantity of the input. In this paper, the yield is expressed in terms of mass (kg seed) and the input is represented by the quantity of energy (MJ) or water (ML) used for crop production. Energy productivity is expressed as kg seed/MJ, while water productivity is expressed as kg seed/ML.

3 Results

The energy and water consumption patterns change depending on the irrigation method used and the soil water deficit for any given year. Results are shown for the period of the last six years, with the net crop water requirement changing each year according to climatic conditions which drive the soil moisture deficit. The amount of water applied is reliant on net crop water requirement and the efficiency of the system. As the order of assumed efficiency is SSD (0.95), CP (0.75), F (0.5), this is reflected in the quantity of water applied for each method, as illustrated by Figure 1.

As shown by Figure 2, energy consumption follows a similar pattern to water application; however, the ranking between the systems is reversed in terms of the quantity of energy consumed. Water productivity is an index of the yield and the amount of water applied (Figure 3). In this situation, the SSD system is more water productive than both CP and F. Energy productivity is an index of the yield and the amount of energy consumed (Figure 4). F is the most energy productive, since it uses the least amount of energy, followed by SSD and CP. Figure 5 shows emissions, which closely follow water use, as energy for groundwater pumping is the biggest energy input on-farm. The F system has higher emissions followed by the SSD and CP systems.

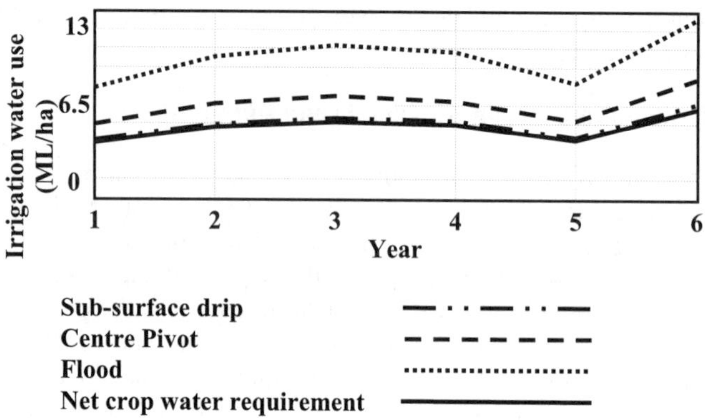

Figure 1: Irrigation water use based on soil water deficit (ML/ha).

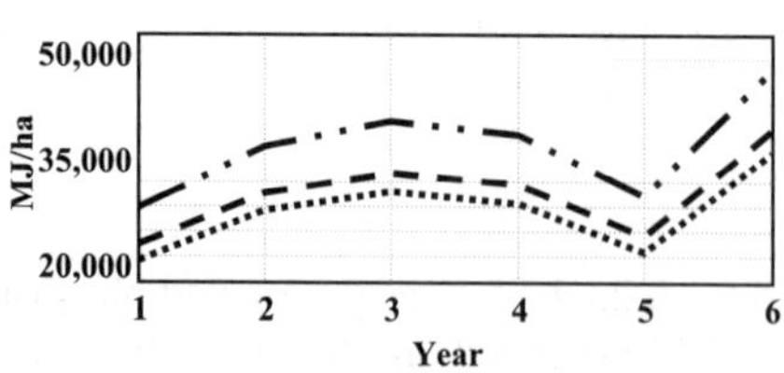

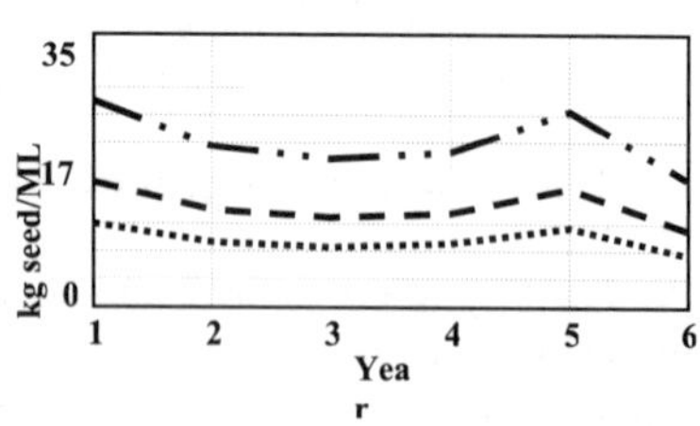

Figure 2: Energy consumption (MJ/ha).

Figure 3: Water productivity. (kg seed/ML).

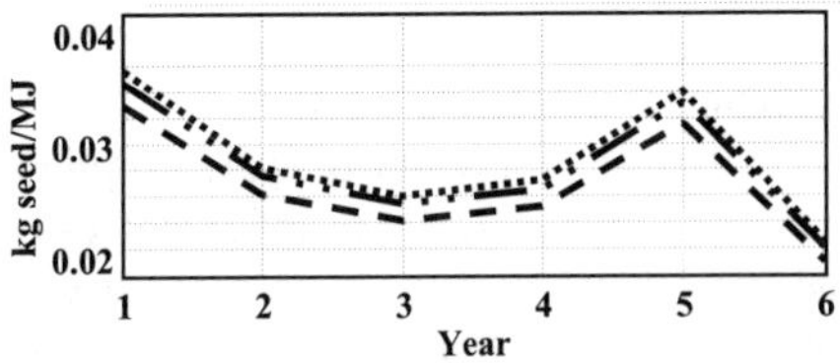

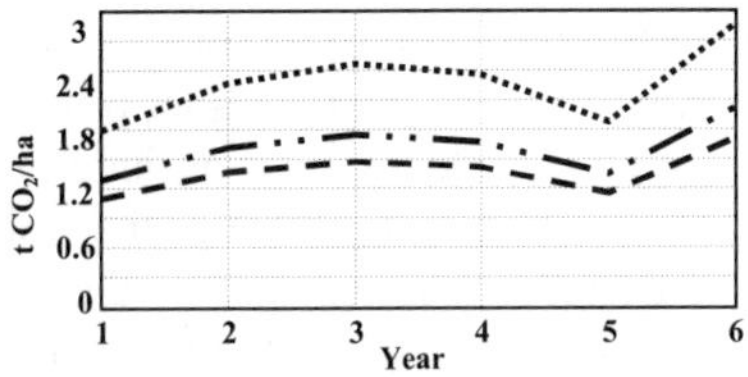

Legend
Sub-surface Drip_ . . _ . . _ Centre Pivot_ _ _ _ _ _ _ Flood.......................

Figure 4: Energy productivity (kg seed/MJ).

Figure 5: CO_2 emissions (t CO_2/ha).

Table 2: Changes to groundwater under each irrigation method.

Method	Ave water table change (m)	Ave soil salt concentration change (dS/m)
Flood	-5.65	0.16
Centre Pivot	-1.75	0.72
Sub surface drip	-0.25	1.05

Table 2 shows changes to groundwater under each irrigation method. The extent of the change in soil salinity increases with decreasing water application, while the change in depth to groundwater is greater when more water is removed by pumping.

4 Discussion

The results from the Vensim and Swagman models show that the choice of irrigation method and pump fuel source can impact on on-farm productivity and the environment. The amount of water applied is reliant on the soil water deficit and the efficiency of the system. As the order of system efficiency in terms of water delivery is SSD, CP, F, this is reflected in the quantity of water applied for each method, as illustrated by Figure 1. In terms of water use, SSD is the most efficient, followed by CP and F.

However, when the same systems are assessed from an energy perspective, the opposite is true. As shown by Figure 2, energy consumption follows a similar trend to water application, since groundwater pumping is the largest component of energy consumption. Other inputs are low and remain relatively constant due to the fact that this is a perennial legume that requires no nitrogen fertiliser and land preparation only once every six years. When energy consumption is considered, the ranking between the systems is reversed, due to the fact that SSD and CP are operated under pressure, which requires a large amount of energy.

Due to low pressure bubblers being used with the CP, in this situation the SSD system is operated at a higher pressure and hence uses more energy.

Water productivity is an index of the yield and the amount of water applied (Figure 3). In this situation, the SSD system has higher yields (+25%) and uses significantly less water than the other systems; consequently, it is more water productive than both CP and F. The F system is the least water productive due to its significantly higher water use for no yield gain.

Energy productivity is an index of the yield and the amount of energy consumed (Figure 4). As with energy consumption, energy productivity follows a similar trend to water productivity, as it has been established that energy consumption in this situation is heavily influenced by groundwater pumping. F is the most energy productive, since it uses the least amount of energy. SSD is the next most energy productive despite its higher energy use due to increased yields, while CP is the least energy productive; despite using less energy than SSD, it has similar yields to the F system.

Greenhouse gas emissions are associated with any energy that is consumed by a system, both direct and indirect. In this situation, emissions closely follow water use, as energy for groundwater pumping is the biggest energy input on-farm (Figure 5). The difference between the systems in this case is that the F system has higher emissions despite lower energy consumption rates due to the fact that it is supplied by a diesel pump, which has higher associated emissions than electric pumps, which serve the SSD and CP systems.

Groundwater pumping and the subsequent re-application of this water for irrigation impact on the concentration of salt in the rootzone and on the depth to watertable. Based on one average climatic year simulation, the modelled results show that soil salinity increases with decreasing water application. This is due to a reduction in the proportion of water applied that is available for flushing salts from the rootzone. The removal of groundwater also impacts on the depth to watertable. The modelled results show the depth to watertable increasing with higher amounts of water removed, with the largest increase in depth associated with the F system. In reality, the watertable in this region does not change as dramatically as illustrated here. This model does not incorporate lateral or upward flow into the region.

In order to obtain a view of the overall impact on consumption and the environment for the three irrigation methods, their ranking for a number of indicators have been summarized in Table 3. The results from this study show that the choice of irrigation method and pumping fuel source can greatly impact on resource consumption and environmental impact.

Table 3 shows that the F system operated by a diesel pump ranks lowest in terms of water use, water productivity, carbon emissions and change in depth to watertable. This system may be enhanced by improving flood irrigation layout to increase efficiency and converting to an electric pump using a "green power" source. This would have a number of effects, namely reducing the water applied, improving water productivity, reducing the change in water table depth and reducing emissions associated with pumping.

In this situation, the CP system ranks lowest in terms of energy productivity and net recharge. Improving yield and further improving the efficiency of the system may improve the ranking of this system against these indicators. The use of low pressure "bubbler" emitters in this case makes this CP system more energy efficient than CP systems which use high pressure sprays.

Table 3: Ranking of the three irrigation methods.

	Irrigation method		
Indicator	*Flood*	*Centre Pivot*	*Sub-surface drip*
Water use	3	2	1
Energy use	1	2	3
Water productivity	3	2	1
Energy productivity	1	3	2
Emissions	3	1	2
Water table change	3	2	1
Soil salinity change	2	1	2

The SSD system ranked worst in terms of energy use only. This could be improved by reducing the operating pressure of the system; however, this may result in compromising distribution uniformity if non-pressure-compensating drippers are installed in this system.

5 Summary and conclusions

The results of this study show that when an irrigated system is evaluated by more than just water consumption, its ranking among other methods can change. Given that current local and global concerns associated with agricultural production extend beyond just water consumption by a system, it has been demonstrated that by assessing systems against a range of indicators, a better understanding of the wide-spread consumptive and environmental impacts can be determined. In addition to this, areas for improvement can be identified and the impacts of these improvements across different categories explored. The decisions made at the paddock level can have wider implications on the environment. The appropriate selection of irrigation method and fuel source can lead to efficient and environmentally sound production systems at the local and regional levels.

Acknowledgements

The authors wish to acknowledge funding from Charles Sturt University and the Co-operative Research Centre for Irrigation Futures. On-farm data provided by Damian Smart is greatly appreciated. Help with determining appropriate parameters for SWAGMAN Farm by Tariq Rana is acknowledged.

References

[1] van der Werf, H.M.G., Tzilivakis, J.T., Lewis, K. & Basset-Mens, C., Environmental impacts of farm scenarios according to five assessment methods. *Agriculture, Ecosystems and Environment*, **118**, pp. 327–338, 2007.

[2] Entry, J.A., Sojka, R.E. & Shewmaker, G.E., Management of irrigated agriculture to increase organic carbon storage in soils. *Soil Science Society of America Journal,* **66 (6),** pp. 1957–1964, 2002.

[3] Lal, R., Carbon emission from farm operations. *Environment International,* **30**, pp. 981–990, 2004.

[4] Khan, S., Rana, T. & Hanjra, M., A cross disciplinary framework for linking farms with regional groundwater and salinity management targets, *Agricultural Water Management*, **95,** pp. 35–47, 2008a.

[5] Ozkan, B., Akcaoz, H., & Fert, C., Energy input-output analysis in Turkish agriculture. *Renewable Energy,* **29**, pp. 39–51, 2004.

[6] McInnes, K.L., Suppiah, R., Whetton, P.H., Hennessy, K.J. & Jones, R.N. *Climate Change in South Australia. Assessment of climate change, impacts and possible adaptation strategies relevant to South Australia*, CSIRO Atmospheric Research: Victoria, 2003.

[7] Ozkan, B., Fert, C., Feyza Karadeniz, C., Energy and cost analysis for greenhouse and open-field grape production. *Energy,* 32**,** pp. 1500–1504, 2007.

[8] Hatirli, S., Ozkan, B., & Fert, C., Energy inputs and crop yield relationship in greenhouse tomato production. *Renewable Energy,* **31**, pp. 427–438, 2006.

[9] Mandal, K.G., Saha, K.P., Ghosh, P.K., Hati, K.M. & Bandyopahyay, K.K., Bioenergy & economic analysis of soybean-based crop production systems in Central India. *Biomass & Bioenergy*, **23**, pp. 337–345, 2002.

[10] FAO, *The Energy and Agriculture Nexus*, Food and Agricultural Organisation of the United Nations: Rome, Italy, 2000.

[11] Wells, C.M., *Total Energy Indicators of Agricultural Sustainability: Dairy Farming Case Study FINAL REPORT*, Report to MAF Policy, Dept. of Physics: University of Otago, 2001.

[12] Canakci, M., Topakci, M., Akinci, I., & Ozmerzi, A., Energy use pattern of some field crops and vegetable production: Case study for Antalya region, Turkey. *Energy Conversion and Management,* **46(4),** pp. 655–666, 2005.

[13] Barber, A., *Seven Case Study Farms: Total Energy and Carbon Indicators for New Zealand Arable and Outdoor Vegetable Production*, AgriLINK: New Zealand, 2004.

[14] Skewes, M., Treloar, N. & Bailey, G., *Irrigation Innovations in the South East of South Australia*, Rural Solutions SA: Loxton, South Australia, No date.

[15] Jackson, T.M., Khan, S. & Ahmad, A., Exploring energy productivity for a groundwater dependent irrigated farm using a system dynamics approach. In Oxley, L. and Kulasiri, D. (eds) *MODSIM 2007 International Congress on Modelling and Simulation.* Modelling and Simulation Society of Australia and New Zealand, pp. 156–162, 2007.

[16] Khan, S., O'Connel, N., Rana, T. & Xevi, E., Hydrologic–Economic Model for Managing Irrigation Intensity in Irrigation Areas under Watertable and Soil Salinity Targets. *Environmental Modeling and Assessment*, **13 (1),** pp. 115–120, 2008a.

Computer application for optimization of turn assignment on a tree-structured irrigation network

A. Almiñana[1], L. F. Escudero[2], M. Landete[1], J. F. Monge[1], A. Rabasa[1] & J. Sánchez-Soriano[1]

[1]*Instituto Centro de Investigación Operativa, Universidad Miguel Hernández, Spain*

[2]*Departamento de Estadística e Investigación Operativa, Universidad Rey Juan Carlos, Spain*

Abstract

Most of the easier structural solutions for greater utilization of water resources have already been implemented. The principal objective of this paper is to develop a methodology for helping to produce irrigation scheduling policies for the deployment of water resources for agricultural use. We have, therefore, developed WISCHE (Water Irrigation Scheduling), a decision support system for irrigation scheduling in an arborescent network. Computational experience has been implemented with satisfactory results in "La Comunidad de Regantes, Riegos de Levante, Canal 2nd", eastern Spain. Its irrigation area covers 2188 Hectares, with 2831 nodes distributed in 20 sectors. Irrigation is needed on a daily basis in five time periods. Network topology, periodical demands and historical consumptions are the inputs and scheduling the hydrants and turns are the output. Assignments must satisfy two types of constraints: the first one based on physical network limitations and the other one based on previous consumptions and demands. In order to deal with the first limitations, a mixed 0–1 separable quadratic program is proposed. Statistical information about previous consumptions on turns demanded and non-attended demands is used to assign a priority to hydrants. Thus, the system favours hydrants which used what they applied for and hydrants which could not be well attended previously.

Keywords: water resource scheduling, agriculture irrigation, mixed 0–1 separable quadratic programs.

WIT Transactions on Ecology and the Environment, Vol 112, © 2008 WIT Press
www.witpress.com, ISSN 1743-3541 (on-line)
doi:10.2495/SI080251

1 Introduction

Southern Spain is going through a period of serious drought. Due to these circumstances and as a result of the impact of the changes in the general climatic conditions, there is increasing demand on the use of water resources and the need for rational planning and scheduling of water resources is becoming stronger than ever, as is justified by Gomez-Limon and Martinez [1]. We should differentiate between water resource planning (usually long-term) and water distribution scheduling (usually daily basis). In the case of planning, we distinguish between the deterministic environment as proposed by Andreu et al. [2] and the stochastic environment which consider uncertainty among its main parameters such as water inflow and needs as presented by Escudero [3]. We focus on the latter case.

To make an effective use of irrigation water, "La Comunidad General de Regantes, Riegos de Levante, Canal 2nd", in Elche (Spain), is introducing a complete plan for the modernization and improvement of its irrigation infrastructures. A telemetric and monitoring system for the whole irrigation network is being implemented as part of the whole project. Thus, it has become necessary for hydrants to be assigned turns efficiently.

The application we propose generates the turn assignment, by considering physical network constraints and previous usages by farmers.

This paper is organized as follows: the problem definition and the main objective are outlined in section two. The application functionality with a brief description of its modules is presented in the third section. The fourth section offers the results of the computational experiment through a step by step example execution. Finally, in the fifth section some conclusions derived from this software application are presented along with the future research lines to be carried out by the group.

2 Problem definition and main objective

The tree-structured irrigation network consists of 2405 hydrants over a surface of 2188 Hectares. There are five four hour turns a day for irrigation. A hydrant can demand one, two, three or four of them, depending on its irrigable surface. Because of network topology constraints and the quantity of water available, some users' demands cannot be attended. So, an assignment generation system becomes necessary.

Thus, our objective is to develop an application for the optimization of turn assignment, which considers both network physical restrictions (flow and pressure) and the priority demand based on previous demand usage in such a way that farmers can be equally attended.

3 Application functionality

3.1 Developing tools

We chose C++ language for application coding in order to obtain the most efficient program; and the chosen developing tool was Borland Builder® C++

v6.0 because of its wide range of facilities in Graphic User Interfaces development. The mathematical optimization model for final assignments was written using CPLEX. The application runs under Microsoft Windows® Operating System.

3.2 Input data files

-*Network_Topology*: This file includes a list of hydrants with their code, coordinates, antecedent node, type and other information. This file is read automatically when application starts and it represents the real tree-structured irrigation network.
-*Weekly_Telemetry*: This file includes date, hydrant code and daily water consumption in a given week period. These values are provided by the SCADA system and the file is loaded by the user in *Telemetría* form. If the user wants to a *Weekly_Telemetry* file is added to the *Log_Telemetry* file (if it is possible).
-*Daily_Applications*: This file is generated from a template filled in by the user, and it contains the turn when each hydrant wants to start irrigation. This planning is repeated for every day of the week. This file is loaded by the user in the *Solicitudes (Applications)* form. Every *Daily_Applications* file is added to *Log_Applications* file after processing (if it is possible).

3.3 Module functionality

The WISCHE application starts loading the *Network Topology* file and then the Control Panel brings to users three modules: *Telemetrias (Telemetries)*, *Asignación de Prioridades (Priority Assignment)*, *Asignación de Turnos (Turn Assignment)*. From the *Telemetrías* module, the user can load the *Weekly_Telemetry* file and it can be automatically added to the *Log_Telemetry* file (which was automatically loaded when this form started). This module, optionally, allows the user to show the loaded files, data pre-processing facilities and customizable graphs generation with telemetric data. Customizable graphs generation is described in section 3.3.3.

Asignación de Prioridades (The Priority Allocation) module offers a frame to fill in and change an application form template, the weekly loading application, and the priority generation procedure, described in detail in section 3.3.1. Optionally, the user can show generated priorities and create new customizable graphs with previously applied data.

From *Asignación de Turnos (Turn Allocation)* module the user defines the previous conditions: maximum speed, minimum pressure, and pressure on head-nodes. Optionally, the user can require a maximum speed minimization and/or maximum pressure minimization. Finally the form brings a heuristic use option that generates really good but non-optimal solutions with a very low execution time. It is described in detail in section 3.3.2.

3.3.1 Priority generation procedure

The priority associated to a hydrant in a certain turn is calculated as an addition of two factors: the first of which is the previous usage of turns demanded (real consumption divided by maximum consumption in a period), and the second factor is the proportion of turns that could not be attended previously. Both of them can be weighed by the user through its impact values. The calculated priorities are returned in a bi-dimensional structure (hydrant, turn). Figure 1 shows the application window used to customize and calculate the priorities.

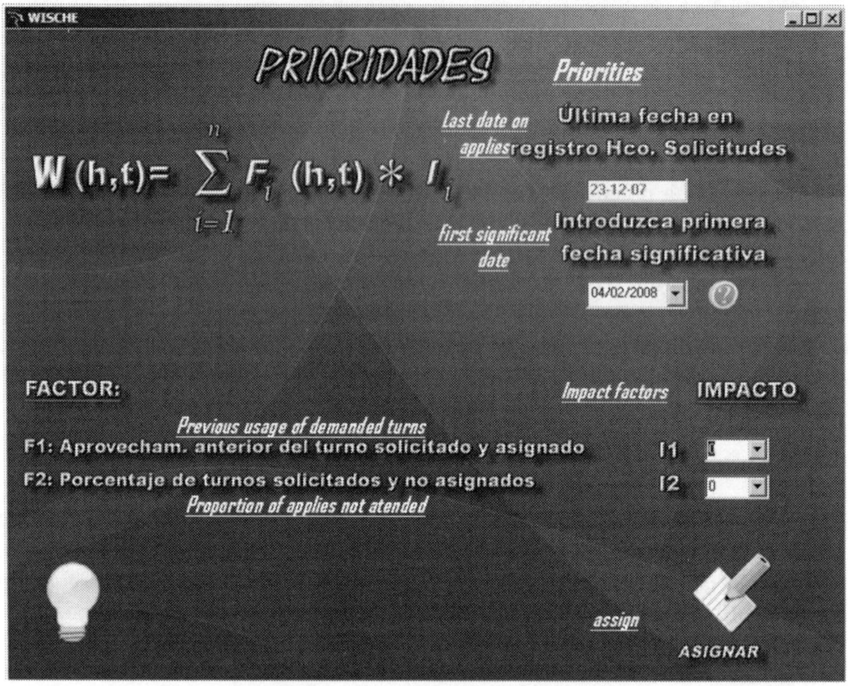

Figure 1: Priorities form.

3.3.2 Mathematical model for turn assignment

The objective is to turn on the maximum number of hydrants with the highest priority factor. The pressure head at any hydrant must not fall below the specific threshold. This constraint is formulated by employing the Darcy-Weisbach equation. The discharge of water through each node during each time period is defined too and the water flow velocity allowed along the immediate upstream pipe segment of any node is enforced as shown in Figure 2.

Hydrant irrigation must not be interrupted, in which case it can only be turned on once and will be kept open during consecutive time periods. The irrigation schedule for some hydrants must be fixed, due to logistical considerations imposed by the system operator. All this is considered in the proposed mixed 0–1 program by Almiñana et al. [4]. The computation of the friction factor is calculated by using the Colebrook-White [5] equation. An explicit calculation of the friction factor in a set of special pipes is presented by Yoo et al. [6]. Final turn scheduling is given by iterating this mixed 0–1 linear problem. Turn assignment results must be filtered as shown in Figure 3.

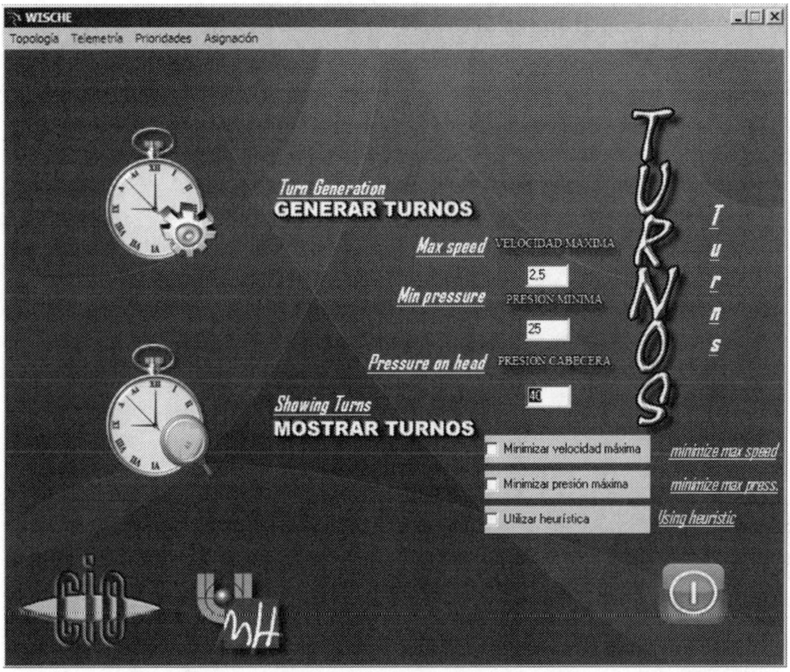

Figure 2: Turn assignment form.

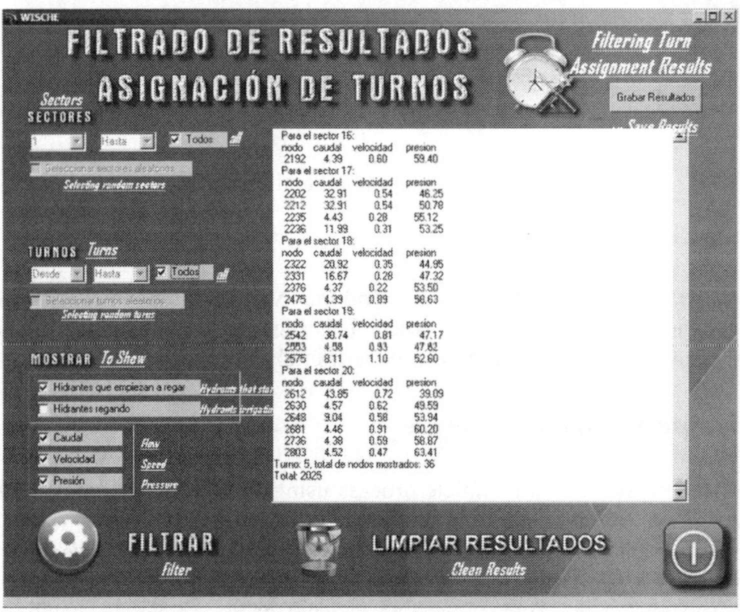

Figure 3: Turn assignment visualization form.

3.3.3 Customizable graph generation

Applications allow the user to create customizable graphs over two data sets: telemetric values and previous applications. The user must select a time period and a type of graph. Also, he can select the parameters he wants to show. There is 3D-visualization and clustering possibilities. Figure 4 shows an example of graph generation with telemetric data where maximum flow is compared.

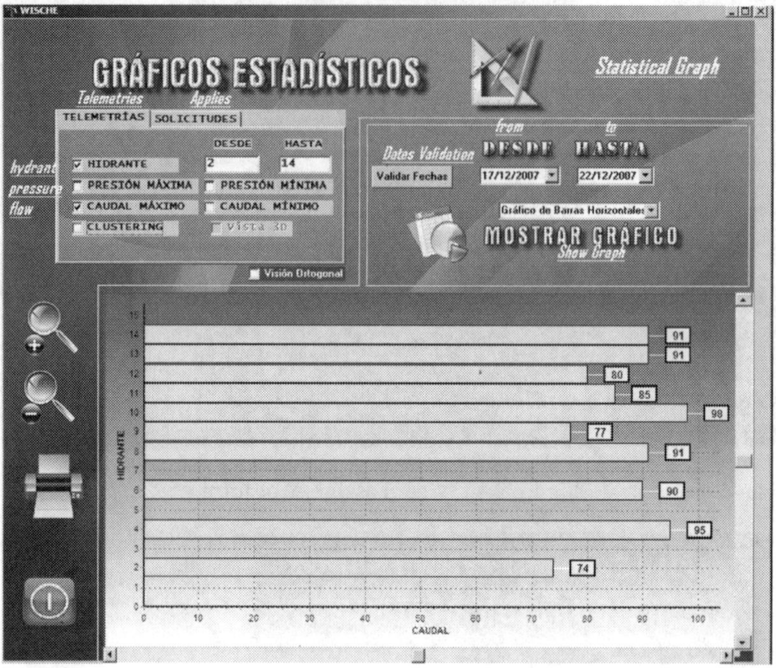

Figure 4: Graph generation form.

3.4 Output data file.

-Assignments: This file contains a list of all the hydrants on the network with the turns assigned by the system, for a week. This file is generated by *Asignación de Turnos (Turn Allocation)* form when the user clicks on this option.

4 Computational experience

In this section we show a complete process using WISCHE. The computational experience was executed on an Intel Core Duo 1,66 GHz. Processor, with 2 GB. RAM running under Microsoft Windows® XP Operating System. Topology was a real data file, but the rest of files were simulated because the network infrastructures and the SCADA system had not been completely installed yet. Below, Table 1 shows an example of a complete application trace, where the actions in brackets are optional.

Table 1: Example of complete application trace.

Action	Form	Option	Time /Size
Loading topology	Starting	(auto)	1 sec. /321 Kb.
Loading telemetric log file	*Telemetries*	(auto)	1 sec. 360 Kb.
Loading weekly telemetries file	*Telemetries*	*Read Weekly Telem. .or Record*	9 sec. 7360 Kb.
[Data pre-processing]	*Telemetries*	*Data Preprocesing*	-
[Showing telemetric]	*Telemetries*	*Show Telemetry*	-
[Graph-generation]	*Telemetries*	*Graphs*	-
Filling in application form	*Pr. Assig.*	*Application Forms*	-
Loading weekly applications	*Pr. Assig.*	*Read weekly Application Forms*	1,5 sec. / 369 Kb.
Priorities generation	*Pr. Assig.*	*Generate Priorities*	4 sec.
[Showing priorities]	*Pr. Assig.*	*Show Priorities*	-
[Graphs generation]	*Pr. Assig.*	*Graphs*	-
Turns generation	*Asig. Tur.*	*Generate Turns*	2,5 - 12 min.
[Showing turns]	*Asig. Tur.*	*Show Turns*	-

5 Conclusions

WISCHE is an application built in collaboration with Riegos de Levante, an irrigation community which has a real need for irrigation scheduling because water demands usually exceed water availability. This software allows irrigation community managers to schedule hydrant turns one week in advance. Scheduling is as flexible as possible and the priority system presented allows for all farmers to be attended according to their previous usages, so farmers who make early applications and usually make reasonable use of their demands are benefited.

Because all types of problem restrictions are considered, the turn assignment model brings optimal solutions in a really short time (12 min). If heuristics are used, this time is reduced to 2 or 3 min. and turn scheduling is produced, which managers consider as quite good and network operation is safe even in critical conditions of water demand.

Also, by using the WISCHE graphic facilities, the farmers' behaviour becomes more understandable, and network telemetric log files can be examined in detail highlighting flow and pressure conditions in the most demanded turns.

At present, the authors are researching new heuristics to reduce the application running time, and therefore obtain satisfactory scheduling. Also, Data Mining techniques, such as Classification Rules, are being studied to be incorporated in a new predictive module.

References

[1] Gomez-Limon, J.A., Martinez, Y. Multi-criteria modelling of irrigation water market at basin level: A Spanish case study. *European Journal of Operational Research*, 173, pp. 313–336, 2006.

[2] Andreu, J., Capilla, J., Sanchos, E. AQUATOOL, a generalizad decisión-support system for water-resources planning and operational management. *Journal of Hydrology*, 177, pp. 269–291, 1996.

[3] Escudero, L.F., WARSYP, a robust approach for water resources planning under uncertainty. *Annals of Operations Research*, 95, pp. 313–339, 2000.
[4] Almiñana, M., Escudero, L.F., Landete, M., Monge, J.F., Rabasa, A., Sanchez-Soriano, J., On a mixed 0-1 separable quadratic program for water irrigation scheduling, *IIE Transactions*, 40:4, pp. 398–405, 2008
[5] Colebrook, C.F., Turbulent flow in pipes, with particular reference to the transition region between the smooth and rough pipe laws. *Journal of the Institute of Civil Engineers*, 11, pp. 133–156, 1938
[6] Yoo, D.H., Singh, V.P., Explicit design of commercial pipes with no secondary losses. *Journal of Irrigation and Drainage Engineering*, 130, pp. 437–440, 2004.

Cost-benefit analysis of different solutions for sustainable irrigation in Fucino Plain (Italy)

E. Frank[1], A. Gallina[2], M. Baldacci[2] & M. Miolo[1]
[1]*BETA Studio srl, Italy*
[2]*HR Wallingford, UK*

Abstract

The Fucino Plain (200 km²) was the largest lake in Central Italy prior to the 1800s when it was reclaimed for agricultural use. In the past 15 years the original crops, mostly wheat, maize and sugar beet, have been progressively replaced with much more profitable (but more water demanding) horticultural crops. Increasing demand for water together with the actual irrigation techniques and the changes in the climate (decreasing precipitation and increasing temperatures) has caused a number of environmental and social-economical problems. These are mainly related to groundwater resource depletion, spring exhaustion, a decline in summer and annual surface water discharges, scarcity of irrigation water and conflict between different users. Based on a comprehensive study of water supply and demand, several measures for a sustainable use of water were considered. These were mainly related to the implementation of more efficient irrigation techniques, the increase of storage capacity and, subsequently, a reduction in the use of ground water. An accurate cost-benefit analysis showed a positive environmental impact (decline of 70% of ground water exploitation, reduction of energy consumption) and social benefits (increasing of quantity and quality of agricultural production) for the chosen solution.

Keywords: sustainable irrigation, irrigation systems and planning, cost-benefit analysis, water balance models, water resource management.

1 Introduction

In the recent past water availability in the Fucino Plain was not a problem: both agriculture and public water supply needs were met using surface water from canals and springs. Since the 1950s, the increasing water demand was satisfied

WIT Transactions on Ecology and the Environment, Vol 112, © 2008 WIT Press
www.witpress.com, ISSN 1743-3541 (on-line)
doi:10.2495/SI080261

by pumping groundwater from wells on the boundaries of the Plain where the limestone aquifers, considered as a virtually inexhaustible resource, are located. Over the past two decades the rising demand for water for agriculture and human uses (civic and industrial) has exceeded supply creating a negative supply-demand balance in dry years and during specific high demand periods in 'normal' years. This situation has caused various environmental and socio-economic problems mainly related with groundwater resource depletion, spring exhaustion, a decline in surface water discharges, scarcity of irrigation water and conflict between different users (Burri and Petitta [1]).

A comprehensive study (BETA Studio and HR Wallingford [2]) financed by Liri-Garigliano and Volturno River Basin Authority and Abruzzo Region, was carried out to identify different solutions to the problem and to suggest the best solution from a technical, environmental and economic point of view.

The first part of the study completed a detailed analysis of all the different aspects connected to the water balance: climate, hydrology, hydrogeology, social and economics issues and infrastructure, making use of knowledge, experience and initiatives of the different stakeholders. The accurate knowledge of all those aspects made it possible to implement and validate a monthly water balance model of the study area. The analysis of the results of the water balance highlight all the different problems (present state and future) related to water availability and water uses in the basin, taking into account both quantity and quality issues. To solve these problems a number of measures (structural and non-structural) were proposed for all the different sectors analysed (civic and industrial water supply, water collection and treatment, irrigation). This paper considers the problem and the results obtained from an analysis of the irrigation sector. It shows that measures proposed are able to close the supply-demand balance and reduce environmental and social problems with a positive cost-benefit balance.

2 Study area

The study area is the Fucino basin (900 km^2) located in the carbonate Apennine range in central Italy (Figure 1). In the centre of the basin is a 200 km^2 alluvial plain that used to be the location of the largest lake of central Italy prior to the late 1800s when a tunnel was made to drain the lake so that the area could be used as farmland. The drainage was completed in 1875 by Prince Torlonia and the necessary infrastructure was built including 210 km of roads, over 100 km of canals and 618 km of drainage ditches, in addition of farm-houses and stables (Brisse and De Rotrou [3]). The disappearance of the lake caused important social changes due to the transition of the local economy from fisheries to farming but also relevant environmental impacts caused by the disappearance of a balanced biological ecosystem.

The Fucino Plain is encircled by limestone ridges, bounded by extensional or thrust belt faults (Ciotoli *et al* [4]). Since Pliocene times, the endorheic depression has been filled by terrigenous, detrital and lacustrine alluvial deposits. The Plain is drained by artificial canals that collect water and direct it to the tunnel outflow.

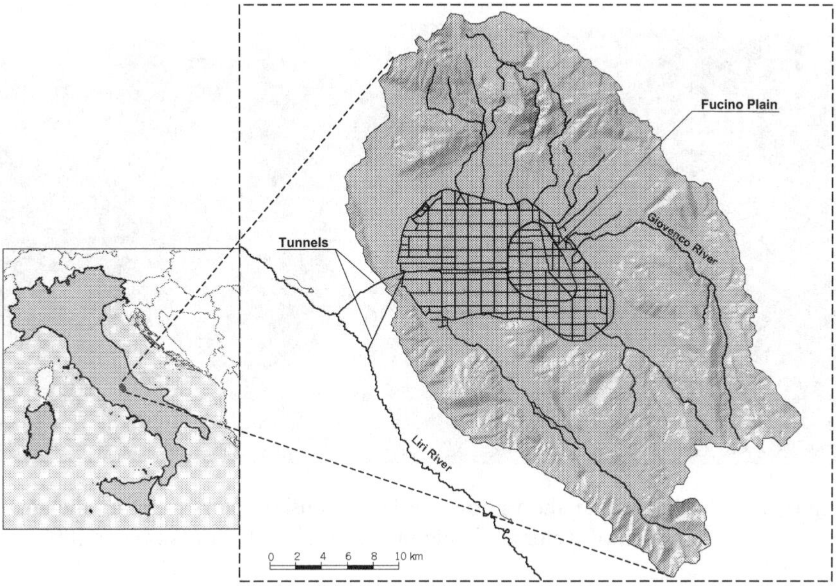

Figure 1: Location of the study area.

The fractured and karstified hills around the Plain are drained at their boundaries by high-discharge springs, which ensure steady discharges even during the dry season (Boni *et al* [5]; Burri and Petitta [6]). The climate of the area is characterised by an average yearly rainfall of 883 mm, evapotranspiration of 609 mm and an average water surplus (in the Plain) equal to 220 mm/year. Water surpluses only occur in the November-March period. From April through October, the low amount of rainfall is not able to satisfy crop water demands and thus large amounts of water are used to irrigate farm crops.

In the past, farmers grew three main crops (wheat, potatoes and sugar beet) on the basis of a three-year rotation. These crops required minimal amount of water and the demand was satisfied by the natural flows in the canals. In the last decade, however, farmers have gradually switched to vegetable crops, such as carrots, salads, Belgian endive, fennel and celery. Vegetable crops are field grown, thanks to favourable circumstances, including the option of repeating two to three growing cycles on the same field. To be able to satisfy this large water requirement several well fields (with a total capacity of 2.6 m^3/s) have been drilled in the limestone aquifers surrounding the Plain. The water taken from the wells is discharged into the canal network, where is exploited by individual using mobile pumps via tractor 'power take offs' (PTOs) (Figure 2).

3 Methodology for problem assessment

To analyse the main problems of the study area and to assess the benefits of the identified measures two models were implemented: a monthly water balance

Figure 2: Picture that shows how the farmers use tractors to pump the small amount of water available in the canals network (summer 2007).

model for the whole catchment and a hydraulic model of the Plain channel network.

The water balance model implemented was RIBASIM (RIver BAsin SIMulation) developed by WL|Delft Hydraulics. RIBASIM simulates the behaviour of river basins under various hydrological conditions, representing through nodes and links hydrological water inputs (surface and underground), natural and artificial waterways and the water users in the basin. The implementation of the model enabled the evaluation of the quantity and flow composition of water available for the user in the basin under different hydrological condition and operational scenarios.

Considerable work was completed to develop hydrological scenarios (monthly series of surface runoff and spring flows) used as inputs in RIBASIM. This included implementing a rainfall-runoff model based on the model developed by Ibrahim and Cordery [7]. The area was subdivided in 21 sub-catchment and 20 acquifers: the intersection of those identified 72 calculation elements with unique surface and underground characteristics. Each of these elements was schematised by means of two storage components: a soil storage, which represents the unsaturated zone and plays an important part in determining runoff and the amount of actual evapotranspiration, and groundwater storage, which represents saturated zones and determinates the baseflow. Input to the model consists of monthly rainfall and temperatures and soil parameters (such as soil moisture storage at wilting point and field capacity) for each calculation element. The model was successfully calibrated and validated in different locations using measured data series of rainfall, temperature and discharges from 1921 to 2002.

The hydrological model was used to develop two runoff scenarios: average and 1 in 10 year drought. These runoff scenarios were used in RIBASIM as input together with data and rules typical of the different water users (civil, industrial, hydropower, agriculture), and the characteristic of the main structures in the basin (wells, rivers, pipelines, sewage treatment plants, …). A total number of 552 nodes and 679 links have been used to schematise the whole study area.

The RIBASIM model results were judged to be sufficient to identify problems and evaluate solutions for the several users in the basin except for the irrigation users in the Plain: the complexity of the hydraulic behaviour of the canal network used by farmers for irrigation meant that this required the implementation of a more detailed hydraulic model and InfoWorks was chosen for this purpose.

InfoWorks CS (Wallingford Software) allows water utilities to complete hydrological and hydraulic modelling of the complete urban and non-urban water cycle. The applications include flood risk management, pollution prediction and the modelling of water quality and sediment transport throughout networks. As well as supporting network modelling, this software is sufficiently flexible to model subcatchment abstractions and infiltration characteristics.

The Fucino Plain network (340 km) was implemented in InfoWorks using a ‘nodes’ and ‘links’ hydraulic representation. Moreover, to obtain an accurate model of the real situation, all gates, siphons, weirs and pumps present in the Plain were included in the model. The results obtained with RIBASIM were used as boundary conditions in the hydraulic model giving the quantity and the quality of water inflow (springs, wells, rivers, urban discharges). The water demand, drawn by the farmers, was implemented using additional orifices to represent specific outflows. InfoWorks CS model has permit to obtain accurate results considering quantity and quality balance and network propagation for the average condition of each month for average and 1 in 10 year drought.

4 Problem assessment

The current and future state analysis, with the use of the models described above, has defined the main irrigation and environmental problems. The major criticalities in the irrigation sector are related to the current system organization. A large amount of irrigated surfaces are not provided with a distribution system and irrigation is achieved by direct abstraction from the channels. In summary the problems are:

- water deficits for irrigation;
- poor management of the irrigation system;
- high-level of energy consumption;
- poor quality of water used for irrigation;
- use of potable quality groundwater.

In particular there are many zones in the Plain with a water deficit during summer periods when the water demand from crops is at its highest. This supply-demand deficit is related to:

- insufficient water availability for the basin as a whole and
- non-uniform distribution of water resources within the Plain.

The unsatisfied water demand is equal to 3 300 000 m^3 for the ‘average year’ and 12 400 000 m^3 for the ‘critical year’ (one in ten year return period).

With regard to the irrigation system, there is an absence of coordination and planning of water use. Moreover, there is no formal irrigational scheduling to satisfy the crop life cycle. Irrigation practices are generally reactive and focused on providing water to satisfy periods of crop stress. Another important problem is related to the use and maintenance of the gates controlling the hydraulic profile in the channels with the intent to permit the water withdrawal by farmers. Detailed management across the whole area is difficult to achieve and is exacerbated by farm fragmentation (a typical irrigation plot is equal to 1.5 ha), which has led to the growth of low efficiency irrigation technologies.

Current irrigation operations entail a high-level of energy consumption due to well withdrawals to fill the channels and the use of tractor motor pumps (very low efficiency) to abstract water from the canals. The average water supplied yearly by underground aquifers is equal to 11 300 000 m^3 that entails an energy consumption of 2 300 MWh. Abstraction from the channels is 30 720 000 m^3 per year with a gas oil consumption of 3 millions of litres.

The modelling results, in agreement with the analysis lead by Local Authority, show that the water quality present in the Fucino Plain channels is, in general, not appropriate for irrigation according to the present legislature. In particular in the main channels, for the medium year, Escherichia Coli is present at with counts of 100 per 100 ml of water (UFC/100 ml) and of BOD_5 concentration is approximately 20 (mg O_2/l). The poor water quality affects product quality and therefore the commercial value of the crops grown.

The use of water resources from underground aquifers for irrigation is not sustainable and these water resources should be prioritised for human use and be preserved for future generations.

5 Proposed measures for sustainable irrigation

The methodology used to define the interventions was based on the study of the 'critical year' (1 in 10 dry year) for the baseline and future scenarios, the analysis of the Local Authority proposals and the identification and description of new interventions.

To choose the best solutions, in terms of economic benefit, appropriate location and environmental functions, a cost-benefit analysis was implemented. As part of this process interventions were refined to provide greater benefits, for example by reducing their energy use. The last step of the method is an interactive procedure to compare the various solutions.

The economic evaluation of the various cash flows permits a cost-benefit analysis that considers all aspects of a proposed solution. The evaluation methods used in this study consider also the actualization of the investments values – the Net Present Value (NPV) and the Investment Return Period, both considering a 5% of discount rate.

The study of 'current state' problems clearly shows that the structural solutions must consider the storage of surface water during winter and its use for irrigation, the distribution of water with a pressurised network and the adoption of more efficient irrigation techniques.

The hydrological analysis demonstrated that the Giovenco River provided the only feasible option for a water intake. The study considered a reservoir design based on the maximum water volume that could be abstracted during a 'scarce year' (1 in 10 dry year).

The first step was to identify ten possible combinations of storage and transfer according to the hydrologic and hydrogeology characteristics of the basin, different abstraction rates and storage locations. The cost-benefit analysis approach was used to define the best solution. Taking into account this storage and conveyance solution two interventions for water distribution – rain-gun or drip – were considered. These hypotheses have been compared, with a cost-benefit analysis, with two other interventions proposed by the local Authorities.

At the end of the process, the optimal solution identified was the "*Project for the construction of a Giovenco river intake near Pescina, of an artificial lake near Arciprete and of a storage basin in the Tristeri valley – Drip irrigation network*", coded IR54-P2 in the study (Figure 3). This project aims to realize an intake in the Giovenco river up to Pescina (East part of the basin) at 740 m above the sea level, a transfer pipe of 1 200 mm of diameter and 5 300 m long from the intake to the irrigation network in the Plain. The project also includes an artificial lake of 500 000 m³ near Arciprete (in the south respect the Plain) at 710 m a.s.l. This elevation was indicated by the model as the optimal to have the right pressure for drip irrigation in every point of the system.

The irrigation network could transport an average annual flow of 1.06 m³/s from the intake in Giovenco river to the Arciprete Lake. In the Tristeri valley, which is a natural karstic depression up to the artificial lake (South part of the Basin), a reservoir of 9 260 000 m³ (maximum amount of water that can be withdraw and stored during the winter period of a scarce water year) will be realized, with a small dam (11.5 m high) and grouting to waterproof the base

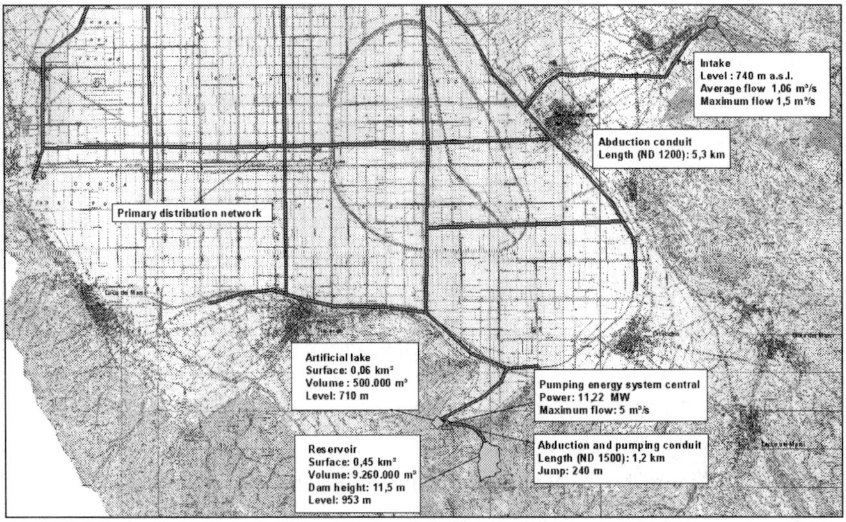

Figure 3: Schematisation of the solution IR54-P2.

of the reservoir. A 1 200 m pipe with 1 500 mm of diameter can connect the reservoir with the lake. The head between the two lakes (more than 200 m) is utilized to produce electricity with a turbine during the high energy price hours. The water is pumping up during the low price energy hours.

The project also includes a pressurised distribution network in the Plain, completely a small network which is already in place. The network can reduce the water deficit in the all Plain, level the water availability for every farmer, permit good control and improve the water quality used for irrigation as well as decrease energy consumption.

In the cost-benefit analysis the environmental and social benefits analysis have not been monetized. The economics benefits that have been taken into account for the evaluation could be absolute or relative compared with the present state. An absolute benefit is the use of hydro-power that includes selling excess hydro-energy, providing energy for the pumping system and the benefit of a Renewable Energy Certificate. Relative benefits are:

- a reduction in the costs of water pumping from underground aquifers;
- a reduction or elimination of water pumping from the channels to the fields;
- a decrease in water deficits that increases crop yield;
- improvements in product quality due to use of an unpolluted water use.

The chart below (Figure 4) compares the last four project solutions analysed using the NPV index, the black line is the IR54-P2 project, and the dark grey line just under the black one is the IR54-P1 project that differs from the IR54-P2 because it uses a rain-gun irrigation network. The two other projects have been proposed by the local Authorities. The IR01 considers to construct a surface water storage lake near the Plain and to distribute the water with a low pressure rain-gun irrigation network; the IR03 considers to use just the water from aquifers and to construct a high pressure rain-gun irrigation network supplied with four small daily storage units.

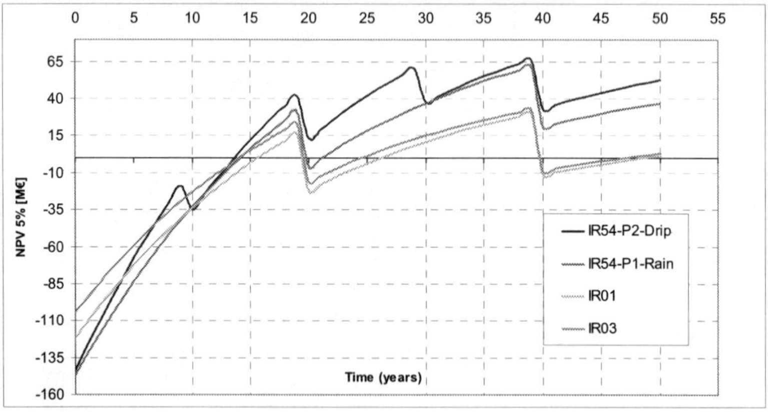

Figure 4: Net present value of the analysed measures.

The results of the economic analysis show that project IR54-P2 provides the highest cost-benefit if a long term perspective of 50 years is used. This considers the periodic replacement of the irrigation supply infrastructure that is more expensive for drip irrigation than for rain irrigation systems.

The benefits of the proposed measures could be summarized as follows:

- 70% reduction of precious water resources using (underground water) for the average year;
- reduction in leakage that entails a diminution of the water needed in the medium year from 46 000 000 m³ to 24 000 000 m³;
- complete satisfaction of the water demand for the medium year and reduction of the water deficit from 12 000 000 m³ to 2 400 000 m³ (-81%) for the critical year. Note that complete water demand satisfaction entails an additional cash input of € 7.5 millions per year;
- improved quality agricultural products due to the use of better quality water that causes a significant increase in commercial value (estimated to be equal to € 7.6 millions per year);
- reduction of energy consumption due to the aquifer withdrawal for an amount of € 120 000 per year, and due to the cancellation of the motor pump withdrawal from canals that has been quantified to be equal to € 2.2 millions per year;
- energy production benefit of the pumping system for on amount of 1.4 millions of €/year;
- better user coordination, management and irrigation scheduling.

6 Conclusion

The results obtained through a cost-benefit analysis of possible solutions to overcome irrigation water scarcity and poor quality problems in the Fucino Plain show that it is possible to find measures that are not only effective in reducing the current environmental problems of the area but are also beneficial from an economic point of view.

The proposals require active involvement and 'buy-in' from the farming community that will have to adopt new irrigation techniques. Their participation in the planning and decision-making processes is crucial so that widely acceptable solutions for irrigation and more general for river basin planning can be promoted in the catchment.

The proposed measures are based on shared knowledge, experiences and a scientific approach including the use of a validated water balance models, hydraulic modelling tools and cost-benefit analysis, which is fundamental to improve decision-making.

The potential benefits of completing this study, which combined considerations, related to public water supply, irrigation and water quality include:

- an increased public awareness of environmental problems in Fucino basin;
- more transparent decision making by considering a wide range of interventions and cost-benefit methods that included environmental considerations;

- less misunderstanding about the costs and benefits associated with particular environmental problems and possible solutions;
- public acceptance, commitment and support with regard to decision taking processes.

Acknowledgements

This paper presents part of the activities promoted by the Liri-Garigliano and Volturno River Basin Authority and the Abruzzo Region awarded to BETA Studio srl and HR Wallingford Ltd: "*Design of a program for the protection, use and management of the surface and ground water resources of the Fucino plain (Abruzzo region), combining structural and non-structural measures*" in 2007. The authors gratefully acknowledge Dr. Vera Corbelli (Technical Director and General Secretary of the Liri-Garigliano and Volturno River Basin Authority) and the staff involved in the study that provided us with useful and helpful assistance.

References

[1] Burri, E. & Petitta, M., Agricultural changes affecting water availability: from abundance to scarcity (Fucino Plain, Central Italy). *Irrigation and Drainage*, **53**, pp. 287–299, 2004.

[2] BETA Studio srl & HR Wallingford, *Design of a program for the protection, use and management of the surface and ground water resources of the Fucino plain (Abruzzo region), combining structural and non-structural measures*, Liri-Garigliano and Volturno River Basin Authority, Abruzzo Region, 2007.

[3] Brisse, A., & De Rotrou, L., *Dessechément du Lac Fucino executé par S.E. le Prince Alexandre Torlonia. Precis historiques*, Impremerie de la Propagande, Roma, 1876.

[4] Ciotoli, G., Guerra, M., Lombardi, S. & Vittori, E., Soil gas survey for tracing seismogenic faults: a case study in the Fucino basin, Central Italy. *Journal of Geophysical Research,* **103**, pp. 23781–23794, 1998.

[5] Boni, C., Bono, P. & Capelli, G., Schema Idrogeologico dell'Italia centrale. *Memorie Società Geologica Italiana*, **36**, pp. 991–1012, 1986.

[6] Burri, E. & Petitta, M., Farming and water management in the Fucino Plain in the last century. *Proc. of the XVII Int. Congr. ICID*, Granada, **1D**, pp. 257–268, 1999.

[7] Ibrahim, A.B. & Cordery, I., Estimation of recharge and runoff volumes from ungauged catchments in eastern Australia. *Hydrological Sciences Journal*, **40(4)**, pp. 499–515, 1995.

Determining the climatologically suitable areas for wheat production using MODIS-NDVI in Mashhad, Iran

S. H. Sanaeinejad, S. M. Hasheminiya & S. Khojasteh
Water Engr. Dept., Ferdowsi University of Mashhad, Iran

Abstract

One of the most important factors in sustainable irrigation is the adaptability of crops to climate. Vegetative vigor or "greenness" of wheat could be considered as an appropriate index to measure water availability and deficiency stress and also plant health, plant density and quality. The index is called Normalized Difference Vegetation Index (NDVI). In this study MODIS-NDVI values were compared with climatological parameters to assess the relations between vegetative vigor and climatological parameters. The NDVI values for three selected wheat farms in Mashhad area were calculated using MODIS images for 2003 and 2004 growing seasons. The data of four climatological parameters including air temperature, precipitation, relative humidity and sunshine hours were also collected from the nearest weather stations. Then a multi-regression statistical analysis was performed to find the relation between wheat NDVI and climatological parameters in the study area. Pertaining statistical methods including Mixed, and Stepwise (Forward and Backward) were used in the analysis. Scattering matrix was used to determine the data scattering of the models and NDVI values for comparison. The results showed that backward method was more appropriate than the other two methods for predicting NDVI values of the study area. After finalizing this model the results were statistically tested using 20% of the samples for the test purpose and the remaining 80% for running the model. The results showed that there was no significant difference between Backward, Testing Backward and Training Backward models. The results from the latter method showed that the NDVI of the pixels could be estimated for 79% of the cases. It can be stated that the rest of NDVI values could be affected by other environmental parameters such as soil type and characteristics, topographical conditions, agronomical practices, plant diseases and other unknown factors. Finally, some maps were developed showing the potential wheat farming in the area according to the model results.
Keywords: NDVI, wheat, GIS, multi-regression model, MODIS.

WIT Transactions on Ecology and the Environment, Vol 112, © 2008 WIT Press
www.witpress.com, ISSN 1743-3541 (on-line)
doi:10.2495/SI080271

1 Introduction

Sustainable irrigation and crop production is closely related to climate adaptability of crops. On the other hand vegetative vigor or "greenness" of wheat could be considered as an appropriate index to measure water deficiency, and also plant health, density and quality. This index can be addressed by NDVI (Normalized Difference Vegetation Index). Climate adaptability of crops is also essential for the maximum climatological potential of farming and food production [1]. Therefore, if a model could be developed a model to use spatial distribution of climatological parameters along with NDVI one could determine the potential area for planting a specific crop such as wheat.

Many researches have been done in different areas in different parts of the world in this field of study. Vento [2] used a GIS model to collect environmental, economical and agricultural data for an area located in Northern Italy and developed a model for yield production and making a decision support system for farm management. Wright et al. [3] also used a spatial analysis with Ikonos and Quickbird satellite images for wheat farms to investigate yield production under water deficiency conditions.

Wheat is an important and strategic cereal which covers most of the agricultural plantation areas in the world [8] and in Iran as well. Vegetative vigor or "greenness" could be considered as an index for wheat farm conditions. This index can be measured from satellite datas and processed into Normalized Difference Vegetation Index (NDVI) composites [4]. It is obvious that wheat production as well as other plants is influenced by climatological parameters. In this study we considered the most four important climatological parameters influencing wheat crop and used developed multi-regression analysis models to identify the potential areas which are climatologically suitable for wheat cultivation.

2 Materials and methods

Mashhad county was selected as the study area, which is located between 59.01 to 60.31 Longitude and 35.40 to 37.06 Latitude. The mean height of the area above sea level is 1812 meters with maximum of 3100 and minimum of 800 meters. This area is located in North East of Iran. Three farms were selected in Mashhad area, which cultivated to wheat during 2004 to 2005 growing seasons and their geographical location was determined by a GPS. Normalized Difference Vegetation Index (NDVI) was used to extract vegetation cover in the area from the satellite images. The index is defined as [5]:

$$NDVI = (NIR - Red) / (NIR + Red) \quad (1)$$

In which NIR is near infra red band and Red is red band of the images. Theoretically, the index value varies between -1 and +1[5]. The index value varies from 0.3 to 0.6 for dense vegetation cover [6].

MODIS (Moderate-resolution Imaging Spectroradiometer) images from Terra satellite were selected to calculate NDVI. "MODIS makes it possible for continuous monitoring (16 days returning period) of the environment by measuring atmospheric trace gases and aerosol density, and mapping the surface of clouds, land and sea in a variety of spectral ranges from the blue to the thermal infra-red (4 to 15 micrometers)" [7]. The Red band (0.62 to 0.67 micrometers) and NIR (0.841 to 0.876 micrometers) were selected to calculate NDVI using the above equation. The images were taken by the satellite in 49 and 289 Julian days in 2004. Figure 1 shows the calculated NDVI in a map for the area. ArcMap software was used to calculate NDVI.

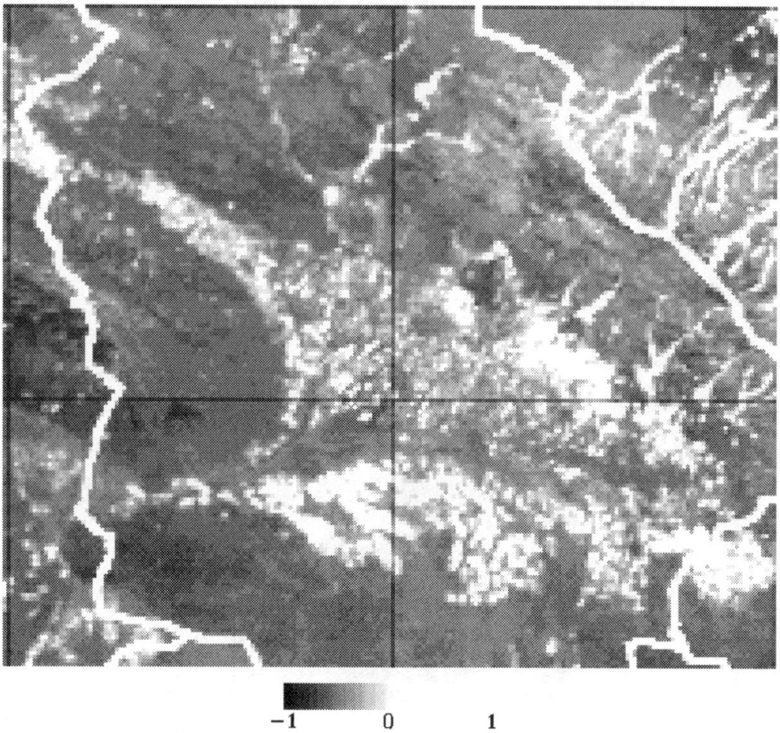

Figure 1: NDVI, calculated by using MODIS images for Mashhad area.

The image processing and also geo-processing procedures were performed as it was reported by Sanaeinejad and Shahtahmasebi [8]. Twelve locations were selected across the image to generalize the NDVI values which were derived for the reference farms. The locations were considered to match the following criteria if:

- They were uniformly distributed across the images and
- Their NDVI values were in the same range of the reference farms.

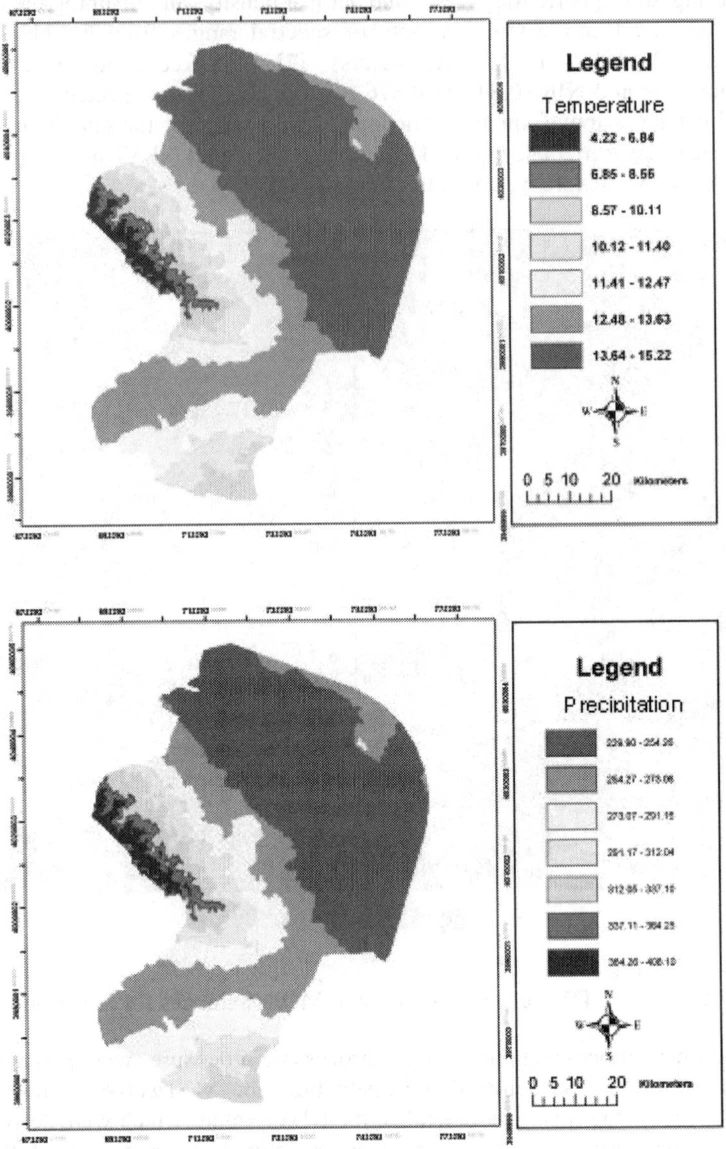

Figure 2: Spatial distribution maps of mean annual air temperature (top) and total Precipitation (bottom) in the study area.

Four climatological parameters including air temperature, precipitation, relative humidity and sunshine hours were considered and their relations with the calculated NDVI for each area were determined. The climatological data were collected from the nearest weather stations. The value of climatological parameters were determined by using some regression equations derived from the climatological data and elevation values of the area from a DEM (Digital Elevation Model). Fig. 1 shows spatial distribution of air temperature and precipitation in the study area.

Stepwise regression method was used to perform regression analysis between the climatological parameters and NDVI values all over the study area.

Stepwise regression is a semi-automated process of building a model by successively adding or removing variables based solely on the t-statistics of their estimated coefficients. The three stepwise methods including Forward, Backward and Enter were examined in the statistical analysis and the following equations were derived for each method:

Forward method

$$Y = -5231.37 + (9.321\ X_1) + (89.081\ X_2) + (53.725\ X_3) \tag{2}$$

In which y is the wheat NDVI, X_1 is sunshine hours, X_2 is relative humidity and X_3 is the mean air temperature.

Backward method

$$Y = 37.8373 + (84.77\ X_1) + (3.036\ X_2) \tag{3}$$

In which y is the wheat NDVI, X_1 is air the mean temperature and X_2 is total annual precipitation.

Enter method

$$Y = -4182.75 + (6.763X1) + (68.093\ X2) + (1.832\ X3) + (56.608\ X4) \tag{4}$$

In which is the wheat NDVI, X_1 is the total sunshine hours, X_2 is the relative humidity, X_3 is the total annual precipitation and X_4 is the mean air temperature.

3 The results

The three statistical methods were used to model NDVI values over the study area based on the data derived from the reference wheat farms. The three statistical models were compared and it was concluded that Backward method had the highest correlation (0.78) with the NDVI values over the study area. Table (1) shows the comparing results.

The p-values from Wilkakson test showed that zero hypothesis for equality of the means were acceptable for Backward and Enter methods according to NDVI values, however, the best results were derived frrom Backward method. Therefore, backward method was selected for modeling wheat NDVI over the area using 80% of the collected data.

The results derived from the model were tested using the remaining 20% of the data. The correlation coefficient for training the model was 0.89 while for testing the results was 0.91 which shows a good correlation between real NDVI values derived from the images and the values calculated from the model.

Using the model, the NDVI for wheat crop was estimated over the area. Figure 3 shows the mapping of the results. Actually, this map shows the areas where in practice the climatological conditions are suitable for wheat cultivation.

Table 1: The comparing results of correlation analysis for the three Backward, Enter and Stepwise methods.

Model	R^2 Adjusted	R2	p-value
Forward	0.047	5%	0.001
Enter	0.36	38%	0.229
Backward	0.78	79%	0.389

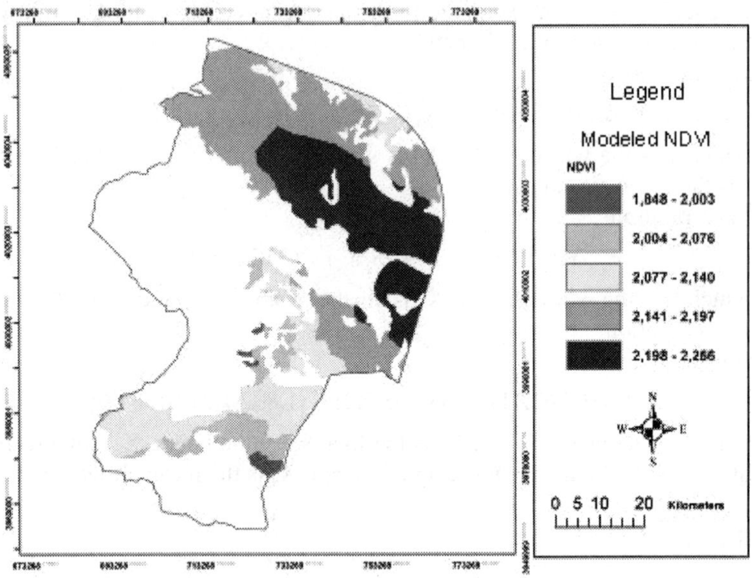

Figure 3: The map showing NDVI derived from the Backward model for Mashhad area.

4 Conclusions

This study showed that a multi-regression model (here Backward method) of climatological parameters with NDVI can be used to develop potentially suitable areas for wheat production. Since MODIS images are easily available, this method can be used to extend the results for remote areas all over the world for predicting the potentially suitable areas for wheat production and also for managing wheat fields under different climatic condition. In addition, the same analysis can be used to determine crop water needs for irrigation planning.

References

[1] Ministry of Agriculture, http://www.agri-jahad.org/introduce.asp?page=importantplan.htm
[2] Vento, D.
[3] Wright, D. L., D. Ramsey, R. Baker, D. V. Rasmussen, 2003 "A Comparison of two Geospatial Technologies in Non-Uniform Wheat Fields: Yield Monitors and Remote Sensing ". ASPRS 2003 Annual Conference Proceedings.
[4] FAOonline, 2002, http://www.fas.usda.gov/pecad2/highlights/2002/04/syria/
[5] Rouse, J. W., R. H. Haas, J. A. Schell and D. W. Deering, 1973. Monitoring vegetation Systems in the Great Plains with ERTS. Third ERTS Symposium, NASA SP - 351, vol. 1, pp. 309.
[6] Simpson J. and J. A. Stitt. 1998. Procedure for the detection and removal of clouds shadow from AVHRR data over land. IEEE Transactions on Geoscience and Remote Sensing, vol. 36, no 3, pp: 880–897.
[7] MODIS, 2005, http://eoweb.dlr.de:8080/short_guide/D-MODIS.html
[8] Sanaeinejad, S. H. and A. R. Shahtahmasebi, 2007, Estimating Wheat Planted area in North east of Iran by MODIS sensor, Proceedings, Map Asia 2007, Kualalumpor, Malaysia.

Mass mixing approach for assessing effects of irrigation on water quality

I. Fraser[1], S. Douglass[1], B. Ellwood[2] & N. Norton[3]
[1]*URS New Zealand Limited, New Zealand*
[2]*Meridian Energy Limited, New Zealand*
[3]*National Institute of Water and Atmospheric Research, New Zealand*

Abstract

Predicting the magnitude of any increase in nutrient loads on surface water and groundwater resources under future irrigation scenarios is a complex matter that necessarily relies on modelling approaches to simplify and approximate the key elements of the catchment(s).

Our water quality assessment for the proposed Hunter Downs Irrigation Scheme (HDIS) in New Zealand utilised a mass mixing model to predict nitrate concentrations within groundwater (and surface water) individual catchments. This approach was selected due to the hydrogeology of the scheme area, which is characterised by poorly draining downland hill country intersected by permeable alluvial gravel aquifers which are interconnected with intermittent or perennial streams. This approach requires a robust understanding of the current hydrology, hydrogeology, existing and proposed landuse distribution and associated nutrient concentrations in soil drainage water. Our modelling was calibrated by comparing the output of the existing landuse simulations with the available current day groundwater and surface water quality data, providing further confidence in the modelling output. The approach also allowed the effects of different landuse scenarios and management processes to be considered.

The modelled effects on water quality varied across the scheme area due to differences in existing landuse and interactions between groundwater and surface water that affected the dilution of irrigation drainage. The predicted changes in nutrient concentrations for each catchment were used to assess the risk of increased algae blooms in rivers and coastal lagoon that is of significant ecological and cultural value to the local indigenous Maori people.

In this paper we show how the use of a mass mixing model has provided a realistic basis for quantifying the potential adverse effects of irrigation on groundwater and surface water quality, and evaluating the potential effectiveness of mitigation measures such as nitrification inhibitors.

Keywords: nutrients, assessment of environmental effects, surface water, groundwater, modelling, New Zealand.

WIT Transactions on Ecology and the Environment, Vol 112, © 2008 WIT Press
www.witpress.com, ISSN 1743-3541 (on-line)
doi:10.2495/SI080281

1 Introduction

The Canterbury region encompasses approximately 42,200 square kilometres of land east of the Southern Alps, between the Waitaki River in the South to the Kaikoura Ranges in the North. The landscape can be generally characterised as extensive alluvial plains extending from the Alps to the coast, with braided alpine rivers cutting across the plains at regular intervals.

The Southern Alps create a rain-shadow effect over the plains, which has historically restricted the intensity of the land use and more recently created demand for water irrigation schemes. Irrigation has enabled Canterbury farmers to accommodate for the climate variability, and to intensify and diversify farming operations. To that end, Canterbury currently accounts for approximately 70% of the irrigated land in New Zealand.

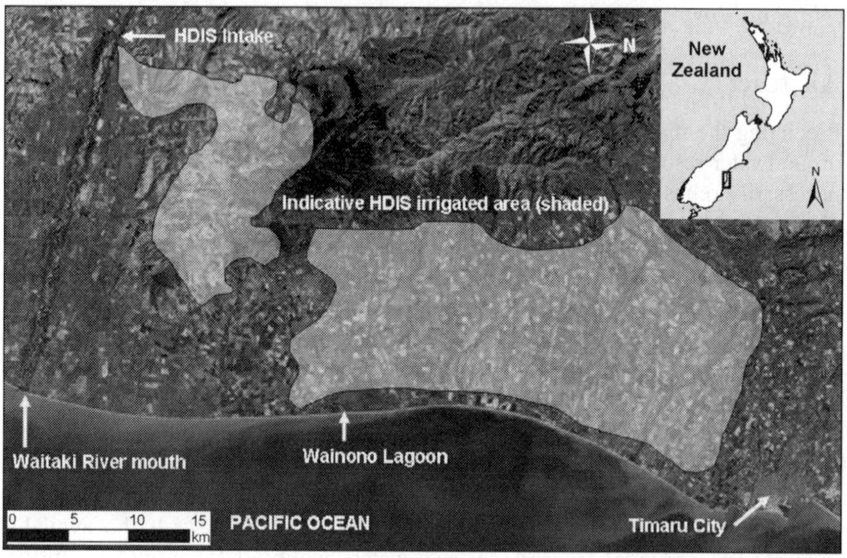

Figure 1: Satellite image showing location of the proposed HDIS (see [14]).

The Hunter Downs Irrigation Scheme (HDIS) is proposed to provide irrigation to approximately 40,000 hectares of South Canterbury that currently has limited groundwater and surface water resources available in-catchment (Fig. 1). Surface water resources in this area are typically ephemeral, with only two catchments in the Scheme area being classified as perennial. Existing utilisation of the surface water and groundwater resources is considered to be very high, with many catchments considered to be over allocated [1].

The HDIS would provide water from an out-of-catchment surface water source (the Waitaki River). Under the New Zealand the Resource Management (RM) Act (1991), the application for such consent must include an assessment of environmental effects (AEE) and consultation with stakeholders and the affected public. This is particularly important due to the presence of a highly valued coastal lagoon located at the eastern margin of the HDIS area, as well as the reliance of the community on high quality groundwater for drinking purposes.

2 Environmental setting

2.1 Geology

The geology of the HDIS Scheme area is dominated by Quaternary deposits of alluvial fans, terraces and floodplains, beach, estuarine and swamp deposits, and extensive loess layers [2]. Terraces of Pleistocene age have been incised by hill catchment rivers to create alluvial valleys of late Quaternary age. The terraced 'Downlands' comprise streams and fan alluvium with loess ridges up to 20 metres thick, and have been created by an accumulation of loess on interfluves [3]. The western boundary of the downlands is formed by the Hunter Hills, a north-south trending fault controlled basement rock range [3].

2.2 Groundwater / surface water interaction

The HDIS area incorporates parts of the river catchments from the Pareora River in the north to the Waihao River in the south. Typically, the rivers carry water from the hill country into the alluvial valleys, where surface flow is lost to groundwater. As the rivers approach their lower reaches they typically regain flow from groundwater. For the purposes of the water quality investigations, flow statistics presented by Aitchison-Earl *et al.* [2] were utilised to determine baseline conditions.

As a result of the low permeability clay and loess immediately below the soil profile, infiltration of drainage water below the root zone is likely to be limited, with 10% of mean annual rainfall estimated to recharge groundwater on the downlands [4].

The Wainono Lagoon is located in the southern part of the scheme area, a coastal lagoon which has been created by the natural formation of a beach barrier arm that has restricted river flow out to the sea. The lagoon has been influenced significantly by the past actions of humans, who have drained the marginal land around the lagoon and created an outflow arm which takes flow to the Waihao River mouth to the South. The water levels in the lagoon are influenced by storm events, with sea water over-topping the barrier arm frequently.

3 Contaminants of concern

A key environmental concern with land use intensification is the generation of water-borne contaminants such as nutrients (various forms of nitrogen and phosphorus), suspended sediment and micro-organisms. Our study focused on the key contaminants nitrate nitrogen (nitrate-N) and phosphorus, although increased sediment and micro-organisms were assessed by others in the project. The environmental consequences of increased nitrogen and phosphorus loads may be divided into two types: i) the potential toxicity effects of nitrate-N on human health and aquatic fauna ii) the nutrient enrichment effects of both nitrogen and phosphorus.

3.1 Nitrate and toxicity effects

Nitrate-N is susceptible to leaching and the potential for this to occur can be increased through a variety of agricultural practices, including ploughing, urine from stock and fertiliser application.

Excessive consumption of nitrate in drinking water has been associated with the risk of methaemoglobinaemia or 'blue baby syndrome' [5]. For this reason, the Drinking Water Standards for New Zealand (2005) has specified the standard for nitrate in potable water at 11.3 g nitrate-N/m^3 (50 g NO_3/m^3) [6]. The World Health Organisation (WHO) recommends the same limit [7]. In order to protect aquatic ecosystems in New Zealand from toxic effects of nitrate-N, the Australian and New Zealand Guidelines for Fresh and Marine Water Quality (ANZECC, 2000 [8]) provide trigger levels that, if exceeded, indicate a potential environmental problem. A trigger value for nitrate toxicity of 7.2 g N/m^3 is currently used in New Zealand [9].

3.2 Nutrient enrichment effects of nitrogen and phosphorus

Nitrogen and phosphorus are essential plant nutrients and increases in either nutrient can enrich waterways, leading to nuisance algae growth and associated eutrophication effects. Such effects include daily fluctuations in water pH and dissolved oxygen concentrations, potentially to levels that are lethal for fish and other aquatic fauna. Relationships between nutrient concentrations, hydrological disturbance and algal biomass have been developed for New Zealand rivers [10]. National guidelines set maximum algal cover and biomass thresholds to protect aquatic ecosystems, and also recommend nitrogen and phosphorus concentration limits necessary to achieve those thresholds, depending on other factors such as hydrological disturbance [11]. Nitrate can cause ecological problems as a nutrient at concentrations an order of magnitude lower than the guideline trigger values for toxicity.

3.3 Water quality status

Existing surface water sample data showed that water quality varied greatly across the HDIS area. Nitrate concentrations were generally low relative to human and aquatic fauna toxicity guidelines but were high enough to cause breaches in algal cover and biomass guidelines in some rivers during summer [12]. Groundwater quality records were relatively limited but did show a trend of better water quality in wells closer to the streams and rivers, and lower concentrations of nitrates with depth. Overall, nitrate-N concentrations in groundwater were shown to be at low to moderate concentrations relative to toxicity guidelines, reflecting the existing predominantly dryland farming practices.

Based on the available nutrient data and investigations undertaken by the National Institute of Water and Atmospheric Research (NIWA) during the 2006/2007 summer, the lagoon was classified as varying from eutrophic to hypertrophic under the existing landuse conditions [12].

4 Assessment methodology

The hydrogeology of the scheme area is dominated by large areas of downland with limited drainage potential, dissected by valleys containing recent permeable alluvial gravel and interconnected river systems. Due to these characteristics the majority of current rainfall on the downlands currently drains laterally to intermittent streams before discharging to the valley bound alluvial gravels.

Following implementation of the HDIS, excess drainage water is also expected to predominantly drain laterally from the downland areas to the alluvial valleys mixing with groundwater and/or discharging directly to the rivers. Drainage water affected by landuse intensification will therefore be mixed with the existing surface water and groundwater flowing though the alluvial valleys, which is sourced predominantly from dryland areas upgradient of the Scheme area. The resulting groundwater and surface water quality is dependent on a number of variables including the relative size of catchment upgradient of the scheme area, the post irrigation landuse mix and hydrogeology / surface water flow characteristics of the particular catchment.

4.1 Nitrate nitrogen

Due to the number of variables influencing groundwater and surface water quality, predicting the magnitude of any increase in nutrient loads necessarily relies on modelling to simplify and approximate the key elements of the catchment(s). The water quality assessment for the HDIS undertaken has utilised a three step process that necessarily involved coupling models between disciplines i.e. 1) an agricultural nutrient budget model 2) a mass mixing hydrogeological model 3) an empirical algae biomass model. The thrust of this paper is on the mass mixing model that has been used to predict nutrient concentrations within groundwater and surface water.

Equation 1 provides the analytical solution which forms the basis of the mass-mixing model used.

$$C_f = \frac{((V_d \times C_d) + (V_i \times C_i))}{V_f} \tag{1}$$

Where C_f is the final concentration, V_d is the volume of drainage water, C_d is the concentration of nitrate nitrogen in drainage water, V_i is the volume of groundwater influx, C_i is the concentration of nitrate nitrogen in the groundwater influx, and V_f is the combined total volume of groundwater flow.

The mass mixing model requires the flux of contaminants from the range of expected landuses to be quantified for each catchment. Drainage concentration were calculated using the OVERSEER® model [13], which estimates the mass of nitrogen lost from the rooting zone under various land use scenarios and mixes it with the calculated mean annual rainfall recharge or rainfall and irrigation recharge. The extent of each land use type, combined with the modelled drainage flux of nitrate-N enables the left hand side of the primary equation to be solved. Calculating the groundwater flow volume within the alluvial aquifers required information on the aquifer thickness, width, hydraulic conductivity etc to be obtained for each catchment. In addition, the background concentration of nitrate-N was established for each catchment. This information enabled the right hand side of the primary equation to be solved.

Figure 2 schematically illustrates the expected pathways for drainage water within the scheme area. The following provides a summary of the principal assumptions used in the modelling.

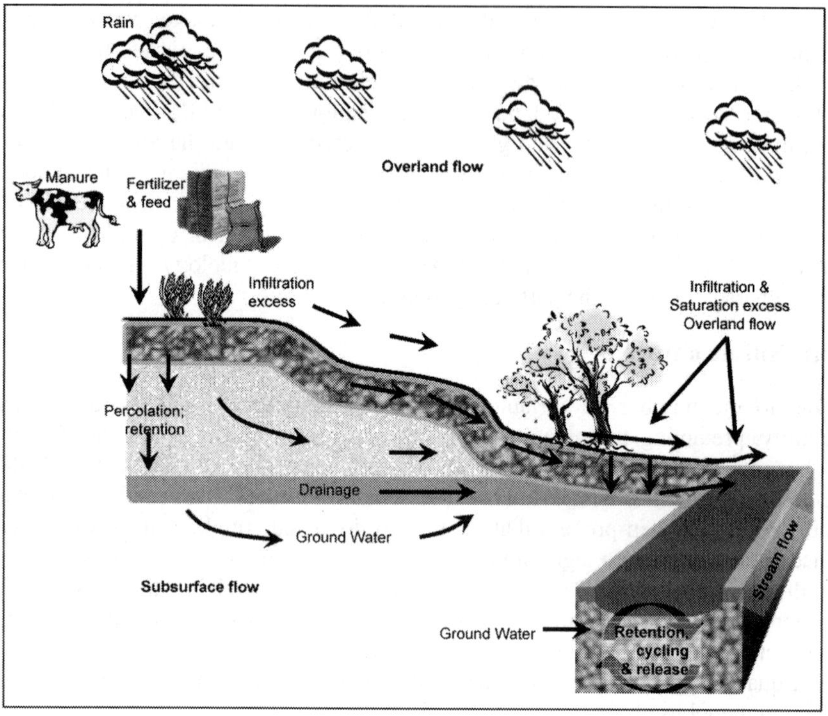

Figure 2: Drainage water pathways.

Drainage Pathways:

- 100% of drainage from the alluvial flats enters groundwater
- 80% of drainage from the downlands enters the ephemeral streams and flows to the alluvial flats
- 20% of drainage/runoff not entering the ephemeral streams either infiltrates through the downlands into the deeper groundwater system and/or is captured in non-irrigated areas

- 80% of water entering the ephemeral streams is lost to groundwater on the alluvial flats

The mass-mixing model was calibrated by comparing the modelled output for the existing landuse distribution with the available groundwater quality data at the down gradient end of each catchment. The model assumptions were modified in an iterative manner until the model output showed a reasonable agreement to the existing background environment. However, this calibration of the model also required the model assumptions to be consistent with the conceptual understanding of the behaviour of drainage on the downlands and lowlands.

4.2 Modelling phosphorus losses

The mass-mixing model approach was also utilised to assess the potential impacts of increased phosphorus concentration in surface water on the Wainono Lagoon. Current understanding of P losses to the environment suggests that P enters surface waterways via overland runoff. Studies in New Zealand show that within a catchment approximately 20% of the land area will contribute 90% of the total P losses. Whilst this is a generalisation, it provides a basis for determining the area of land that may contribute to P loadings within the lagoon catchment. The mass mixing methodology utilised the flux of P calculated for runoff from the catchment contributing to the Wainono Lagoon and baseline surface water flow statistics to determine P concentrations in the lagoon under existing and proposed landuses. The mass balance approach estimates a potential increase of 0.03 g/m^3 following irrigation development within the Wainono inflows area, which is more than 1.5 times the modelled average existing Total P concentration.

4.3 Best management practices

The modelling was also used to demonstrate the effectiveness of best management landuse practices (BMPs) in reducing the contaminant loading on groundwater and surface water resources. The mass-mixing model enables the benefits of the BMPs to be quantified for the wider catchment and provides a benchmark against which HDIS can assess the usefulness of promoting BPMs.

The two mitigation measures that have been assessed and the benefits they provide in terms of nitrate nitrogen generation are: nitrification inhibitors (net 30% reduction in nitrate leaching assumed) and the use of cover crops over the winter period that are reported to reduce nitrate leaching by up to 50%, (28% reduction utilised in modelling).

5 Results

Table 1 summarises the results of the mass balance modelling for groundwater and surface water resources within the scheme area. The mass balance modelling indicates that utilising current landuse practices, landuse intensification under the HDIS has the potential to cause increases in nitrate-N in groundwater and surface water. Within the Wainono Lagoon peak nitrate-N concentrations are anticipated to increase on the order of 1.8 g/m^3 and P by around 0.03 g/m^3 above the existing mean concentrations.

Using the mass balance modelling approach indicates that the implementation of the two BMPs evaluated to reduce derived post irrigation nitrate-N groundwater concentrations on average by 18% and as a result surface water quality improvements are achievable.

Table 1: Mass mixing model output.

	Existing environment (average)	Modelled existing	Future under HD1	Net increase
Nitrite nitrogen (NO_x-N) - g/m^3				
Pareora G/W	5.0-5.5	4.9	8.5	3.6
Hook River G/W	2.0-2.5	2.8	5.9	3.1
Hook River S/W	0.9	1.8	5.1	3.3
Wainono area G/W	2.5-3.5	3.4	7.3	3.9
Phosphorus (P) – g/m^3				
Hook River P	0.01-0.037 (0.025)	0.019	0.050	0.031
Wainono lowland streams P	0.076-0.26 (0.156)	0.155	0.346	0.191
Wainono Lagoon – P	0.015-1.7 (0.27)	0.019	0.049	0.03

5.1 Nitrate and toxicity effects

The results in Table 1 can be directly compared with guidelines to assess the toxicity effects of nitrate-N. With regard to toxicity effects, it is considered that the increases predicted for the HDIS are unlikely to result in widespread breaches of guideline thresholds for nitrate-N toxicity; i.e., the New Zealand Ministry of Health's human drinking Maximum Acceptable Value (MAV) of 11.3 g nitrate-N/m^3, and the ANZECC trigger value for toxicity to aquatic species of 7.2 g nitrate-N/m^3. The predicted increase in nitrate-N loads will obviously push the environment closer to these thresholds and therefore increase the risk of breaches.

5.2 Nutrient enrichment effects

It is more difficult to assess thresholds for nutrient enrichment effects in rivers and lakes because factors other than nutrients(e.g. hydrogeological regime, light, temperature and bed substrate) are important for determining the biological response to enrichment. However the results in Table 1 can be used, in conjunction with modelled relationships between these other factors, to predict the increase in algae growth associated with predicted increases in nutrients. While the method is not described here, such an assessment was undertaken [12] and resulted in the prediction that algae biomass could increase by 60% in rivers and 50% in the Wainono Lagoon if BMPs were not employed. The quantification of this effect allowed other flow-on effects to be considered, such as the likely extent of aquatic habitat smothered by algae in rivers, the visual impact and risk of toxic blooms in the Wainono Lagoon.

6 Discussion

The mass mixing approach utilised is a simplification of the actual conditions, which introduces a number of conservatisms into the assessment of effects. For example nitrate-N has been conserved throughout the modelling with no allowance being made for natural denitrification processes and/or nutrient uptake which will tend to reduce nitrate-N concentrations in groundwater and surface water. In addition, the dilution effect that is experienced during runoff events when drainage water is flushed from downland areas is not allowed for, with the mass mixing utilising mean annual surface water flow conditions. Furthermore, in all cases the modelling has assumed that the entire catchment within the command area is mixing with the flux of groundwater moving down the alluvial valley(s). In reality this is only the case once nearing the bottom of the catchment, therefore nutrient concentrations are expected to reduce from that calculated with distance upstream.

7 Conclusion

The mass mixing modelling approach is a relatively straightforward and appropriately conservative tool for assessing water quality effects associated with the introduction of irrigation into a catchment. However, the approach requires a robust understanding of the current hydrology, hydrogeology, existing and proposed landuse and existing water quality such that the inputs can be appropriately allocated. Due to the outcome of the mass balance approach being highly dependent on the nature of the receiving aquifer, it is most useful when the system can be reasonably well defined in terms of aquifer width, saturated thickness etc. Knowledge of the current landuse distribution and existing water quality status is required in order to calibrate the model and confirm the appropriateness of the principal assumptions used in the modelling process.

The approach is useful because:

i.) It allowed quantitative predictions of nutrient concentrations in surface waterways can be directly compared to toxicity guidelines.

ii.) The predicted nutrient concentrations can also applied to established biological models to predict biological effects. Such effects include quantitative predictions of increased algae growth and associated negative visual and ecological impacts.

The mass mixing approach has provided a realistic basis for quantifying the potential adverse effects of irrigation on groundwater and surface water quality, and evaluating the potential effectiveness of mitigation measures such as nitrification inhibitors.

References

[1] Thorley, M. and Ettema, M., 2007. Review of Water Allocation Limits for the South Canterbury Downlands. Unpublished Technical Report U07/09. Environment Canterbury.

[2] Aitchison-Earl, P., Ettema, M., Horrell, G., McKerchar, A., Smith, E., 2006. Pareora-Waihao River: Water Resource Summary. Unpublished Technical Report U06/20. Environment Canterbury. ISBN 1-86937-602-1.

[3] Forsyth, P.J., 2001, Geology of the Waitaki Area: 1:250,000 Geological Map. Institute of Geological and Nuclear Sciences, Lower Hutt, 64pp.

[4] Aitchison – Earl, P., 2005: *Deep Aquifer Review: Timaru – Waitaki River.* Environment Canterbury unpublished Technical Report U05/02.

[5] Fan, A.M. and Steinberg, V.E., 1996: *Health implications if nitrate and nitrite in drinking water: an update on methemoglobinemia occurrence and reproductive and developmental toxicity.* Regulatory Toxicology and Pharmacology 23, 35–43.

[6] Ministry of Health, 2005: *Drinking Water Standards New Zealand 2005.* Ministry of Health, Wellington.

[7] World Heath Organisation, 2004: *Guidelines for drinking water quality, 3rd Edition.* WHO, Geneva.

[8] ANZECC, 2000: *Australian and New Zealand guidelines for fresh and marine water quality.* National Water Quality Management Strategy Paper No.4, Australian and New Zealand Environment and Conservation Council & Agriculture and Resources Management Council of Australia and New Zealand, Canberra, Australia.

[9] Hickey, C., 2002: *Nitrate guidelines values in ANZECC 2000.* Unpublished memo from Chris Hickey, NIWA, to the Ministry for the Environment dated 30 September 2002.

[10] Biggs, B.J.F., New Zealand periphyton guideline: detecting, monitoring and managing enrichment of streams. Ministry for Environment, Wellington, 2000.

[11] Biggs, B.J.F., Eutrophication of streams and rivers: Dissolved nutrient-chlorophyll relationships. *Journal of the North American Benthological Society* 19(1), pp. 17–31.

[12] Norton, N., Floeder, S., Drake, D., Hunter Downs Irrigation Scheme: Assessment of potential effects of increased nutrients on aquatic ecology values in rivers and Wainono Lagoon. NIWA Client Report CHC2007-057, Christchurch, 2007.
[13] AgResearch Limited, OVERSEER Nutrient Budgets, www.agresearch.co.nz/overseerweb/default.aspx
[14] Google Maps New Zealand, http://maps.google.co.nz/maps

Designs for the future: the role of sustainable irrigation in northern Australia

J. K. Camkin[1,2], K. L. Bristow[1,2], C. Petheram[1], Z. Paydar[1], F. J. Cook[1,2] & J. Story[1]
[1]*CSIRO Land and Water, Australia*
[2]*Cooperative Research Centre for Irrigation Futures, Australia*

Abstract

Northern Australia (NA) accounts for 40% of the Australian landmass, is largely undeveloped and is one of the few large natural areas remaining on earth. The interplay between the landscapes, rivers, groundwater and strongly monsoonal weather pattern has resulted in unique, diverse and iconic ecological systems that will need special attention to ensure their integrity is retained as development pressure increases. With 60 to 70% of Australia's fresh water discharging from tropical rivers, and reduced water availability in much of southern Australia due to drought, climate change and increasing water demand, there is rapidly growing interest in the land and water resources of the north.

The Australian Government has recognised this growing pressure and that there is a unique and historic opportunity to ensure the management and use of Australia's northern land and water resources takes place within a strategic, sustainable framework. Taking this opportunity requires forethought about the future of NA and what role irrigation should play in that future. Key questions are whether irrigation should play a significant role, and if so, where should it be located, what should it look like and how should it be managed? Established in 2003, Northern Australia Irrigation Futures (NAIF) is a collaboration of four governments responsible for NA, research organisations and industry developing new knowledge, tools and processes to support debate and decision-making about the future of irrigation in NA.

NAIF exhibits many of the characteristics of the emerging discipline of implementation and integration science. It has been a catalyst for further work to help understand and deal with the many complexities and uncertainties relating to decisions about the future of irrigation in NA.

Keywords: Northern Australia Irrigation Futures, irrigation, hydrology, mosaics, decision-making, implementation and integration science.

1 Introduction

There is a unique and historic opportunity to ensure that management of Australia's northern water resources takes place within a strategic, ecologically, culturally and economically sustainable framework (Government of Australia [1]). Northern Australia (NA) is defined here as that area north of the Tropic of Capricorn - approximately 40% of Australia's land mass (Figure 1).

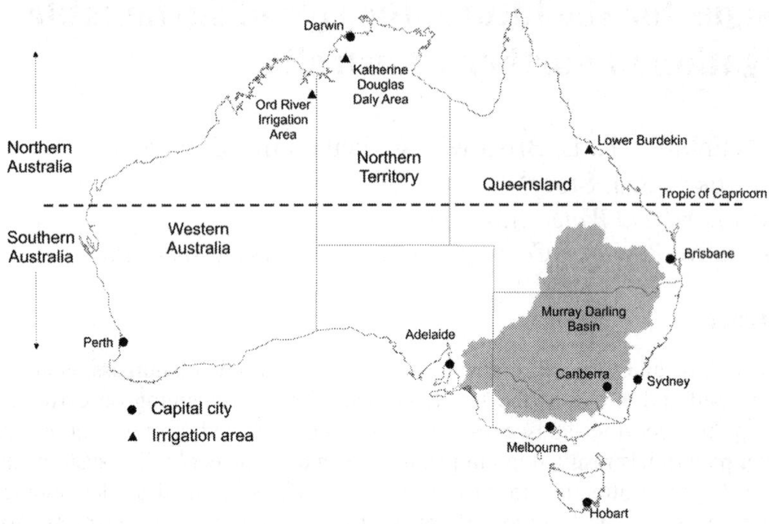

Figure 1: Map of Australia.

The 2nd World Water Development Report (UNESCO [2]) predicted the world will need 55% more food by 2030. Nationally, there is mismatch between water availability and demand. 60 to 70% of Australia's freshwater runoff occurs in the north but greater than 80% of the population lives in five expanding southern coastal cities where the increasing urban footprint, drought, climate change and increasing demand are reducing the water available for irrigation. Together, these trends are increasing pressure for development of the water and land resources of NA, and increasing public and political debate on this important national issue.

The uniqueness of northern landscapes and ecosystems has been strongly influenced by the quantity, quality and timing of water flows through the catchments. For Indigenous Australians, the surface and groundwater systems of the north have a strong cultural significance (Jackson [3]) and for many other Australian's the ecosystems of the north also hold an iconic status. There is growing recognition of these values and of the potential for irrigation and other water use to impact on them. Policy initiatives reflect these changing values and irrigation development proponents face unprecedented scrutiny of their proposals.

Three recent initiatives are recognition of the increase in public, scientific and political focus on NA and are of major significance to the question of the future role of irrigation in NA. The Northern Australia Irrigation Futures project (NAIF) was established in 2003 to develop new knowledge, tools and processes to support debate and decision-making about irrigation in NA and to encourage strategic thinking (see http://www.clw.csiro.au/naif/). The Tropical Rivers and Coastal Knowledge (TRaCK) hub (see http://www.track.gov.au/) was established in 2007 to provide the science and knowledge that government, communities and industries need for the sustainable use and management of Australia's tropical rivers and estuaries. In January 2007 Prime Minister Howard released a National Plan for Water Security (Australian Government [1]). The new Rudd Government has confirmed its commitment to a Northern Australia Land and Water Taskforce to examine the range of development opportunities in NA that impact on land and water resources (Wong [4]). The Taskforce will be informed by Northern Australia Land and Water Futures Assessments. The remainder of this paper focuses on NAIF.

2 Northern Australia Irrigation Futures

2.1 Background

NAIF is a collaboration of the Australian, Queensland, Northern Territory and Western Australian Governments, the Cooperative Research Centre for Irrigation Futures, the National Program for Sustainable Irrigation and Australia's Commonwealth Scientific and Industrial Research Organisation (CSIRO). It involves targeted research focussing on improving understanding of the links between irrigation and the quantity and quality of downstream water systems (particularly groundwater), and the relationship between irrigation and the ecological, economic and social systems within which it takes place. Working with northern governments, and guided by a representative and skills based Steering Committee, researchers are developing frameworks that support strategic thinking and decision-making about irrigation.

2.2 The hydrology, context and history of irrigation in northern Australia

Cultivated agriculture and irrigation developments are not new to NA. Several irrigation schemes, such as the Lower Burdekin, Pioneer and Mareeba schemes in north Queensland are over 100 years old, and irrigation at the Ord River in Western Australia has been in operation for over 40 years (Petheram et al. [5]). Other irrigation developments collapsed very soon after they commenced operation (e.g. Humpty Doo; Mollah [7]). Numerous studies and reports have examined the reasons for European development either failing or never starting in the north of Australia. A much cited factor is a failure to understand the northern environment (Woinarski and Dawson [6]).

NAIF has laid foundations for understanding the hydrology of NA by providing a broad overview of the surface and groundwater resources with respect to irrigation development. The literature on the hydrology of NA was reviewed and some key biophysical issues, opportunities and constraints for irrigation identified. Particular emphasis was placed on illustrating the differences between water systems in NA (tropical) and those of the Mediterranean and temperate southern Australia – the latter being more familiar to most Australians (Petheram and Bristow [8]).

Because of its position and the orientation of the Australian continent within the global circulatory system, NA is characterised by high year-round temperatures, a distinct seasonal rainfall pattern, some of the greatest rainfall intensities in the world, large inter-annual variability in rainfall and large evaporation rates. Due to of a lack of rainfall irrigation is essential for cultivated agriculture or perennial horticulture during the dry season (May to November). High evaporation rates during the dry season mean that a greater volume of water is required to irrigate a given area than in the south (Petheram and Bristow [8]). Twelve major drainage basins characterise the Australian continent. Half of these are partly or entirely located within NA, and approximately 60% of Australia's runoff is generated in the north (NLWRA [9]). Streamflow in NA is strongly seasonal and has a large inter-annual variability compared with rivers of similar climate elsewhere in the world (Petheram et al. [10]). This means that permanent settlements and irrigation during the dry season requires surface water storage structures, unless suitable groundwater resources are available (Petheram and Bristow [8]).

NAIF examined three well known irrigation areas to draw out experiences and lessons relevant to northern Australia. The three focus areas were the well established Lower Burdekin (LB), the expanding Ord River Irrigation Area (ORIA); and the emerging Katherine-Douglas-Daly Area (KDDA). The LB is NA's largest irrigation area with approximately 80 000 ha of land under irrigation, predominantly sugarcane. The ORIA is a 13,000 ha surface water scheme with plans for further expansion. The KDDA is an evolving irrigated area of 2,200 ha located within the Daly River catchment. The Daly River is one of the few perennial rivers in the Northern Territory and has the largest dry season flows. These flows are central to the variety of unique ecosystems that inhabit the Daly river catchment (Erskine et al. [11]). Unlike the other two irrigation areas, the KDDA has no in-stream storages. The perennial flows and groundwater yields make the region attractive to irrigators and of high ecological and cultural significance (Petheram et al. [10]). Considering past experiences in both northern and southern Australia, and from other tropical regions of the world, will be important to avoiding a repeat of problems that have plagued irrigation developments around the world. Success will require transparency and accountability to local communities, ongoing monitoring and early identification of emerging problems, and funding to deal with unexpected problems that invariably occur well after development (Petheram et al. [10]).

2.3 Irrigation mosaics: a possible alternative approach to irrigation

Most irrigation areas in Australia are characterised by large-scale contiguous irrigation systems. Irrigation mosaics, discrete patches of irrigated land dispersed across the landscape, may offer an alternative and could be particularly attractive for delivering improved social and economic opportunities for rural and remote communities in NA. However, the longer-term environmental impacts of irrigation mosaics are still largely unknown (Paydar et al. [12]). NAIF examined some of the issues associated with irrigation mosaics, focussing on their bio-physical effects compared to large scale contiguous irrigation systems (Figure 2).

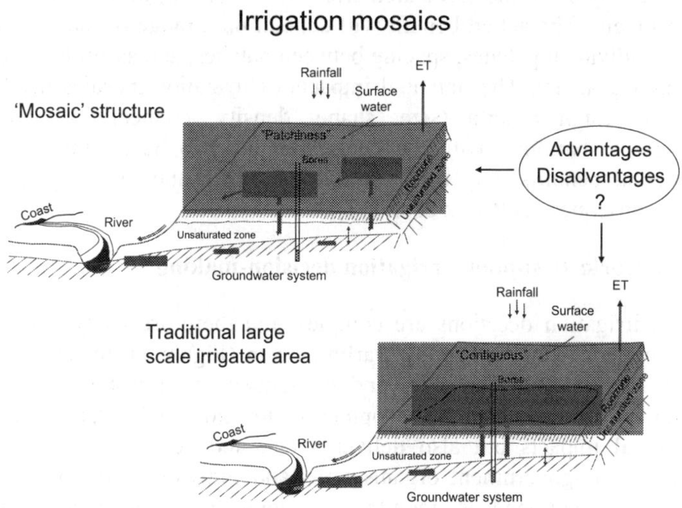

Figure 2: Irrigation mosaics v contiguous areas (Cook et al. [13]).

Documented knowledge on irrigation mosaics and implications within the context of sustainable development was found to be very limited. However, there are some findings from studies of other systems, with spatial patterns in the landscape, which can be used to help with analysis of irrigation mosaics (Paydar et al. [12]).

From ecological research it appears that patch size, shape and spatial arrangement are important characteristics in landscape analysis. For conservation planning, the bigger the reserves, the closer together, the more circular, and the more they are linked by habitat corridors, the better they serve the purpose of nature conservation. Irrigation mosaics could be used to create or enhance ecotones in the landscape for greater biodiversity, adjusting the microclimate, minimising erosion, and in absorption of surplus material (nutrients, sediments, solutes) flowing from the surrounding fields, thus decreasing the discharge out of an irrigation area, a possible environmental off-site effect. Conversely, fragmentation, involving discontinuity of patches, can increase the vulnerability of patches to external disturbance (e.g. wind storms, droughts; Paydar et al. [12]).

Based on the literature review it appears that irrigation mosaics could have both negative (more lateral recharge, salinisation, increased operational losses) and positive (filtering nutrient surplus, enhanced biodiversity, preventing erosion, reduced area of impact around the irrigation area, lower rate of watertable rise) effects on the environment (Paydar et al. [12]).

New analytical and numerical solutions and programs that considerably reduce computation time were also developed to help analysis of spatial and temporal issues associated with irrigation mosaics. This research also suggests that irrigation mosaics could have both negative (e.g. higher evapotranspiration, increased operational losses) and positive (e.g. reduced water-table height, but noting there may be an increased area affected by a rise in water table) effects on the environment. The actual benefit will depend on a range of factors including the size of individual patches, spacing between patches, and assimilative capacity of surrounding areas). The potential impacts of irrigation mosaics need careful study, and design criteria (size, shape, density, connectivity and spatial arrangement in harmony with the landscape) need to be established because environmental benefits may be short lived if space and time lags just delay unwanted consequences (Cook et al. [13]).

2.4 Frameworks to support irrigation decision-making

Water and irrigation decisions are complex and there are many uncertainties. Recent experiences in the Murray Darling Basin (Figure 1), for example, have increased awareness of the risks and consequences of water use decisions. Communities now expect developments to not only have acceptable environmental impacts but also deliver social and economic benefits to the community. Non-government organisations and individuals are better trained, connected and equipped to monitor decisions. Together, these trends are increasing pressure on decision-makers and dealing with complexity and uncertainty emerges as a shared need and responsibility for government, developers and the community so that good decisions can continue to be made (Camkin et al. [14]).

NAIF initially aimed to *"...deliver a framework based on sustainability indicators and management criteria at a range of scales..."*. Thinking about the framework changed as issues of complexity, uncertainty, resilience, risk and adaptive management emerged through the research. Focus shifted away from a simple set of biophysical indicators to frameworks to support communities and decision-makers deal with complexity and uncertainty in a comprehensive, transparent and inclusive way (Camkin et al. [15]).

In complex decisions there is a need to understand and address the difference between uncertainty and risk. The matrix created by considering at every scale (e.g. on-site, local, catchment) all of the potential environmental, social, economic and external factors that may be relevant to a decision is very large. However, not all of those potential factors are relevant to a particular decision and not all of those that are relevant are of equal importance. NAIF has worked with governments and stakeholders to develop tools to aid understanding of these differences so that managers and the community can be confident effort is focused on the most important issues. A prototype framework was developed for the LB consisting of an ESD Component Tree system, a web-based catchment knowledge platform and processes for improving the integration of science, policy, stakeholders and industry (Figure 3).

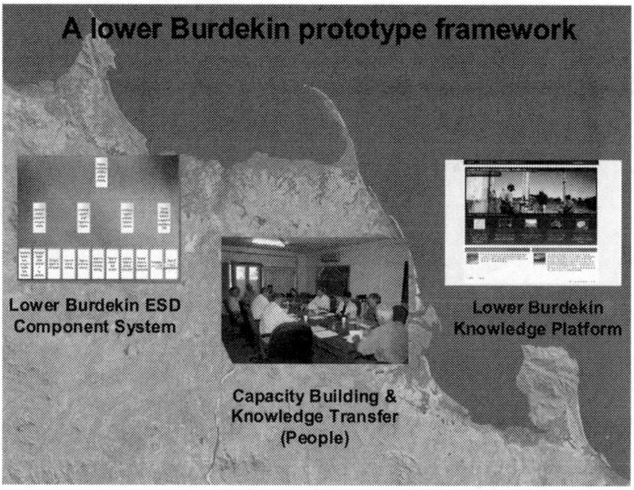

Figure 3: A lower Burdekin prototype framework (Camkin et al. [14]).

Economic activity is part of and takes place within the social system, which in turn is part of and takes place within the ecological system. The key question is not whether an industry or individual development is sustainable but rather what positive and negative contributions it makes across the full range of ESD (Chesson et al. [16]). ESD Component Tree systems that identify the factors relevant to a particular industry, proposal, or location, have been successfully applied in fisheries, aquaculture, agriculture, and irrigation reporting. NAIF has developed an ESD Component Tree approach to support irrigation decision-making in NA (Figure 4; Camkin et al. [14]).

There is rapid growth in the development of on-line technology that utilises new approaches to learning and supports communities of practice in the resolution of complex problems. NAIF has applied these approaches to develop a prototype catchment knowledge platform for the LB. The platform will help the community progressively build and tell the story of how the catchment operates in a biophysical, social, economic sense and governance sense.

3 Conclusion

The range of concerns expressed about irrigated agriculture in NA highlights the need for irrigation design to be part of a much broader development planning process that considers physical and economic aspects of resources as well as cultural, social and ethical values. The relatively small number of players in NA

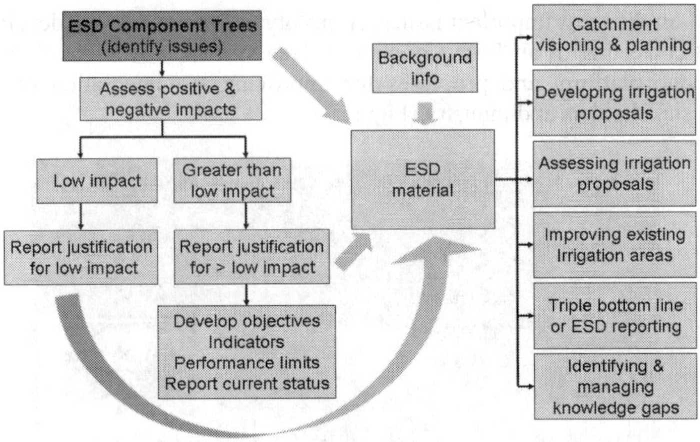

Figure 4: Applications for ESD Component Trees (Camkin et al. [14]).

provides a unique opportunity for collaboration and strategic planning. NAIF exhibits features consistent with those described by Bammer [17] as characteristic of integration and implementation science. Firstly, the project aimed to find better ways to deal with the complexity, uncertainty, change and imperfections in understanding of irrigation in NA. Secondly, the project was conducted consistent with the three theoretical and methodological pillars of integration and implementation science, namely (i) systems thinking and complexity science; (ii) participatory methods; and (iii) knowledge management, exchange and implementation. Thirdly, NAIF is grounded in practical implementation, with a focus on developing practical tools for decision-makers and the community, and involved collaboration with a wide range of government, industry, community, and research stakeholders relevant to NA.

References

[1] Australian Government (2007). *A National Plan for Water Security*. 26 January 2007.

[2] UNESCO (2006). Water a shared responsibility. The United Nations World Water Development Report 2.

[3] Jackson, S. (2004). Preliminary report on Aboriginal perspectives on land-use and water management in the Daly River region, northern Territory. A report to the northern Land Council.

[4] Wong, P. (2008). Elevated role and more expertise for Northern Australia Land and Water Taskforce. Media Release. Senator the Hon Penny Wong, Minister for Climate Change and Water Resources, Australian Government. 25 January 2008.

[5] Petheram, C., Tickell, S., O'Gara, F., Bristow, K.L., Smith, A. and Jolly, P. (2008a). Analysis of the Lower Burdekin, Ord and Katherine-Douglas-Daly Irrigation Areas: Implications to future design and management of tropical

irrigation. CSIRO Land and Water Science Report 2008, CRC for Irrigation Futures Technical Report No. 04/07 101 pp.

[6] Woinarski, J. and Dawson, F. (1997). Limitless lands and limited knowledge: coping with uncertainty and ignorance in northern Australia. In Ecology, Uncertainty and Policy: managing ecosystems for sustainability. (eds.) Norton, T.W., Handmer, J.W. & Dovers, S.R. Cambridge University Press.

[7] Mollah, W.S. (1982). Humpty Doo – rice in the Northern territory. North Australia Research unit. Monograph, Darwin. ANU Press Canberra.

[8] Petheram, C. and Bristow, K.L. (2008). Towards an understanding of the hydrological factors, constraints and opportunities for irrigation in northern Australia: A review. CSIRO Land and Water Science Report 2008, CRC for Irrigation Futures Technical Report No. 07/07 111 pp.

[9] NLWRA (2000). Australian water resources assessment 2000. Retrieved from http://audit.ea.gov.au/anra/water/docs/national/Water_Contents.html on 12 February 2006.

[10] Petheram C., McMahon, T.A. and Peel, M.C. (2008b). Flow characteristics of rivers in northern Australia: implications for development. Submitted to Journal of Hydrology

[11] Erskine, W.E., Begg, G.W., Jolly, P., Georges, A., O'Grady, A., Eamus, D., Rea, N., Dostine. P., Townsend, S. and Padovan, A. (2003). Recommended environmental water requirements for the Daly River, Northern Territory, based on ecological, hydrological and biological principles. Supervising Scientist Report 175. (National Riverhealth Program, Environmental Flow Initiative, Technical Report 4), Supervising Scientist, Darwin NT.

[12] Paydar, Z., Cook, F.J., Xevei, E. and Bristow, K.L. (2007). Review of the current understanding of irrigation mosaics. CSIRO Land and Water Science Report No. 40/07, CRC for Irrigation Futures Technical Report No. 08/07. 38 pp.

[13] Cook, F.J., Xevi, E., Knight, J.H., Paydar, Z. and Bristow, K.L. (2008). Analysis of biophysical processes with regard to advantages and disadvantages of irrigation mosaics. CSIRO Land and Water Science Report 2008, CRC for Irrigation Futures Technical Report No. 09/07 64 pp.

[14] Camkin, J.K., Story, J. and Bristow, K.L. (2008a). An Ecologically Sustainable Development Component System to Support Irrigation Decision-Making in Northern Australia. CSIRO Land and Water Science Report No. 17/08, CRC for Irrigation Futures Technical Report No. 11/07 19 pp.

[15] Camkin, J.K., Kellett, B.M. and Bristow, K.L. (2008b). Northern Australia Irrigation Futures: Origin, evolution and future directions for the development of a sustainability framework. CSIRO Land and Water Science Report No. 73/07, CRC for Irrigation Futures Technical Report No. 10/07 37 pp.

[16] Chesson, J., Whitworth, B., Stewart, J. and Yapp, G. (2005). Signposts for Australian Agriculture, Stage 2a: Refinement of preliminary framework

and industry profiles to include pathways to ESD. Final report to the National Land and Water Resources Audit, September 2005.

[17] Bammer, G. (2005). Integration and Implementation Sciences: building a new specialisation. Ecology and Society 10(2): 6. (online) URL: http://www.ecologyandsociety.org/vol10/iss2/art6/

Definition of ecological flows downstream of dams located in the South of Portugal: a new method

M. M. Portela
Department of Civil Engineering and Architecture, Environment and Water Resources Division, IST, Technical University of Lisbon, Portugal

Abstract

One of the environmental issues related with the numerous dams of the Alqueva irrigation system (South of Portugal) was, and still is, the definition of the ecological regimes. Besides other features, those regimes must account for the extreme water scarcity that characterizes the region.

For the 12 dams of the Alqueva system (either existing or new ones) several methods aimed at defining the ecological flows were applied. The comparison of the ecological flows thus predicted revealed the inadequacy of the different methods, suggesting the need for a new approach capable of providing comparable ecological flows under similar hydrologic constraints. In the previous scope a new method supported by hydrologic and hydraulic criteria was developed. The hydrologic criteria account for the water scarcity and for the temporal irregularity of the natural hydrologic regime and the hydraulic criteria, for the geometry of the cross sections and of the river reaches. The method is briefly described and the ecological flows achieved for 12 case studies are presented.
Keywords: ecological flows, hydrologic-hydraulic method, hydrologic regime, hydraulic characteristics, cross section, mean flow velocity.

1 Introduction: scope of the study

In the next few years an intense agricultural development is expected in Alentejo (South of Portugal) as result of the construction of the Alqueva dam, located in the Guadiana River and providing a huge reservoir, in fact, the largest artificial

WIT Transactions on Ecology and the Environment, Vol 112, © 2008 WIT Press
www.witpress.com, ISSN 1743-3541 (on-line)
doi:10.2495/SI080301

lake in Europe, with a gross and a net storage capacity of 4500 and 3150 million cubic meters, respectively. Alqueva is the "heart" of an irrigation system that will supply water to 115 thousand hectares, by means of 15 dams spread over the region (existing and new ones), more than 300 km of open channels and more than 2000 km of buried conduits [3].

Figure 1: Alqueva dam: view from downstream to upstream.

One of the environmental issues related with the Alqueva irrigation system was, and still is, the definition of the ecological flows to be implemented downstream of each dam. In fact, Alentejo has very specific hydrological constraints being one of the driest of even the driest region of Portugal [11], with a mean annual rainfall of about 500 mm and a mean annual flow below 150 mm, these hydrological variables also being characterized by a very pronounced temporal irregularity (within each year and among years): about 75 to 80% of the rainfall and 90 to 95% of the runoff occur during the wet season (from October to March). Most of the rivers are of the torrential type, only having runoff during a few days of the wet season.

The availability of water that became possible through the Alqueva system may suggest that more water could be launched into the rivers during the dry season, by means of artificial ecological flows. This perspective, though somehow tempting, may not be the most correct one as the local river ecosystems are naturally adapted to extreme water scarcity. Also a wise and tight management of the water in Alentejo is crucial, as the price of that essential asset is not expected to be small.

In the previous scope several methods were tested and compared, aimed at defining the ecological flows downstream of the 12 dams (either existing or new ones) of the Alqueva system.

In a broad sense, the ecological flow for a given river reach is the flow that ensures the conservation and maintenance of the natural aquatic ecosystems, including their biodiversity, the production of species with sporting or commercial interests, as well as the conservation and the maintenance of the riparian ecosystems, of the esthetic features of the landscape or of other features of scientific and cultural interest [2]. An ecological flow regime is a temporal sequence of flows, generally defined in a monthly basis. Therefore, any flow or sequence of flows able to preserve the "dynamics" (performance, composition and structure) of the "fluvial-related" ecosystems in natural conditions can be considered an ecological one. This implies that for each river reach there is no such thing as "the ecological flow" but instead a range of ecological flows, varying from minimum ones to maximum ones. With water being a resource that is becoming progressively scarcer, the minimum ecological flows are generally the envisaged ones. It should also be pointed out that nowadays "natural conditions" do not mean pristine or untouchable conditions, which no longer exist, at least in the European rivers. They refer to the conditions that occurred prior to the construction of the infrastructure (such as a dam) that is responsible for changes in the fluvial corridor and, consequently, in the ecosystems connected with that corridor.

The physical organization of each natural fluvial corridor as well as the biologic "performance" of the ecosystems connected with it are deeply dependent on the flow regime as this regime determines the morphologic, the hydraulic, and by extension, the biologic parameters of such a corridor. Consequently, several methodologies and criteria aimed at defining ecological flows utilize the characteristics of the natural river flows, with emphasis on the values of the flows themselves as well as on their temporal variability (within each year and among years). Also, the river flows are most of the time the only easily available data when the definition of a given ecological flow regime is envisaged.

In the previous understanding, three methods of the hydrologic type were applied to 12 dams of the Alqueva system and the ecological flows thus achieved were compared. The methods under consideration were the wet perimeter method [1], a method specifically conceived for Portugal, focusing on the characteristics of the monthly flow series [3], and the basic flow method developed by Palau and Alcázar [5,6]. These methods will be further referred to as the WP method, the INAG method and the QB method, respectively. Their application to each location/dam requires only monthly flow series (INAG and QB methods) or cross sections of the river reach downstream from the dam (WP method).

However, the comparison of the ecological flows thus predicted for the 12 dams showed that: i) *at each* location the different methods led to disparate ecological flows, as they could be either very large or very small; b) the ecological flows estimated by applying the *same methodology to the different cross sections* were also quite dissimilar and totally uncorrelated.

This sort of "anachronism" among ecological flows seemed even more abnormal as the region under consideration presents a very "coherent" hydrologic regime, characterized, as previously mentioned, by a very small mean annual flow depth, with almost the same value in the whole region, and by a very pronounced temporal irregularity. Notwithstanding the differences among locations related with the geometry of the cross sections and with the area of the respective watersheds, it was expected to achieve ecological flows of the same order of magnitude when expressed as a percentage of the modulus, Q_{mod}.

Besides the values of the ecological flows, some hydraulic features of the flow regimes were also compared, namely the flow heights and the mean flow velocities (to simplify the presentation, the mean flow velocity in a given cross section will be referred as the flow velocity). This comparison showed that pronounced differences among ecological flows did not necessarily mean differences equally pronounced among the previous hydraulic parameters. In fact, the flow heights and especially the flow velocities were much closer than the differences among ecological flows could indicate. These results suggested that to recommend an ecological regime based only on the values of the natural flows may not be the most correct decision as only part of the features of the flow regime are taken into account.

In the previous scope, research was carried out in order to develop a method capable of providing comparable ecological flows under similar hydrologic constraints. The method thus achieved is supported by hydrologic and hydraulic criteria [7,10]. The hydrologic criteria account for the water scarcity and for the temporal irregularity (within each year and among years) of the natural hydrologic regime and the hydraulic criteria, for the geometry of the cross sections and of the river reaches.

2 The hydrologic-hydraulic method: general description

In each cross section and besides its detailed geometry, the application of the hydrologic-hydraulic method (HH method) requires a long series of mean daily flows which, for Portugal, does not represent an obstacle as that kind of series can be easily established by applying the procedures developed in [8,9] and widely proved.

By considering only part of the mean daily flows (in accordance with the criteria shortly presented), the flow heights and the flow velocities are computed, as well as the mean values of those hydraulic parameters. The mean monthly ecological flow is such that its velocity is equal to the mean velocity previously achieved [7,10]. Based on that flow, a month-by-month regime is established by applying a kind of "monthly rotation", in accordance with the following equation, which accounts for the temporal variability of the flow regime throughout the year:

$$Q_i = Q_{eco} \times Q_{ave_i} / Q_{mod} \tag{1}$$

In the previous equation Q_{eco} is the mean monthly ecological flow; Q_i the ecological flow in month i; Q_{ave_i} the average of the mean daily flows in month i; and Q_{mod} the modulus (all variables expressed in the same units, usually m^3/s).

The selection of the range of mean daily flows that supports the computation of Q_{eco} takes into account the particular hydrologic features of the hydrologic regime in the region in what concerns the extreme flows.

In fact, most of the time the rivers present extremely small flows and often, for two months or even more, no flows at all. Under these constraints the floods, though rare and restricted to a few days per year, may contribute significantly for the total runoff, as they may present flood discharge exceptionally large, with maximum values often several set of tens bigger than the modulus. As those floods do not really represent the flow regime in terms of water availability along the year, it was decided to discard part of the maximum mean daily flows, namely those flows with a mean annual duration (for a given set of n years, the duration, D, of a given flow/discharge, Q, is the number of days with flows equal or larger than that one. The mean annual duration, $\overline{D}$, is the average number of days per year with flows equal or larger than Q ($\overline{D} = D/n$)) smaller than 5 days (*criterion for the extreme large flows*).

On the other hand, the irregularity of the hydrologic regime combined with the extremely dry conditions that may occur during a significant part of the year, could justify ecological flows very small as those issues suggest that the local ecosystems are adapted to water scarcity. To prevent, somehow, ecological flows essentially influenced by the water scarcity, part of the flows during the dry season were discarded, namely the flows with mean annual durations $\overline{D}$ (days) computed by the following equation (*criterion for the extreme small flows*):

$$D' \geq 365 - (100 - \overline{D}) \tag{2}$$

where $\overline{D}$ (days) is the mean annual duration of the modulus estimated as a function of the mean annual flow depth $\overline{H}$ (mm) by applying the following equation:

$$\overline{D} = 0.2108\ \overline{H} + 15.101 \tag{3}$$

The latter equation is supported by the extensive hydrologic regionalization studies developed in [8,9]. Those studies proved that the mean annual flow depth is a regional parameter capable of "describing" the hydrologic regime and of providing a powerful tool that enables the establishment of flows series at ungauged river sections. Some of the results from those studies are presented in Figure 2 which contains the representation of the relationship then established between the mean annual flow depth, $\overline{H}$, and the mean annual duration of the modulus, $\overline{D}$.

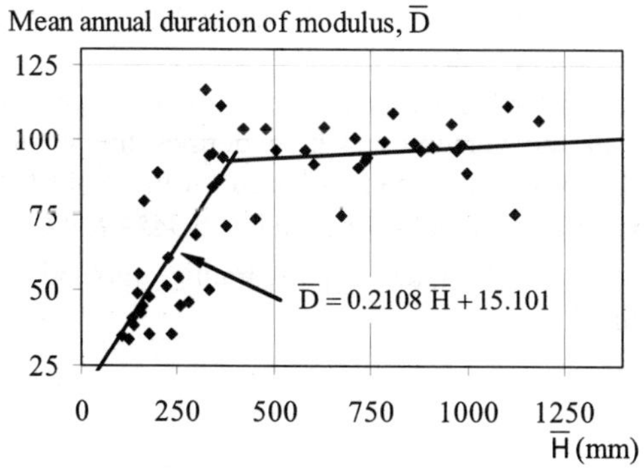

Figure 2: The relationship between $\overline{H}$ and $\overline{D}$, based on the records at 52 Portuguese stream gauging stations.

Figure 2 shows that for flow depths smaller than 400 mm a linear dependency between $\overline{H}$ e $\overline{D}$ is expected. The decrease of $\overline{D}$ as $\overline{H}$ decreases denotes a flow regime progressively more irregular, with only a few days with flows greater than $\overline{H}$. For larger values of $\overline{H}$, $\overline{D}$ becomes more or less constant (approximately equal to 100 days), i.e., independent of $\overline{H}$, which means that the temporal irregularity of the flow regimes is, in relative terms, more or less the same. This kind of "hydrologic behavior" can also be detected in the mean annual flow duration curves, as exemplified in Figure 3, based on the daily records at seven Portuguese stream gauging stations.

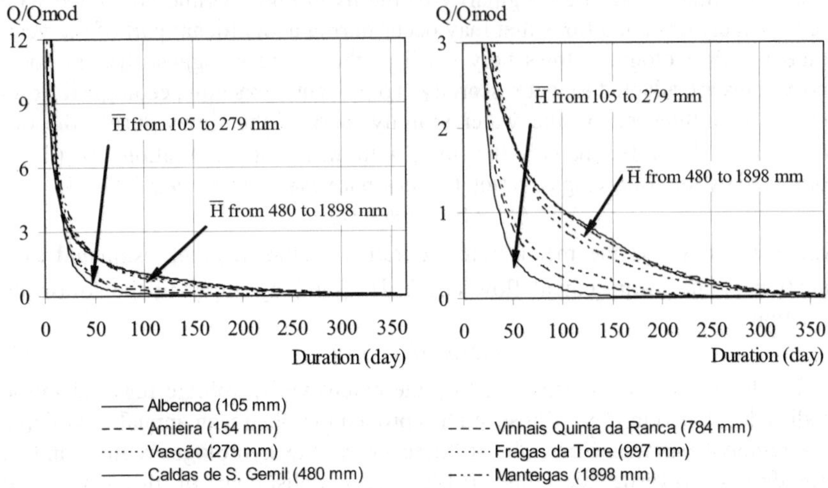

Figure 3: The mean annual flow duration curves at seven Portuguese stream gauging stations (curves and detail of the curves).

The previous results show that in order to increase the ecological flows more days with zero or almost zero flows must be discarded as $\overline{H}$ decreases. To fulfill this criterion eqn. (2) was adopted.

According to the criteria established for extreme large and extreme small flows the average of the velocities (which defines the velocity of the mean monthly ecological flow) was computed based on the daily flows with a mean annual duration comprehended between 5 and $D' \geq 365 - (100 - \overline{D})$ days.

The application of the hydrologic-hydraulic method is schematically represented in Figure 4.

3 Results

Table 1 presents some features of the 12 case studies (watershed areas, mean annual flow depths and modulus) along with the ranges of mean monthly ecological flows (expressed in a non-dimensional form, as percentage of the modulus, Q_{mod}) provided by the hydrologic-hydraulic method (HH method).

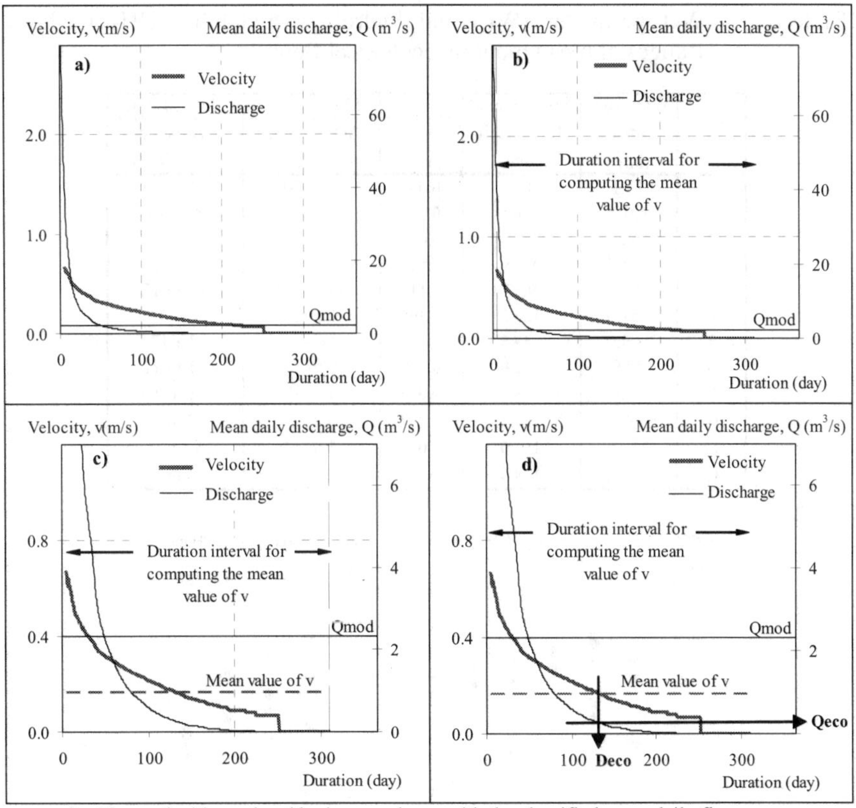

a) Flow velocities ordered by in accordance with the classified mean daily flows.
b) Interval – expressed in durations – considered in the computation of the mean value of the flow velocities.
c) Mean value of the flow velocities.
d) Ecological flow, Qeco, and respective mean annual duration, Deco.

Figure 4: Schematic representation of the application of the hydrologic-hydraulic method.

As more than one section was analyzed for each case, a range of mean monthly ecological flows is presented instead of a specific value of that flow. To allow a brief comparison among methods, the range of mean monthly ecological flows computed by the wet perimeter method was also include (WP method). The mean monthly ecological flows given by the INAG and the basic flow methods are not presented in Table 1 as the former method always resulted in ecological flows too high (for most of the cases of about 24% of Q_{mod}) while latter method proved not to be applicable in the region as it resulted in ecological flows smaller than 1% of Q_{mod}. Figure 5 completes Table 1, by representing schematically the ecological flows of that table (within the range considered for the vertical axis), as well as those from the INAG method. In the figure, the two

Table 1: Wet perimeter (WP) and hydrologic-hydraulic (HH) methods. Ranges of mean monthly ecological flows.

Dam	Watershed area (km^2)	Mean annual flow depth (mm)	Modulus (m^3/s)	Mean monthly ecological flow	
				WP method (% of Q_{mod})	HH method (% of Q_{mod})
1	13.1	83.7	0.035	3 to 30	11 to 13
2	101.8	90.7	0.293	8 to 13	6 to 15
3	6.3	94.5	0.019	8 to 150	12 to 14
4	176.2	95.5	0.534	2 to 10	10 to 13
5	37.6	95.7	0.114	53 to 285	10 to 12
6	351.0	124.2	1.395	1 to 21	9 to 11
7	509.0	143.8	2.321	0.4 to 4	7 a 12
8	38.9	152.0	0.188	2 to 56	11 to 12
9	15.4	153.0	0.076	5 to 37	11 to 16
10	48.0	155.3	0.237	4 to 7	7 to 9
11	212.0	161.0	1.081	6 to 12	7 to 11
12	218.0	178.4	2.432	2 to 3	4 to 5

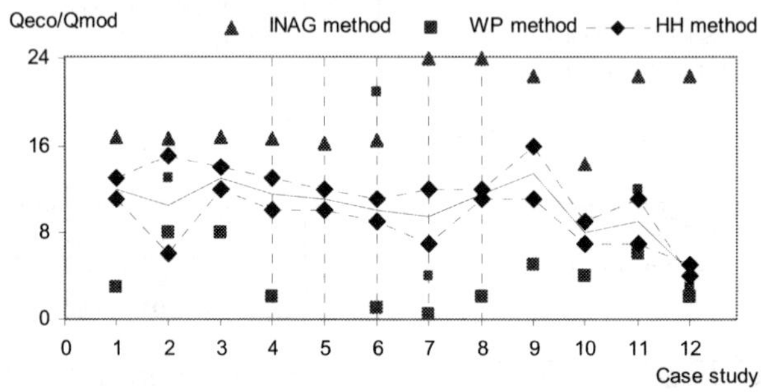

Figure 5: INAG, wet perimeter (WP) and hydrologic-hydraulic (HH) methods. Ranges of mean monthly ecological flows.

dashed lines links the maximum and the minimum ecological flows provided by the HH method while the full line represents the averages of those values for the 12 case studies.

Table 1 and Figure 5 clearly show that:

i) Despite the differences among watershed areas and among mean annual flows depths and excepting, in a certain way, case 2, the hydrologic-hydraulic method applied to each section always resulted in a narrow range of non-dimensional mean monthly ecological flows in clear opposition to the wet perimeter method; this is even more remarkable as both methods utilized the same cross sections; and

ii) Excepting case 12, the ranges of mean monthly ecological flows provided by the hydrologic-hydraulic method are quite similar; in fact, for the others 11 cases the adoption of mean monthly ecological flows comprehended between 9 and 12% of the modulus can always be justified.

4 Conclusions and discussion

A new method to define ecological flows based on hydrologic and hydraulic criteria was developed to Alentejo (South of Portugal) and it is briefly presented.

The hydrologic criteria account for the water scarcity and for the temporal irregularity of the natural hydrologic regime and the hydraulic criteria, for the geometry of the cross sections and of the river reaches. The data required by the application of the method to a given river reach are a series of mean daily flows and, as for the wet perimeter method, cross sections of that reach. In order to ensure that the special features of the flow regime are correctly considered, the previous series must be as long as possible (15 years or more). Also more than one cross section must be considered to attend the spatial variability of the geometry of the fluvial corridor.

The results achieved for 12 dams clearly show that the method is able to provide similar non-dimensional ecological flows despite the differences among watershed areas and mean annual flows depths. Mean monthly ecological flows from 9 to 12% of the modulus seem to be appropriate to the regional constraints. Based on each mean monthly ecological flow a monthly regime is established by applying eqn. (1).

Finally, it should be stressed that the validation of any ecological flow regime requires a continuous monitoring of the local ecosystems, which, for the time being, has not yet been done.

References

[1] Alves, M. H., 1993, *Método de determinação do caudal ecológico*, MSc Thesis. IST. Portugal.

[2] Alves, M. H., 2002, "Aproveitamentos hidráulicos, alterações dos regimes naturais e caudal ecológico" *in Caudais ecológicos em Portugal*, INAG, DSP, 2002, pp. 2.1–2.18. Portugal.

[3] Alves, M. H.; Bernardo, J. M., 2002, "Determinação de caudais ecológicos" *in Caudais ecológicos em Portugal,* INAG, DSP, 2002, pp. 3.1–3.27. Portugal.

[4] EDIA, Empresa de Desenvolvimento e Infra-Estruras do ALQUEVA, SA, www.edia.pt.

[5] Palau, T.; Alcazar, J., 1996, "The basic flow: an alternative approach to calculate minimum environmental instream flows", *Procs 2nd Int. Symp. on Habitat Hydraulic*, Vol. A: 547–558. Canada.

[6] Palau, T.; Alcázar, C.; J. Roi, 1998, *El Caudal Básico. Método para la gestión hidrobiológica de rios regulados*, Informe Técnico para CEDEX. Spain.

[7] Portela, M. M. 2005, "Proposta de procedimento hidrológico hidráulico para definir caudais ecológicos em cursos de água do Sul de Portugal Continental", *Recursos Hídricos*, Vol. 26 (1), APRH. Portugal.

[8] Portela, M. M.; Quintela, A. C., 2005, "Regionalization of hydrologic information: establishment of flow series at ungauged watersheds", *Water*

Resources Management III (Carvoeiro, Portugal), Wessex Institute, WIT Press, ISBN 1-84564-007-1, pp. 11–20. Southampton, Boston.

[9] Portela, M. M.; Quintela, A. C., 2006, "Estimação em Portugal Continental de escoamentos e de capacidades úteis de albufeiras de regularização na ausência de informação". *Recursos Hídricos,* Vol. 27(2). APRH. Portugal.

[10] Portela, M. M., 2007, "A new method to define ecological flows in the South of Portugal", 2nd *International Congress. Energy and Environment Engineering and Management*, IICIIEM2007. Badajoz, Spain.

[11] QUINTELA, A. (1967), *Recursos de águas superficiais em Portugal Continental*. PhD Thesis, IST. Portugal.

Study to improve the effective use of water in the Stompdrift-Kamanassie water scheme in the Oudtshoorn area in the Western Cape Province of South Africa

A. S. Roux
Department of Agriculture, Western Cape, South Africa

Abstract

The Stompdrift and Kamanassie Irrigation Scheme provides water for 13 513 ha in the Klein Karoo region of the Western Cape Province of South Africa. Water losses are high in the unlined earth canals, which cover more than 50% of the total length. The two dams can on average supply 22 million cubic meters per annum (Mm^3/a) of the total allocation of 28 Mm^3/a. The fully allocated quantities of water are supplied erratically, with only a fraction of the allocations supplied in some years, with the result that about 40% of the land is irrigated in most years.

Agriculture is the backbone of the economy of the whole region and the socio-economic circumstances of the population, especially the historically disadvantaged individuals, who rely heavily on agriculture. It is thus essential that the available water should be optimally utilised for the benefit of all.

Several studies in the past could not find an economical solution to the problem and a new study was commissioned to investigate water usage and more beneficial uses. The economic feasibility of developing sources of additional water was reviewed in terms of the current economic conditions.

The improved efficiency of water use, the possible augmentation of water resources, the improved management of the water scheme and the possible effects it can have on poverty alleviation in the area, are discussed.

Keywords: water use efficiency, best management practices, irrigation, plant water requirements, water conservation, agriculture.

WIT Transactions on Ecology and the Environment, Vol 112, © 2008 WIT Press
www.witpress.com, ISSN 1743-3541 (on-line)
doi:10.2495/SI080311

1 Introduction

The Stompdrift-Kamanassie Irrigation Scheme is situated in the valley of the Olifants River near the town of Oudtshoorn (see Figure 1). The main sources of irrigation water for the scheme are the Stompdrift and Kamanassie Dams, which were constructed solely for the purpose of providing water for the scheme.

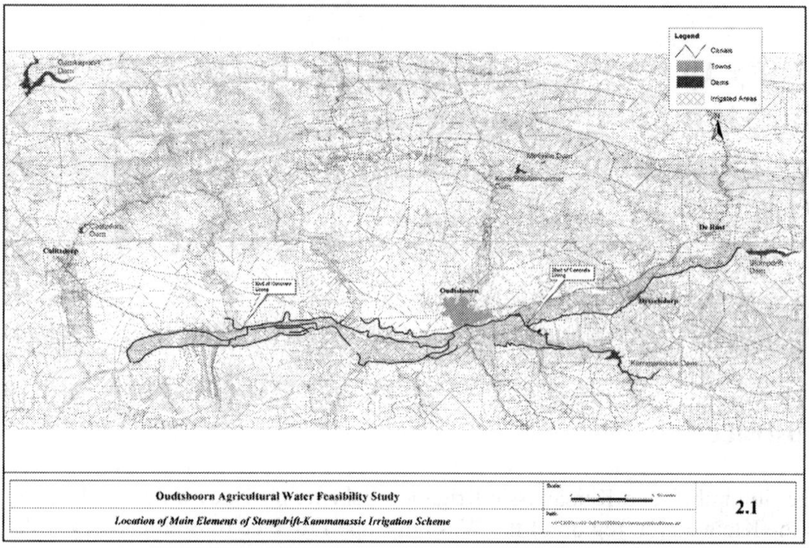

Figure 1: The Stompdrift-Kamanassie irrigation scheme.

The scheme consists of 13 513 ha of irrigated land that extends along the river banks for some 105 km downstream of the Stompdrift Dam and 18 km downstream of the Kamanassie Dam, to its confluence with the Olifants River. Water is conveyed to the irrigated lands by means of a system of canals, which consists of main canals, smaller distribution canals and smaller on-farm furrows. The main canals have a combined length of 137 km. The total length of on-farm furrows is estimated to be in excess of 100 km.

The Kamanassie Dam and most of the canals were constructed between 1919 and 1923. The scheme was extended by the construction of the Stompdrift Dam and Canal in 1963. Since then, some unlined sections of the main canals have been lined with concrete with the result that about 64 km of the 105 km length of main canals are now concrete lined.

2 Approach to the study

After completion of an Inception Phase, the study was carried out in three further phases, the purposes of which were:

- **Phase 1**: to gain a full understanding of availability and utilisation of the water in the two dams, and to assess options for restoring and augmenting their yields or the supply of water.
- **Phase 2**: to determine how effectively the water released from the dams is utilised, both in terms of losses and in terms of crop production, and to suggest improvements.
- **Phase 3**: to evaluate the socio-economic implications of the various options identified in Phases 1 and 2, and to make recommendations on those that are most effective.

3 Irrigation quotas and availability of water

The irrigated area is divided into four zones, each with a different irrigation quota, as indicated in Table 1 below:

Table 1: Irrigation zones, irrigation quotas and areas.

Zone	Scheduled area (ha)	Irrigation quota (mm/ha/a)
A	235	530
B	2 403	263
C	9 500	150
Kamanassie Servitude area	1 375	365
Total area	13 513	

It can be seen from the above that the quotas for Zones B and C are far less than those for Zone A. The reason for this is when the quotas were derived, it was assumed that irrigators further downstream would be able to abstract sufficient water from flow contributed by the tributaries of the Olifants River downstream of the dam to make up the difference. This is not the case in practice because most of the available water is used for irrigation in the tributary catchments before it reaches the main channel of the Olifants River. Consequently, there is normally very little flow in the Olifants River downstream of the dams, and what little flow there is becomes too saline to be used for irrigation, due to the geological conditions.

To obtain maximum yield from the 13 513 ha with the present mix of crops (98% alfalpha and 2% higher value crops), a reliable water supply of 187 Mm^3/a would be required. It is estimated that the average quantity of water released from the two dams is only 22 Mm^3/a, with the actual average quantity supplied at field edge is only 19 Mm^3/a. Therefore, on average, there is a shortfall of 168 Mm^3/a. It appears that, as a consequence of the shortage of water, only about 40% of the land scheduled under the scheme is irrigated in most years.

Higher value crops such as wine grapes, stone fruit and vegetables grown for seed, for which the soils and climate are suitable, require less water than alfalpha does, but at a higher assurance of supply. If water could be provided at high enough assurance for high value crops to be grown on the whole 13 513 ha scheduled under the scheme, the field edge water requirements would decrease from the present 187 Mm^3/a to between 60 Mm^3/a and 80 Mm^3/a, depending on the crop mix.

4 The economy of the Oudtshoorn area

The study highlighted the following aspects:

- The economy of the Oudtshoorn area is based on agriculture with some contribution from the tourism industry, mainly the world renowned Cango Caves.
- There is a high level of unemployment and associated poverty and some 30 000 new jobs are required to alleviate this situation.
- The potential of agriculture to contribute to employment exceeds the potential of the non-agricultural sectors. As the Stompdrift-Kamanassie Irrigation Scheme contains about 40% of the irrigated land in the area, a substantial increase in the output of the scheme would be expected to significantly increase employment opportunities in the area.

5 Possible ways of increasing the output of the scheme

Possibilities for increasing the output of the scheme are:

5.1 Increasing the yields of the Stompdrift and Kamanassie Dams by eliminating any unlawful upstream water use

It is estimated that agricultural development in the catchments of Stompdrift and Kamanassie Dams has reduced the average annual natural inflow to the dams by about 45 Mm^3/a, or 41%. Most of this development occurred before 1992, where after additional water use was severely restricted by law, with only a 5% increase in the storage capacity of farm dams in the catchment of Kamanassie Dam since 1992. This dams only had a further reduction of 0,1 Mm^3/a in the inflow into the dam.

5.2 Increasing the quantity of water available for irrigation by developing water supply augmentation schemes

Possible augmentation schemes in the Olifants River catchment using surface water resources could only provide quantities of water that are negligible in comparison to the shortfall in optimum water requirements of 168 Mm^3/a. Furthermore, the relatively small quantity of water that could be provided is needed either to meet ecological flow requirements or to meet urban requirements. Therefore, it is concluded that there is no significant potential within the Olifants River Catchment to augment supplies to the Stompdrift-Kamanassie Irrigation Scheme from surface water.

The raising of the existing Gamkapoort Dam on the Gamka River might make an additional quantity of about 7 Mm^3/a of water available which could be used to supplement supplies to Zone C at a unit reference value (URV) of R1, 30/m^3. (URVs are an indication of the cost of water from a scheme). However, the existing uncertainties about the reliability of the hydrology, the environmental water requirements downstream of the dam, the needs of other stakeholders for the water supplied from the dam, and the economic viability of the scheme need to be resolved.

The development of local groundwater resources are expected to yield about 70 Mm^3/a, but this need to be confirmed by further investigations. The URVs for groundwater schemes are estimated to range from R0,62/m^3 to R2,58/m^3.

5.3 Reducing canal conveyance losses to make more water available at the field edge

Based on the present average quantity (22 Mm^3/a) of water released from the dams into the canals, water losses in the canals are estimated to be 4,5 Mm^3/a. The estimate of losses is based on published data on canal losses elsewhere, and need to be verified by physical measurements of canal losses. Based on the above estimate it has been found that an average quantity of water of 2 Mm^3/a would be saved at a URV of R6,87/m^3 if the main canals were fully sealed and concrete lined, while 2,8 Mm^3/a could be saved at a URV of R9,78/m^3 if they were replaced with pipelines. See Table 2 below.

5.4 Improving the efficiency of irrigation methods so as to obtain increased yields of crops from the same quantities of water at the field edge

The most practical way of improving the efficiency of irrigation practices appears to be to level, by means of laser controlled methods, the lands that are flood irrigated. This has been shown on those lands where it has already been implemented to increase the efficiency of flood irrigation from 60% or lower to about 80%. The estimated average saving in water if this were done throughout the scheme is 4,4 Mm^3/a at a URV of R0,60/m^3. See Table 2 below.

5.5 Using the available water to produce higher value crops on reduced areas of irrigated lands

Using the available water to produce higher value crops than alfalfa would drastically reduce the area of land irrigated because, even though crops of this type may require less water than alfalfa does to produce optimum yields, the water is required at a considerably higher assurance of supply because the crops are less tolerant of a lack of water than alfalfa is. It would also be necessary to retain water in the dams at the end of an irrigation season to ensure that there would be sufficient water for the next season if the winter rains were poor. This would result in higher evaporation losses.

If all crops were irrigated so as to meet optimum water requirements and the areas of higher value crops remained as at present, the area of alfalfa grown would reduce to the extent that the total area of land irrigated would be 20% of the total scheduled area. In this scenario the higher value crops would receive water at high assurance and the alfalfa would receive water at the current low assurance. The higher value crops would make up approximately 10% of the total area irrigated.

Unless the available water supply was increased significantly, increasing the area of high value crops would further reduce the total area irrigated, and a major reduction in irrigated area would be caused by a policy of irrigating the crops to provide optimum yields instead of the sub-optimum yields that are obtained from the bigger areas of land that are irrigated at present. Whilst it was shown by a financial and economic analysis that such a policy would be likely to benefit the economy of the Oudtshoorn area in general, it would severely disrupt the social structure of the farming community.

Table 2: Potential for water saving.

Description	Capital cost (R million)	Average quantity of water saved (Mm^3/a)	URV (R/m^3)
Fully concrete lining all main canals	206	2,0	R 6,87
Pipelines in place of main canals	412	2,8	R 9,78
Laser levelling of lands	40	4,4	R 0,60

6 Results of the financial and economic analyses

The potential benefits of the possibilities, listed in Section 5 above, for increasing the output of the scheme were compared to the status quo by means of financial and economic analyses of different scenarios of representative agricultural practices that could be implemented on the scheme. The findings may be summarised as follows:

- Levelling by laser controlled methods of cultivated lands that are flood irrigated shows significant financial benefits in comparison to conventional levelling.
- Using the presently available irrigation water supply, the best financial returns from the scheme could be obtained by concentrating the water on an area of land of about 20% of the scheduled area so as to obtain optimum yields from a crop mix of 90% alfalfa and 10% higher value crops.
- If this approach were adopted, none of the lands in Zone C would be irrigated and this would result in the loss of an estimated 155 jobs on the farms. This would probably be more than compensated for by the creation of a greater number of other jobs, spread amongst all sectors of the local economy, as a result of the extra income generated on the remaining irrigated farms.
- If a substantially greater area of high value crops than the 10% of the total irrigated area referred to above was grown with a corresponding reduction in the area of alfalfa, the financial benefits would be less because more irrigation water would need to be provided at a high assurance, with the

result that the annual average quantity of water supplied would be significantly reduced.

- The cost of concrete lining the canals, or replacing the canals with pipelines, in order to reduce water losses, would exceed the income obtained from the additional quantity of water that would be made available. Therefore, this option for improving the efficiency of water use would not be financially viable unless the cost were heavily subsidised from a source outside the Oudtshoorn area.

7 Conclusions

The findings of the study lead to the following conclusions:

- The shortfall in irrigation water has, to some extent, been brought about by substantial increases in water use in the catchments of Stompdrift and Kamanassie Dams and in the tributary catchments of the Olifants River downstream of the dams, since the dams were built.
- The total shortfall in water at field edge with the existing crop mix of 98% alfalfa and 2% other crops is estimated to be 168 Mm^3/a and it is unlikely that there is sufficient water available at justifiable cost to provide the full shortfall. However, it appears that about half of this quantity of water might be available through a combination of developing groundwater resources, raising Gamkapoort Dam, and laser levelling the lands that are flood irrigated. Even this quantity of water would significantly increase the output of the scheme and, with the existing crop mix, could be expected to provide a significant number of new employment opportunities in the local economy.
- If water could be supplied at a 1 in 20 year assurance, higher value crops such as vines, stone fruit, or vegetables for seed, which require less water than alfalfa does, could be grown on a large scale. The field edge water requirements for the scheme would then reduce from 187 Mm^3/a to between about 60 Mm^3/a and 80 Mm^3/a, depending on the mix of crops.
- For the water augmentation schemes to be financially viable, capital and operating costs would have to be heavily subsidised from a source outside the local economy.
- The availability of increased quantities of irrigation water at an affordable cost would improve the financial situations of the farmers on the scheme and put them in a better position for entering into Agricultural Black Economic Empowerment ventures with their farm workers.

8 Recommendations

The recommendations arising from this study fall into three categories, namely:

- Improved efficiency of water use
- Augmentation of water supplies
- Operation of the scheme and poverty alleviation.

The recommendations in each of these categories follow.

8.1 Improved efficiency of water use

- The removal of alien vegetation in the catchments of the Stompdrift and Kamanassie Dams should continue until the alien vegetation has been completely eradicated.
- Laser levelling of flood irrigated lands to improve the efficiency with which the water is applied should be encouraged.
- Reliable measurements of water losses from concrete lined and earth canals should be made and the feasibility of upgrading the canals so as to reduce water losses should be reviewed in the light of these measurements.

8.2 Augmentation of water supplies

- Investigation of potential groundwater schemes should proceed, with particular emphasis on schemes that would use the Table Mountain Sandstone aquifers.
- The Department of Water Affairs and Forestry should be asked to further investigate the feasibility of raising Gamkapoort Dam and supplying some of the additional yield obtained to Zone C of the Stompdrift-Kamanassie Irrigation Scheme.

8.3 Operation of the scheme and poverty alleviation

- If the water supply cannot be increased significantly, the possibility of using in Zone B the irrigation water that is allocated to Zone C at present should be investigated in more detail because there may be economic advantages in using the water in this way. The social impacts and ways of adequately compensating farmers and farm workers and their families who would be disadvantaged by the change should also be investigated, and the apparent potential for the change to benefit the Oudtshoorn community as a whole should be verified.
- The current small-scale farmer activities should be recognised, stabilised and strengthened through actions such as assistance with laser levelling of their irrigated lands, and on-farm skills transfer, especially in matters such as farm management and financial management.
- Zones A and B should be favoured as areas for new opportunities for emerging farmers because of the better water supplies in these zones.
- Mechanisms should be developed to promote workable collaboration between established commercial farmers and emerging farmers so that the latter can avoid commonly made mistakes and learn from successes achieved.
- The job creation potential of tourism on farms should be acknowledged and explored.
- As it is recognized that agriculture alone could not possibly generate 30 000 jobs, efforts should also be made to create employment in the general local economy.

References

[1] Bredenkamp, D B, Botha, L J, Van Tonder, G J and Van Rensburg, H J. 1995. *Manual on quantitative estimation of Groundwater Recharge and Aquifer Storativity.* Water Research Commission (WRC) Report No TT 73/95, 363 pp. & 2 appends.

[2] Davis. 1993. *Davis' Handbook of Applied Hydraulics.* Vincent J Zipparro, Editor-in-chief, Hans Hasen, co-editor. Published by McGraw-Hill, Inc. 1993

[3] DeHayr, R and Gordon, I. 2004. *Irrigation Water Quality* – Salinity and Soil Structure Stability produced by the State of Queensland (Department of Natural Resources, Mines and Energy) Obtained from www.nrme.qld.gov.au/factsheets/pdf/water/w55.pdf

[4] Doorenbos, J. & Pruitt, W.O. 1975. FAO 24. *Guidelines for predicting crop water requirements.*

[5] Green. 1985. *Estimated Irrigation Requirements of Crops in South Africa part 1.* G C Green, editor. Department of Agriculture and Water Supply, Pretoria. 1985

[6] Hewitson, B C. 2003. Developing Perturbations for Climate Change Impact Assessments. Climate System Analysis Group, Department of Environmental and Geographical Sciences, University of Cape Town, EOS, Transactions, *American Geophysical Union,* Vol. 84, No. 35.

[7] Koegelenberg, F.H. & Roux, A.S. 2000. *Water Conservation Campaign in Agriculture: Interim Summarized Report.*

[8] Laker, M.C. 2006. Soil productivity in irrigated agriculture, with special reference to South Africa. *Paper delivered at SARIA meeting, Roode Valley, Pretoria, South Africa.* Southern African Irrigation Association. February 2006.

[9] McDonald, S. and Punt, C. 2004. *Analysis with a social accounting matrix for the Western Cape.* Department of Agriculture, Western Cape. Elsenburg.

[10] Tripp, R. 2001. *Agricultural Technology Policies for Rural Development Policy Review,* Volume 19. December 2001. doi:10.1111/1467-7679.00146

[11] Van Koppen, B., Namara, R., and Safilios-Rothschild, C., 2005, *Reducing poverty through investments in agricultural water management.* Poverty and gender issues and synthesis of sub Saharan Africa case study reports. IWMI, 2005.

[12] Visser, B. 1980. *Soil Survey in Oudtshoorn Area.* Department of Agriculture, Western Cape Province, Elsenburg. 1980.

[13] Umvoto. 2003. *Deep Artesian Groundwater Exploration in the Greater Oudtshoorn District for Bulk Water Supply (DAGEOS).* Phase B Report.

Section 5
Re-use of water

Economic analysis of water reuse in Spain

J. M. Moreno[1,2] & L. D. Chabalina[2]
[1]*Department of Applied Economic Analysis, University of Alicante, Spain*
[2]*Water and Environmental Science Institute, University of Alicante, Spain*

Abstract

The remarkable development of reuse of water in Spain has taken place not only because of the necessity to extend water supplies, but also due to the requirements to improve the management in water treatment. Water supply consumption has increased, along with the increase of population that has taken place in numerous urban zones. As a result, traditional supplying sources are insufficient to satisfy the demand, which is in permanent expansion. This paper will analyze from an economical point of view, the effects of the Directive 91/271/EEC in Spain, as well as the new rule that appears in the Order in Council of December 8, 2007 in which the legal regime of treated water reuse is established. Furthermore, costs of reused and recycled water by means of different treatment systems will be discussed, as well as the environmental costs associated to them.

Keywords: treated water reuse, economic analysis, environmental and economic costs, legal regime.

1 Introduction

The increase reached by water supplies, as well as the increase of population which took place in numerous urban areas, have caused that the traditional supplying sources are insufficient to satisfy the demand, which is in permanent expansion. Environmental limitations and plurianual droughts have led numerous towns to use treated water as an additional water source to take advantage of it, in case drinkable water quality was not necessary. At the same time, sanitary and environmental requirements of continental and marine water quality are rising, in addition to the requirements of location and treatment levels which are more and more strict. For that reason, recycled water has been turned into an alternative

WIT Transactions on Ecology and the Environment, Vol 112, © 2008 WIT Press
www.witpress.com, ISSN 1743-3541 (on-line)
doi:10.2495/SI080321

source of supplying, economic and assured from the sanitary and environmental point of view.

The modification of the National Water Plan, made from the Order in Council 2/2004, has supposed an important change in the hydraulic policy. That has been achieved thanks to the development of activities as desalination and wastewater reuse in basins with lack of water in different fields of water planning. This paper will be focused on water recycling.

Directive 91/271/CEE, in its article 12 establishes: “1. Treated wastewater will be reused when it proceeds. The evacuation methods will reduce up to the minimum adverse effects on the environment”. In Spain, the National Plan of Sanitation and Wastewater Treatment also has established, the promotion of wastewater reuse, as an important management point in the hydraulic domain (published in the Resolution of April 28th, 1995, of the Secretary of State for the Environment, by which is stated in the Cabinet Meeting of February 17th, 1995, of which the National Plan of Sanitation and Wastewater Treatment was approved, BOE nº. 113 of May 12th, 1995.).

The reuse of treated wastewater must not be considered isolated and exclusively in function of the benefit that it could produce on users. Recycled wastewaters must be considered as an unconventional resource, which its management must be included in an integral water resources planning, taking into account economic, social and environmental aspects. Reuse can raise the availability of water resources. As a result, the improvement of effluents quality is the key of the use and water management. This way, recycled water can replace uses that do not need a high quality, avoiding using volumes of better quality for other more demanding uses.

2 Discussion

2.1 State of treatment and water reuse in Spain

From the legal point of view, the promotion of building sites associated to wastewater reuse can be carried out by the state administration through the basin organizations and by the state societies if buildings are considered of general interest, in accordance with the autonomous regions if it is also considered as an autonomic interest. Finally, it is possible that users' regions could take part in this type of action in its condition of public administrations of corporative nature, especially when the actions are related to irrigation.

According to the Ministry of Environment (MIMAM), wastewater treatment in Spain in 2004 covered 2.700 city centres, treating a total load of 70.130.000 equivalent inhabitants (EI). Although the Directive 91/271/CEE was fulfilled, it is not clearly aimed to wastewater reuse, therefore effluents can not be used for many potential uses.

Figure 2 shows the fulfilment degree of the objectives of yield and quality established in the Directive.

With regard to the state of wastewater reuse in Spain, the maximum volume is limited by the amount of urban treated water, by the geographic location with respect to the use of these facilities, by the demand, by the resource acceptance and by its economic and environmental viability. In *El Libro Blanco en España* [1], it was expected that the maximum volume of treated effluents will reach 3,500 hm^3/year, once finalized the National Plan of Sanitation and Treatment in accordance with the Directive 91/271/CEE. Approximately 1,200 hm^3/year of that water could be reused. In 2005, 2,400 hm^3 of wastewater was treated, of which 17% was reused [2]. Although the percentage of reused water with respect to the total of available resources is small (2,67%), this one becomes strategic and even essential in some zones of Spain, mainly in those that have a structural deficit. The majority of the projects already done or in study phase, have an agricultural purpose, showing an increasing trend to reuse water with environmental aims.

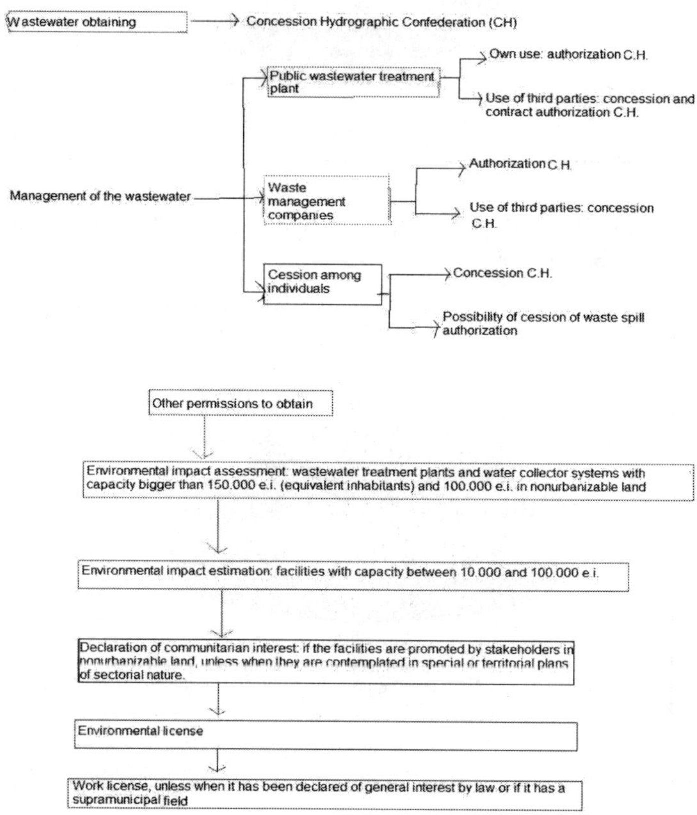

Figure 1: Concessions, authorizations and permissions to request for activities of treatment and water reuse in the different administrations.

In order to make the planned direct reuse in appropriate conditions is necessary to fulfill some requirements: wastewater availability which sometimes will need additional treatments; studies of economic, social and environmental feasibility; rules that define the quality limits based on the possible uses; management and operation systems of recycled wastewater; and, finally, a wastewater prices policy, that contemplates how and who covers the expenses related to infrastructures, facilities and operation costs associated to reuse.

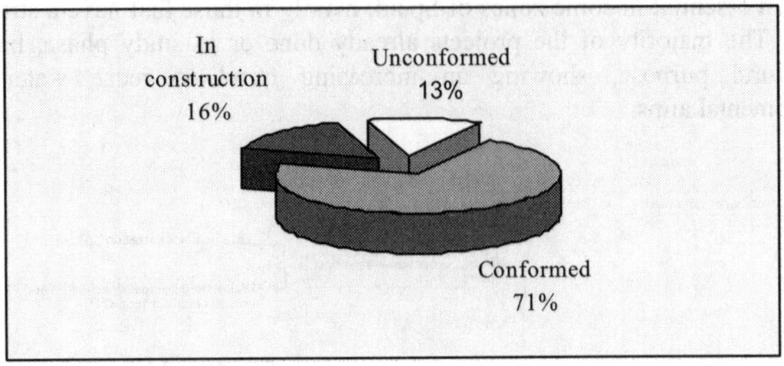

Figure 2: Fulfilment degree of the Directive 91/271/CEE of treatment plants in Spain in 2004.

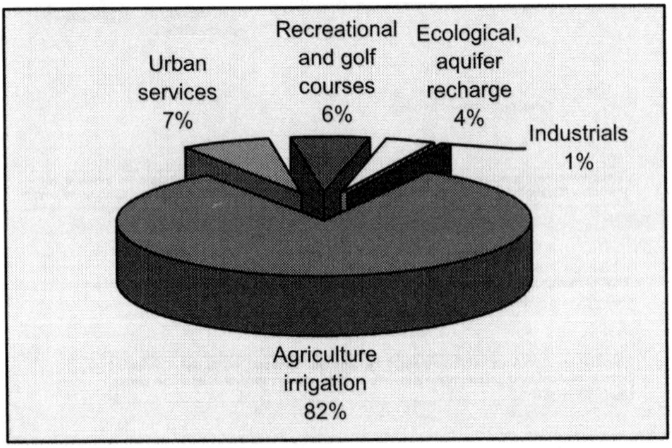

Figure 3: Use percentage of reused wastewaters in Spain in 2002.

Wastewater uses can be incredibly different. It is possible to use them practically for everything, except for human and animal feeding. The use of treated water will depend on its quality. The degree of wastewater treatment(s) will determine the quality of treated water, its cost being variable and increasing as the obtained water has a higher quality. This cost will also depend on the initial characteristics of water received in the plant.

Wastewater management must be included in an integral planning of water resources, where economic, social and environmental aspects should be considered. Thus, reuse can increase water uses of the already used water, replacing those that do not require high quality water. In addition, the availability of water resources increases because water volumes of better quality could be used for other more demanding uses. Therefore, and agreeing with Hernández [3], who indicates that "the reuse of resources obtained from wastewaters would have to be considered that cannot be waived, as much from the social point of view as environmental and sanitary", since "additionally, optimizing the treatment process using these unconventional resources it would be able to reduce the demand pressure on certain conventional water resources".

However, as a treated water market does not exist, it is difficult to obtain a price for this product; for that reason, it is assumed that the cost per cubic meter must be equal to the maximum sale price, this way guaranteeing to cover the costs. A fundamental question, in this sense, is: Who must or can pay the cost that represents the obtaining of a viable project? Only it can be said that at the moment, in the European Union (EU), there are no concrete subventions which promote water reuse [4]. Now the financing mechanisms can be grouped in two categories: 1) Financing of initial costs, 2) Financing of exploitation costs.

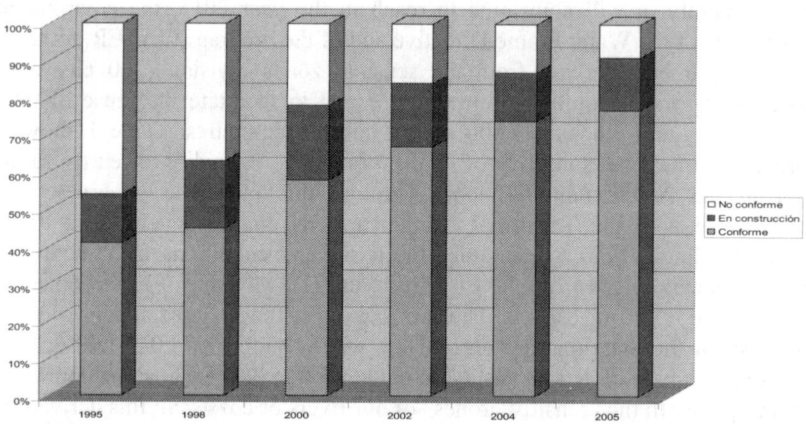

Figure 4: Conformity degree evolution of the polluted load since the publication of the National Plan of Sanitation and Treatment (1995-2005). *Source: MIMAM, 2007.*

Nowadays, the most extended use of wastewater is agriculture, and the prices paid by this use are symbolic or free. In Spain, the price of water for agricultural use ranges between 0.006 and 0.012 €/m^3, the supplying one is 0.77 €/m^3, the desalinated water is around 0.6 €/m^3 and the price of wastewater treatment (including tertiary treatment) ranges between 0.6 and 0.8 €/m^3. In view of this situation, it is possible to ask the following question: how the wastewater price could be assumed? The most advisable solution would be the wastewater subvention to extend its implementation, otherwise it would be necessary to increase the price of water for agricultural and domestic use. Therefore, in Spain it is essential to establish a price policy that distributes the treatment costs and wastewater management according to the total water consumption. Moreover, incentives must be developed to promote wastewater use in all sectors.

The National Plan of Sanitation and Treatment (1995–2005), at the end of 2005, presented a conformity degree of 76%, in construction a value of 13% and 11% of unconformity. The balance of this plan is very satisfactory, because the conformity degree increased from 41% in 1995 to 77% in 2005, in accordance with the EU rule (figure 4). In addition, the total load obtained in December 2005 was 73.3 million of EI, unconformed being 6.5 million of EI (9%). On the other hand, in 2007, around 2533 wastewater treatment plants existed in Spain. They treated more than 3,375 hm^3/year of wastewater, with a reuse rate of 13.25% (450 hm^3/year).

The new National Plan of Water Quality: Sanitation and Treatment 2007–2015 have as the main objective, to satisfy not covered and future necessities related to sanitation and wastewater treatment of the Autonomous Regions and the Local Corporations. Thus, the new plan, approved on June 8th of 2007 by the Cabinet Meeting and which count with an estimated cost of 19,007 million euros, tries to complete the fulfilment of the EU exigencies. Consequently, it will contribute to reach in the year 2015, the environmental objectives of the Water Frame Directive and of the program WATER, to face the new investments derived from the sensible zones revision (200 city centres affected by the Resolution of July 2006) and to facilitate the reuse of treated wastewater until reaching 3,000 annual cubic hectometres. Table 1 shows the provisional national summary of total investments, in million of euros, to carry out in the Autonomous Regions. Through this plan, the government will collaborate with the Territorial Administrations on the development of their corresponding actions to guarantee the terms and conditions fulfillment of the requirements derived from the EU directives. The Ministry of Environment will work in the following aspects. The carrying out of the declared actions of general interest for the State that are unresolved, with a budget of 3,046 million euros. Accomplishment of actions, using an amount of 25% of its cost, to improve the water quality in the "sensitive zones" of our rivers or coasts, all this derived from the declaration made by Spain or Portugal about Sensitive Zones, investing for it 557 million euros. Participation to 50% with the Autonomous Regions in actions that affect the National Parks and in municipalities with territories of the Network Natura 2000, to assure the water quality in these more demanding environments, with a total amount of 1,200 million euros. Financing, without interest, 50% of the actions that will be agreed with the Autonomous Regions, recovering the investment in 45 years, through the State Water Societies, with a maximum amount of 1,430 million euros. In short, the MIMAM will participate with 6,233 million euros, of which 3,046 are inherited of the Plan of Sanitation 1995–2005. The objective, with this National Plan of Water Quality, is to make possible the reuse of water, with its consequential offer increased until reaching approximately 3000 hm^3/year available in 2015.

Table 1: Provisional national summary of total investments, in million of euros, to be carried out by the Autonomous Regions.

1	Declared actions of general interest (actions of Hidrographic Confederations and State Societies are included)	1.114 €	5,7 %
2	Actions without wastewater treatment plant or with wastewater treatment plant unconformed	2.903 €	14,8 %
3	Actions due to the new declaration “Sentitive zones”	4.782 €	24,3 %
4	Actions to cover future necessities	5.620 €	28,6 %
5	Actions to contribute to reach the objectives of the Frame Water Directive (Directiva Marco del Agua DMA)	1.938 €	9,9 %
6	Actions in sanitation (without treatment)	2.741 €	14,0 %
7	Actions to promote I+D+i in sanitation and treatment	547 €	2, 8 %
	TOTAL	19.645 €	100 %

Source: MIMAM, 2006.

Until then, the analysis of treated water reuse focused in technical and legal aspects have had an important development and its methodology in general has been well structured. On the contrary, environmental, social and economic aspects are enormously delayed, and methodologic budgets require to be reinforced. It is necessary to indicate the new legal definition of recycled wastewater: treated wastewater that, in their case, have been treated by an additional or complementary treatment allowing to obtain a required quality according to the use to which they are destined (RD 1620/2007, Dec. 7th). It is a term that every day acquires more strength and it is related with the search of the social acceptance of this water, since, from the technical point of view and in wastewater field, treated effluent, treated water and recycled water are synonymous.

It is necessary to emphasize the potential use of recycled wastewater reuse. In the first place, it is used for agriculture (culture lands, grass, ornamental zones, industrial cultures, aquiculture, nursery...); secondly, for aquifer recharge, with the purpose of solving environmental problems of fight against marine intrusion and regeneration of some lost ecosystems by excessive exploitation; thirdly, for environmental and recreational uses (golf courses, forestry, green spaces inaccessible to the public), urban (irrigation of private gardens, street washing) and industrials (textile sector). So, it is required that the regional administrations regulate these questions of using exclusively treated water forcing to any set going project, as well as the already existing ones.

2.2 Costs of treatment and water reuse in Spain

In Spain, the average cost of treated water by means of membranes for agricultural irrigation is 0.5 €/m^3. In the Valencian Region, in 2006, according to data of the Public Water Management Organisation (EPSAR-Entidad Pública de Saneamiento de Aguas), there are 415 wastewater treatment plants (148 of them are in the province of Alicante). In addition, 117 of them have water reuse with a volume of 154 hm^3, 20 plants have tertiary treatments and 170 hm^3 of capacity. All treatment plants treat an annual volume of 487 hm^3 (132 hm^3/year in the province of Alicante). The yield of Biochemical Oxygen Demand (BOD_5)

elimination is of 92% in the Valencian Region and of 95% in the province of Alicante. For the mentioned year, the Valencian Region reused in agriculture about 175 hm^3, approximately the number that represents half of the total reuse in Spain. In Castellón, the volume of reused water was 16.5 hm^3, in Valencia 88.5 hm^3 and 52.6 hm^3 in Alicante. In the case of Alicante, the annual volume that can be treated is 132 hm^3, of these only 52.6 hm^3 were reused. For that reason, 79.4 hm^3 (39,84%) of total treated water obtained in the province was not reused. According to the EPSAR, in 2010, the region will be able to reuse 350 hm^3/year, which means an increase of double.

Which are the costs to treat and to reuse water? It is necessary to begin indicating that the costs of treatment and water reuse are conditioned by the existence of several factors. In the first place, it has to be considered the type of water to treat, since its origin determines the class and level of polluting agents which have to be eliminated and the type of treatment to apply. Secondly, it must be considered the use (different qualities) that water will have, because different processes will be used according to the application required. There is an enormous variability of costs associated to the different treatments. They increase when the number of processes involved increase. Some uses, like those associated to an industrial reuse and those that are destined to aquifer recharges, also have significant price variations. Special attention has been paid to the subject of power costs, since consumption is very unequal (and, therefore, the cost) according to the chosen technique. Thus, it is necessary to consider clearly the correlation between pollution degree of treated water (measured by the quotient between the EI served and the processed cubic meters) and the energy consumption of the plant.

2.3 Cost examples in the Valencian Region

In the Valencian Region, according to the collected data, treatment cost has a value of 0.220 €/m^3, that are distributed as follows: personnel costs, 0.088 €/m^3 (40%); energy, 0.042 €/m^3 (19%); wastes, 0.035 €/m^3 (16%); maintenance, 0.026 €/m^3 (12%); reagents, 0.015 €/m^3 (7%), and under the denomination of "others" (material of laboratory, vehicles, fuel, gardening, etc.) there are expenses of 0.014 €/m^3 (6%). Also, operation costs differ sensitively according to the different treatments used. For secondary treatment, the cost is 0.26 €/m^3; for tertiary treatment, 0.06 €/m^3; for advanced treatment, 0.14 €/m^3. It is assumed that the distribution cost is 0,1 €/m^3.

In order to determine the price that final users must pay by recycled water consumption, in addition to the price that entails obtaining this resource, it is necessary to consider other additional costs, such as the expenses derived from the use of cooling towers, the update of pipes, etc. On the other hand, it would be beneficial to count with the existence of subventions, financing at low interests or the possibility of reducing the global cost of the system. As a result, it would reduce the price at which recycled water can be offered. In addition, it has to be considered that many companies apply tariffs to recycled water based on a percentage of the drinkable water price. With this measure it is trying to foment its use, but the global cost of the reuse project cannot be recovered, nor either include the cost of distribution systems. The price system of recycled water should consider the costs, including the intrinsic value of water as a resource, its environmental effects and the cost of the opportunity that entails its use.

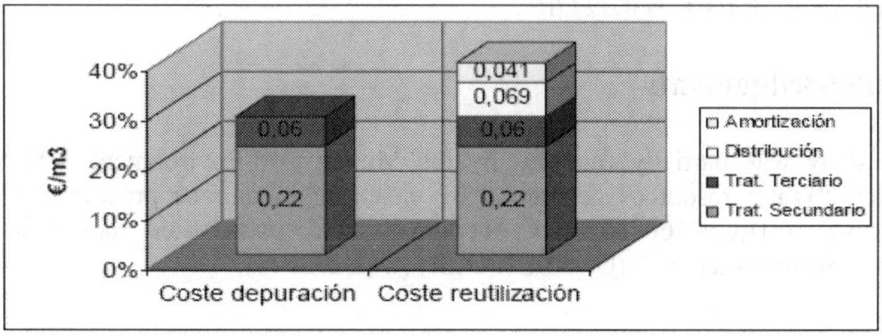

Figure 5: Costs of treated and reused water in the Valencian Region. *Source: EPSAR, 2007.*

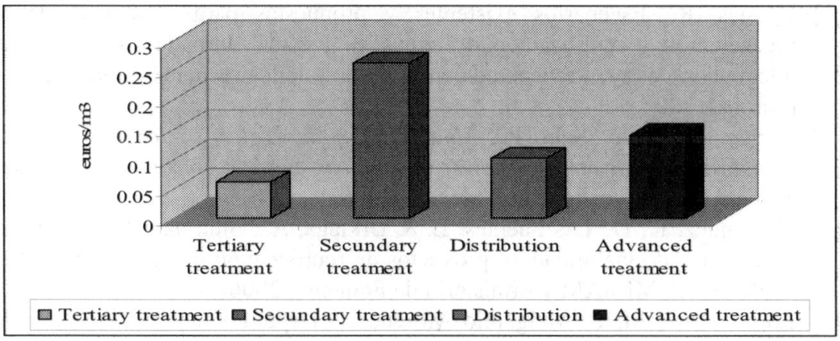

Figure 6: Exploitation costs of recycled water in the Valencian Region. *Source: EPSAR, 2007.*

In the Valencian Region, the prices of recycled water, including treatment costs, increase to 0.28 €/m^3 (0.22 €/m^3 of secondary treatment and 0.06 €/m^3 of the tertiary treatment). Taking into account the reuse cost, the price of recycled water increases to 0.39 €/m^3 (0.22 €/m^3 of secondary treatment, 0.06 €/m^3 of tertiary treatment, 0.069 €/m^3 of distribution and 0.041 €/m^3 of amortization).

3 Conclusions

In countries, where there is a lack of water, treated water with acceptable quality levels becomes essential for later reuse. In some zones of Spain, water treatment is being implemented for its massive reuse. In order to obtain this objective, the

1st National Plan of Sanitation and Wastewater Treatment 1995–2005 has been completed. At the present time, the new National Water Plan: Sanitation and Treatment 2007–2015 is in force, which is expected to reuse more than 3000 hm^3 of water per year. For that, there is a budget of 19 645 million euros, that will be co-financed by the State and Autonomous Regions. In Spain, the average cost of treated water for reuse is 0,39 €/m^3.

Acknowledgements

This study was partially financed by the Ministry of Environment with the projects "Water resources generation by means of membrane processes" (NF MBR/NF) CTQ200766780 and "Treatment and wastewater reuse for a sustainable management" (CONSOLIDER) (CSD200644).

References

[1] LIBRO BLANCO DEL AGUA: Ministerio de Medio Ambiente, Secretaría de Estado de Aguas y Costas, Dirección General de Obras Hidráulicas y Calidad de las Aguas, 2000.

[2] Iglesias, R., Escenarios existentes y propuestas para el avance de la regeneración y reutilización de aguas en España. Jornadas Técnicas: La integración del agua regenerada en la gestión de los recursos, Lloret de Mar (Girona), 2005.

[3] Hernández, F., & Sala, R., *Modelización de costes en los procesos de tratamiento de aguas residuales: un análisis empírico para la Comunidad Valenciana.* Universidad de Valencia, 2006.

[4] Hernández, F., De Las Fuentes, L. & Urkiaga, A., Guía para la realización de estudios de viabilidad en proyectos de reutilización de aguas depuradas. AQUAREC, MIMAM y Ministerio de Fomento, 2006.

[5] López Ortiz, M. & Melgarejo Moreno, J. Depuración y reutilización de aguas en España: los casos de la Comunidad Valenciana y Murcia", en GONZÁLEZ VARAS, S. (Coord.), *Nuevo Derecho de Aguas*. Civitas/Thomson, Navarra, pp. 633–657, 2007.

[6] Mujeriego, R., La reutilización planificada del agua para regadío, XI Congreso Nacional de Regantes, Palma de Mallorca, 2006.

[7] "EPSAR- Entidad Pública de Saneamiento de Aguas" webpage. http://epsar.cop.gva.es/depuradorasv

[8] "MIMAM-Ministry of Environment" webpage. Síntesis del contenido del documento inicial para la evaluación ambiental estratégica". Programa A.G.U.A. Apartado I.S.A: Actuaciones urgentes cuenca mediterránea. www.mma.es

[9] Prats Rico, D. & Melgarejo Moreno, J, Desalación y reutilización de aguas. Situación en la provincia de Alicante, Fundación COEPA, Alicante, 2006.

Treated wastewater reuse for green space irrigation in arid and semiarid regions

G. Badalians Gholikandi, M. Lashgari & P. Kiumarsi
Department of Water and Wastewater Engineering,
Power and Water University of Technology (PWUT), Iran

Abstract

This research investigated the feasibility of reuse of treated wastewater of the existing treatment facilities in Sistan and Baluchestan province of Iran for green space irrigation as a part of waste management in the form of a pilot study. After investigation of the environmental parameters and wastewater treatment conditions, two pilot plants in zahedan and zabol cities were designed and performed. On the basis of the results of qualitative tests of the soil, before and after establishment of the Zahedan Pilot, we can say that the soil quality has been improved. The values of soil EC, pH, SAR, and ESP prior to establishment of the pilot have been 27.8, 8.05, 44.73 and 38.87 respectively. After irrigation of the pilots for a period of 5 months the said values changed to 8.5, 8.34, 13.62 and 15.61 respectively which show the decrease in EC, SAR and ESP for 30, 30.5 and 40.2 percent respectively. After continuation of the irrigation for 12 months, the said values were decreased to 8.34, 7.89, 13.76 and 15.75 percent respectively. The values of soil EC, pH, SAR, and ESP prior to establishment of the Zabol pilot were 9.55, 8.35, 17.46 and 19.39 respectively. After irrigation of the Zabol pilots for a period of 5 months the said values changed to 5.05, 8.59, 7.35 and 7.81 respectively which shows the promotion of soil properties. After continuation of the irrigation for 12 months, the said values were changed to 3.93, 8.02, 7.91 and 21.82 respectively. The results showed a lack of any adverse effect on the soil quality and growth of the selected plants; the majority of the quality factors were approximately at an optimum level and only soil ESP had been significantly increased in the Zabol pilot. Two solutions are suggested for promotion of the treatment plants operation.

Keywords: waste management, treated wastewater, reuse, arid regions, green space irrigation, soil quality.

WIT Transactions on Ecology and the Environment, Vol 112, © 2008 WIT Press
www.witpress.com, ISSN 1743-3541 (on-line)
doi:10.2495/SI080331

1 Introduction

The ever-increasing populations in urban areas and the critical need for infrastructural facilities in developing countries for water resources and waste management have made the optimum use of suitable methods for overcoming the environmental and social crisis an inevitable necessity.

As one of the arid provinces of Iran, Sistan and Baluchestan Province has been faced with water deficit and environmental crisis for a long time.

It is evident that the careful dealing with some issues like safeguarding the public health, soil protection, protection of the irrigation and water supply facilities, use of nutrient content of the wastewater and the public satisfaction is essential towards implementation of such a movement.

The most important chemical parameters of irrigation water, which are effective on the crop growth, soil fertility, and the environment, are: total saline concentration, electrical conductivity, sodium absorption rate (SAR), toxic ions including boron, chlorine, and sodium, trace elements, and heavy metals including aluminium, beryllium, cobalt, fleuron, iron, manganese, molybdenum, lithium, selenium, tin, titanium, tungsten, vanadium, arsenic, cadmium, copper, lead, mercury, selenium, zinc, and also pH value, and nitrogen, bicarbonates and phosphorous contents. In connection with physicochemical quality of the treated wastewater reused in irrigation, the suggestions of FAO should be observed (Ayers and Westcott [2]).

Considering that the high sodium concentration adversely affects the plant growth, this factor has a significant importance.

The quality of the treated wastewater used for irrigation in arid regions (high temperature, low humidity, high evaporation), has a considerable importance. The physical and mechanical properties of the soil and also its strength, porosity and hydraulic conductivity, all are sensitive to the ion contents of the irrigation water. Another area that needs to be considered, is the effects of wastewater soluble salts on plant growth. The water soluble salts increase the osmotic pressure of soil water and the latter in turn results in increases in energy consumption used for water absorption. As a result, the perspiration increases and the growth becomes limited. Most of the plants are sensitive to the active salts in the soil water, which is effective on the osmotic potential.

The activities and reactions of the elements existing in the soil compounds are usually the result of the equilibrium between the soil minerals, organic materials, ferrous hydroxides, magnesium and aluminium and also the soil pH. The concentration of the heavy elements increase proportionally to the increase in soil pH (alkalinity). In acidic soils (pH = 4.2–6.6), some elements like cadmium, mercury, nickel and zinc become movable, hence arsenic, boron and chrome are partially movable and copper, lead, and selenium are hardly moved and/or have a very low movement velocity. In neutral or alkaline soils (pH = 6.7–7.8), arsenic and chrome are highly movable and boron, cadmium, mercury and zinc have a normal movement and copper, lead and nickel have a very low movement velocity. In the case of increase in soil pH from 5 to 8, the absorption of elements like cobalt, copper, manganese and zinc will increase. Soil pH is the most important effective factor on controlling the absorption of some elements like cadmium and zinc existing in the treated wastewater. The cation exchange capacity of the soil mainly depends on soil volume and type, organic materials and oxides of ferrous, manganese and aluminium. The high soil cation exchange capacity is one of factors which are highly affective on absorption of heavy elements by the soil without any adverse effect on the environment. Some heavy elements like cobalt, copper, manganese, nickel, lead, and zinc are capable to be absorbed by organic materials of the soil, as a result, form a resistive soluble and/or an insoluble compound. Ferrous and manganese oxides normally have an important role on absorption of elements and also their fixation in the soil. Other factors like planting method, duration of plant growth, and also the climatic conditions are also effective on absorption of the elements by the plant. Note that very low contents of copper, nickel, and other similar elements are useful for land fertility and plant growth, but they become toxic and/or may prevent growth in high concentrations (Gried [7]).

Different methods may be used for irrigation by reuse of the treated wastewater. The most important factors to be considered in selection of the appropriate irrigation system are labor, the required technology, topography of the region, type of vegetation, and the existing strategic and infrastructural facilities (Petygrove and Asano [15]).

The quantity of the irrigation water differs in different seasons and depends on the climatic condition and also type of vegetation and plant life. The required quality of the treated wastewaters must have some changes because the nutrients needed by the plant change during different seasons. The different sensitivity of plant species to the soil salinity is a very important fact and should be considered. Some plants are capable of absorbing more water in saline soils, and therefore resist well against the salinity. Another important factor is plant toxicity. The toxicity takes place in the plant and does not relate to the water deficit. The toxic ions including chloride, sodium, and boron are absorbed by water and moved in the plant. As a result of perspiration, the water content of the plant is decreased and the concentration of the ions is increased.

The degree of damage to the plant depends on the time, concentration of the toxic ions, plant sensitivity and the quantity of the absorbed water. In sprinkler irrigation, the chlorine and calcium ions are directly absorbed through foliage. The degree of the hazard is specially increased when the temperature is high and humidity is low. In addition to the above-mentioned elements, the micro nutrients (minor metals) may also cause plant toxicity. Also it should be pointed out that the majority of heavy elements are accumulated in surface soil and adversely affect the plant growth.

2 Methods

2.1 Basic study

Sistan and Baluchestan Province is located in southeast Iran between 25°, 3' and 31°, 28' north longitude and 58°, 47' and 63°, 19' latitude.

The province area is about 191,000 km^2 which Sistan Region forms 5% and the remaining 95% is Baluchestan. The provinces area is about one ninth of the country area.

Considering that the study was conducted in a special region, Sistan and Baluchestan Province, and Zahedan and Zabol regions have been selected for pilot studies, the investigation and recognition of the natural specifications of the said regions will have considerable importance [11, 12].

The geological formation of the province mainly includes sedimentary and igneous stones of the 3rd era and alluvial deposits of 4th era.

One of the major problems of the province is the improper soil conditions and its relatively low quality. The majority of soils of the province are alluvial soils with poor organic materials. The main source of these soils is alluvium and other settlements moved by the wind all belonging to 4th era. Marn, gypsum, silt, sand, and salt are the main components forming the soils of the province. The soils of the province are alluvial and, therefore, are fine particles and owe their life to Hirmand and Hamoon rivers. The soil erosion in the province is relatively high and this is the result of seasonal floods and heavy winds of the region [6].

Due to special geographic situations and desert conditions, the region has not got rich vegetation. The existing forest and other plant species mainly include Iranian screw bean, tamarisk, Pakistan screw bean, *Accacia bambulah*, marsh arrow grass, reed, wild cotton, cape tree (Khosravi [9]).

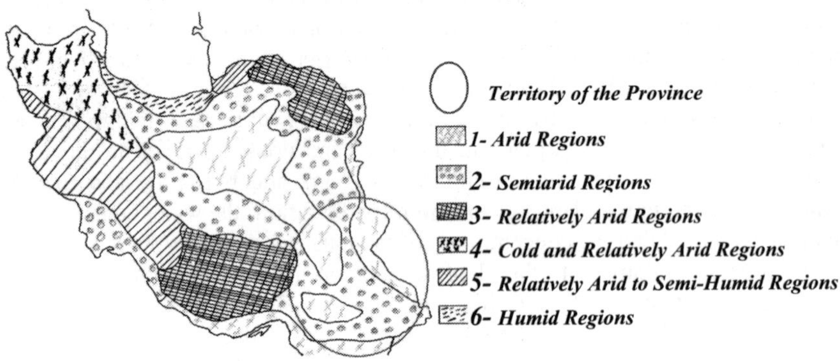

Figure 1: Climatic classification of the country.

As the province is located in a region with an arid and semiarid climate, the annual precipitation is very low and, due to its long distance from humidity resources and west rainfall, does not have rich surface and ground water resources (Bandarian [4]).

Considering the isothermal map of the province, the average annual temperature range of the province is minimum 16 and maximum 26. Only some records of late frost have been recorded in three meteorological stations: Zahedan, Chashmeh Ziarat and Mohammad Abad Koorin and the fruit bearing regions like Khash and Iranshahr are mainly free from frost problems (Parsi [14]).

The calm conditions are lower than all other regions of the country. For example, calm conditions in summer at Zahedan Station occur in less than 8% of cases and in the remaining 92%, the 120-day winds are blowing. Regional winds of the province focusing on 120-day Sistan winds.

From the point of view of the blowing limitation and durability, these winds are unique and are blowing from the northeast mountains of Iran to the southeast region of the country from early May to late September. The wind is very humid and fresh in the south Alborz hills and, after passing through the Kawir and Loot deserts, becomes hot and arid and, especially in the Sistan region, causes serious damage to the vegetation of drastic soil erosion. The accurate blowing duration of the wind is 131 days from May 10th, to September 17 in Sistan and in Zahedan is 117 days from May 11th, to September 5th.

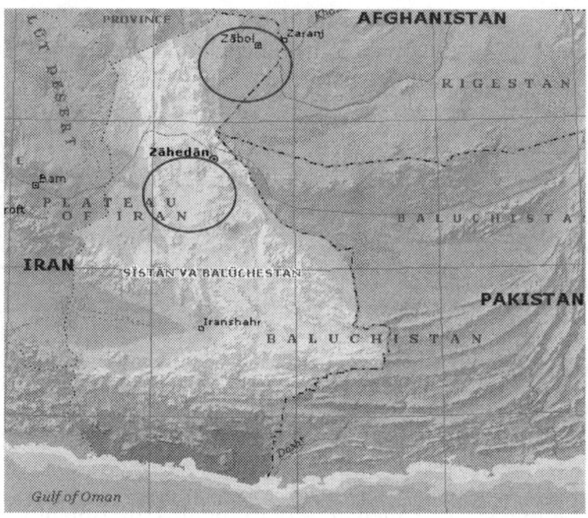

Figure 2: Geographic Situation of Zahedan and Zabol regions.

The number of dusty days in the province is more than almost all other regions in the country, this figure is about 180 days in Zabol.

The winds result in soil erosion and, consequently, destroy the valuable resources of the province. Also the 120-day winds cause some physical injury including 28% of blindness in the Kratit region (Khosravi [9]).

The annual precipitation of the province differs from a minimum of 65mm in Zabol to a maximum of 165mm in Khash Station; considering the average annual precipitation of the country (about 260mm) almost all stations have annual precipitation less than the average of the country.

February and March are the most humid months of the year (20–25mm) and June is the most arid month (0–5mm). The annual radiation in all synoptic stations of the province is more than 3000 hours, where the same in north cities of the country is 2100 hours.

By use of Copen method the province falls into arid desert climates with humid winters.

By use of Demarton method the I_a index of all points of the province were calculated and the province has an arid (I_a <10) climate.

One of the most important specifications of arid regions is little precipitation hence considerable evaporation. The province is not an exception and the evaporation and perspiration is less than the precipitation and in general the crop production is dependant on irrigation.

The lowest evaporation and perspiration occurs in December and January (40–50 mm) and the highest is in July and August (300–400 mm) (Sabeti [17]).

2.2 Pilot studies in Zahedan and Zabol

Considering the results of the investigation and basic information, the place for the pilot and plant species were selected and after preparation of the landscape, the seedlings were planted. The area of each pilot was 2000 m^2 with different dimensions which were established in Zahedan and Zabol wastewater treatment sites.

Figure 3: Zahedan pilot (3 months after planting).

Figure 4: Zabol pilot (3 months after planting).

2.2.1 The chemical quality of Zahedan and Zabol treated wastewater

The wastewater treatment system at Zahedan plant is activated sludge (extended aeration) built in a residential complex, east of Zahedan. The inlet wastewater volume is 8 L/sec and the outlet discharge is approx. 6 L/sec.

Table 1: Average chemical specifications of wastewater treated in Zahedan wastewater treatment plant.

pH	EC	Na	Ca	SAR	Cl (MEC)
7.56	7630	47.82	14.5	17.75	51

The wastewater treatment system of Zabol City is aerated lagoons (first stage) with settlement facultative ponds (second stage) and maturation ponds (third stage). The inlet wastewater volume is 12300 m^3/day and the discharge at the outlet is 9840 m^3/day.

Table 2: Average chemical specifications of wastewater treated in Zabol wastewater treatment plant.

pH	EC	BOD (mg/lit)	COD (mg/lit)	TDS (mg/lit)	SS (mg/lit)
8	4370	82	182	2578	118

Table 3: Chemical specifications of fresh water supplies for the homes.

pH	EC	Na	Ca	SAR	Cl (MEC)
7.3	6660	47	13.5	16.5	51

The comparison between the chemical specifications of the fresh water supplied to the homes and the treated wastewater show that the alkalinity, as well as calcium and magnesium contents of the treated wastewater is high.

2.2.2 Pilot plants

Generally, the soil used for landscapes and green spaces should be light soil with enough depth and suitable EC and pH; otherwise the soil amendment, fertilization and even soil replacement would be required. Prior to pilot establishment, the qualitative structure and classification of the soil was examined through drilling some holes in 0–30 cm, 30–60 cm and also in 1.5–2 m depth. Figure 5 shows the used irrigation systems in Zahedan and Zabol pilots.

The economic issues, climatic conditions, soil and water quality, the results of tests on the inlet and outlet wastewater and soil specifications of the regions are among important factors which should be considered in the selection of the plant species. Therefore, the selected species were those which adapted to the climatic conditions. The planted species are outlined in Figure 6.

Prior to irrigation with wastewater, the irrigation was done by the use of fresh water. In the first week, the irrigation took place on a daily basis and in the second week and thereafter the irrigation took place once every two days and the trend was continued up to one month considering the temperature and evaporation and perspiration rate. The irrigation by wastewater in Zahedan Pilot commenced two weeks after planting.

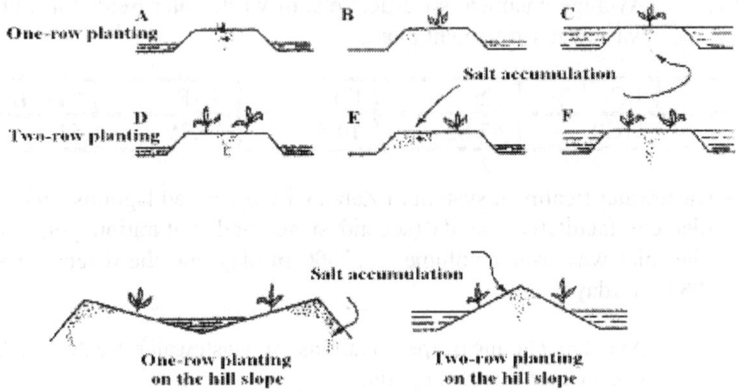

Figure 5: The schematic one and two-row planting in pilots.

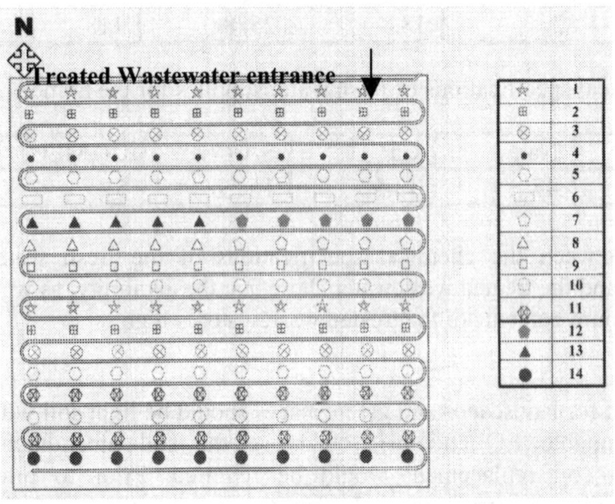

1-Eucalyptus
2-Olive Tree
3-Palm
4-Egypt Silk
5-Oleaster
6-Pistachio
7-Osh Tree
8-Pomegrante
9-Oleander Rosebay
10-Type of Pine
11-Ornamentary Palm
12-Marshmallow
13-Sour Orange
14-Juniper

Figure 6: Schematic plan of Zahedan and Zabol pilots.

3 Results

The results of our periodical investigations show the optimum growth and development of most of the planted species. From among the 200 plants only the growth of some pistachio plants was inappropriate; however, the phenomenon was observed only in 4 plants out of the total 20 pistachio plants.

Pistachio is among plants that need little water and in the case of continuous irrigation the plant will be damaged as a result of root rot. Some other factors were also effective in this regard including the failure in appropriate observation of the planting practice principles and root damage at the time of transplanting.

After renewal transplanting and change of the irrigation regime, the problem was removed. The results of soil tests of pilots prior and after operation are summarized in Figures 7 and 15.

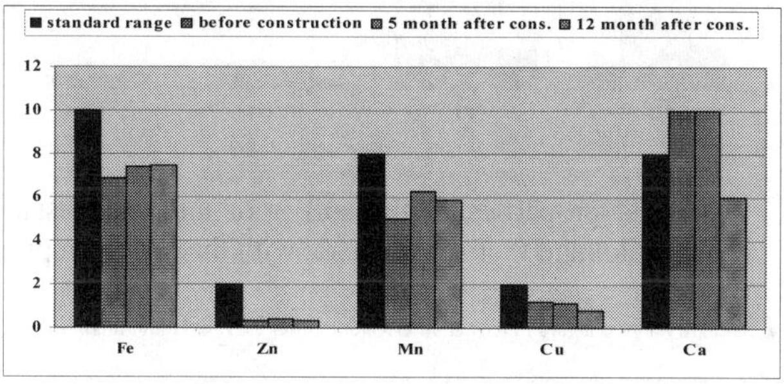

Figure 7: Physical properties of the soil prior and after establishment of Zabol Pilot and their comparison with the optimum properties.

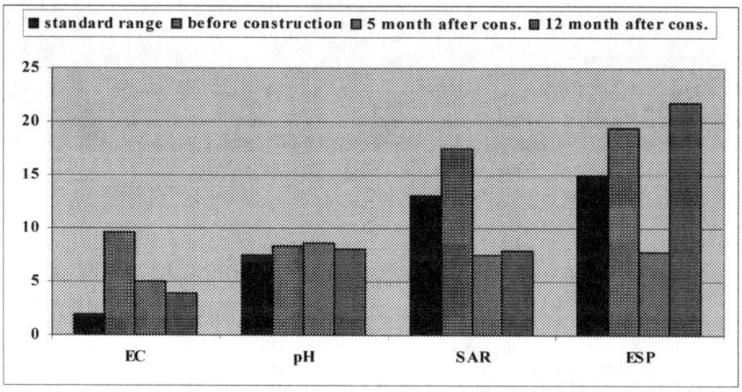

Figure 8: Chemical properties of the soil prior and after establishment of Zabol Pilot and their comparison with the optimum properties.

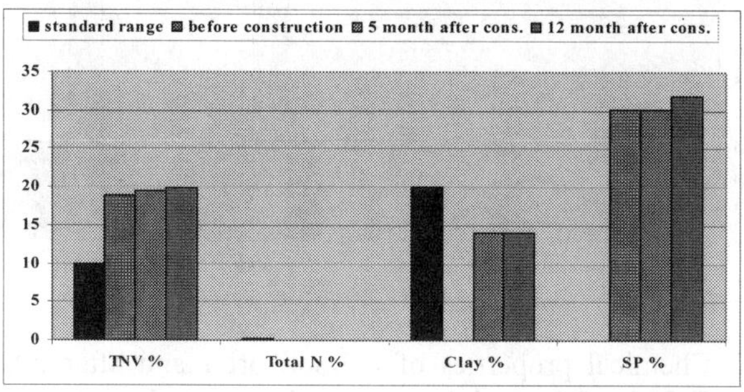

Figure 9: Physical properties of the soil prior and after establishment of abol Pilot and their comparison with the optimum properties.

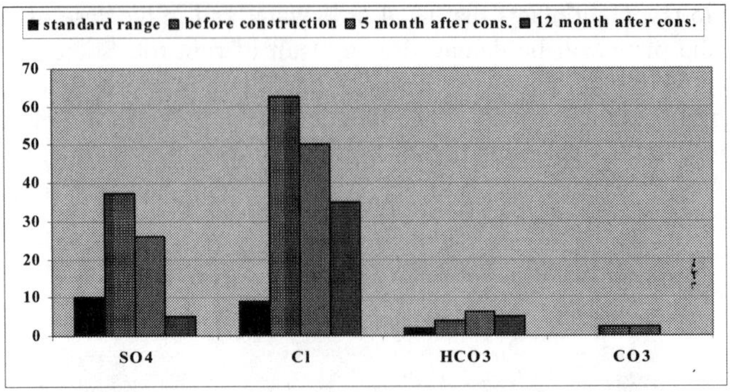

Figure 10: Chemical properties of the soil prior and after establishment of Zabol Pilot and their comparison with the optimum properties.

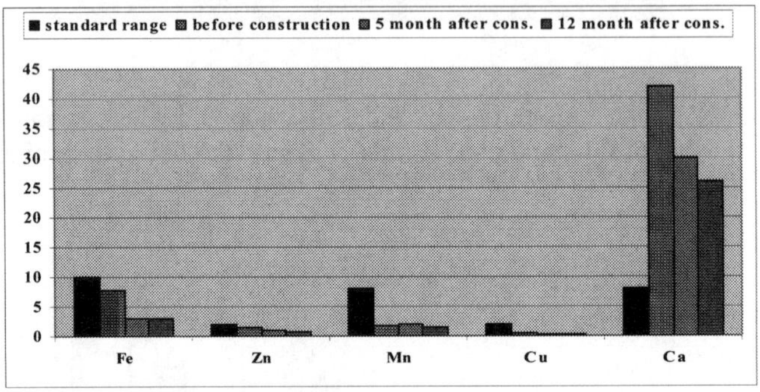

Figure 11: Physical properties of the soil prior and after establishment of Zahedan Pilot and their comparison with the optimum properties.

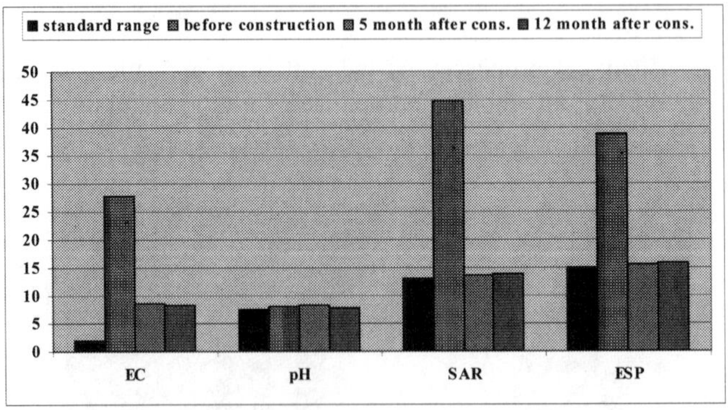

Figure 12: Chemical properties of the soil prior and after establishment of Zahedan Pilot and their comparison with the optimum properties.

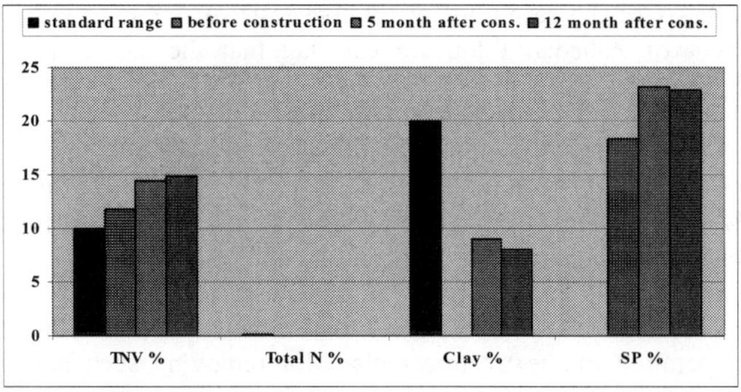

Figure 13: Physical properties of the soil prior and after establishment of Zahedan Pilot and their comparison with the optimum properties.

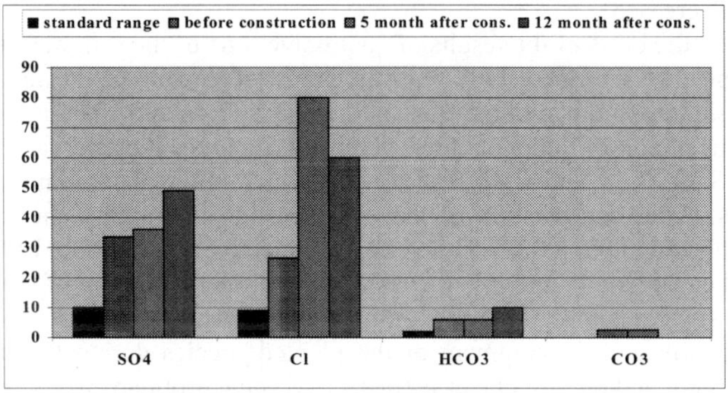

Figure 14: Chemical properties of the soil prior and after establishment of Zahedan Pilot and their comparison with the optimum properties.

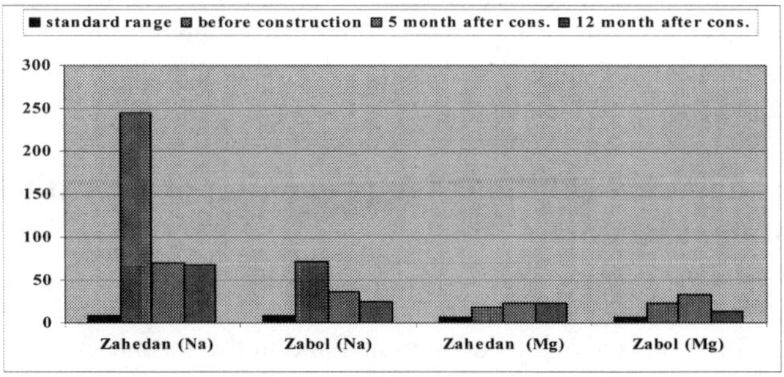

Figure 15: Chemical properties of the soil prior and after establishment of Zahedan and Zabol Pilot and their comparison with the optimum properties.

On the basis of the results of qualitative tests of the soil, before and after establishment of Zahedan Pilot, we can say that the soil quality has been improved. The values of soil EC, pH, SAR, and ESP prior to establishment of the pilot have been 27.8, 8.05, 44.73 and 38.87 respectively. After irrigation of the pilots for a period of 5 months the said values changed to 8.5, 8.34, 13.62 and 15.61 respectively which show the decrease in EC, SAR and ESP for 30, 30.5 and 40.2% respectively. It should be added that soil pH has not followed such trend and has been increased for 3.6% and it can be said that the increase in soil pH is the result of high alkalinity of the treated wastewater which is in turn the result of residuals of alkaline detergents used in households and this implies the improper operation of the treatment plant for removing such residuals. After continuation of the irrigation for 12 months, the said values were decreased to 8.34, 7.89, 13.76 and 15.75 respectively which, when compared with the standard ranges (EC < 2, pH $= 7–7.5$, SAR < 13, and ESP < 15), the decrease trend seems optimum (the pH value, however, has been increased after 5 months and then decreased).

Also, on the basis of the results of qualitative tests of the soil, before and after establishment of the Zabol Pilot, the values of soil EC, pH, SAR, and ESP prior to establishment of the pilot have been 9.55, 8.35, 17.46 and 19.39 respectively.

After irrigation of the pilots for a period of 5 months the said values changed to 5.05, 8.59, 7.35 and 7.81 respectively which show the promotion of soil properties. After continuation of the irrigation for 12 months, the said values were changed to 3.93, 8.02, 7.91 and 21.82 respectively; here, only ESP has been increased considerably and it is the result of improper operation of the treatment plant and sever qualitative changes of the treated wastewater.

The growth and development of the planted species during the pilots have been optimum and only in one case (pistachio) some problems were found which were removed after changing the irrigation regime (decrease of irrigation).

Considering the results of the 12 month study, we suggest soil amendment through promotion of its quality by use of gypsum and sulphuric acid together with leaching. Also some corrective actions should be done towards promotion of the wastewater quality at the outlet of the wastewater treatment plants either through promotion of quality of the fresh water supplies to household or upgrading the quality of the plants.

4 The suggested solutions for promotion of operations of the treatment facilities

Two solutions are suggested for promotion of the treatment plants operation; in the first solution the activated sludge model and in the second one the settlement pond is considered (Poepel [16]).

4.1 Solution I: Activated sludge facilities

4.1.1 First model

The solution is suitable for regions with hot summers and relatively cold winters. During the cold months, the nutrients must be removed and in the hot months the protection of the nutrients takes priority. In this solution the treatment facilities have some parallel routes which the volume of the aeration pond in the cold months is calculated at average temperature of 12°C, and the denitrification share of 40. For the hot months the calculations are done at 25°C. The value of the needed oxygen in the hot months is almost 40% of the said value than in the winter months.

We cannot use the procedure under which one of the routes removes the carbon products and other ponds act as reserve and clarification facilities, could not be used with this solution because the nitrification bacteria are not existent in the active sludge. Therefore, the ordinary procedure together with facilities for the removal of the nutrients in one of the routes should be maintained and in the other routes only the removal of carbon compounds must be done. During the growth season, the best way is to use the routes as follows:

Route I: Remove the nutrients (P, N, C),
Route II: Protection of the nutrients,
Route II: Reserve.

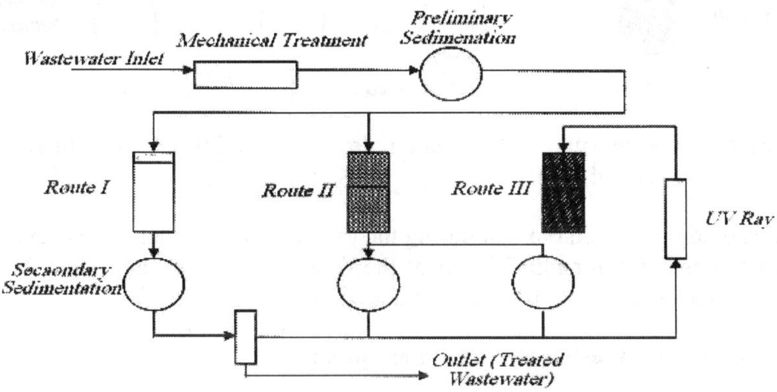

Figure 16: The schematic of exploitation from facilities in growth season – first model.

Among other advantages of this method is attainment to different concentrations of nitrogen and phosphorus by mixing the two flows.

4.1.2 Second model

In this method, the calculations of the needed facilities are done at a temperature of 12°C for cold months. By increasing the temperature up to about 20°C about 50% of the preliminary volume will be needed. As the decrease in oxygen content is relatively less, the aeration facilities must be used in all ponds. In this exploitation method shifting from conditions for protection of the nutrients to removal of the nutrients, due to low propagation of nitrification bacteria, has some problems. The water content of the returned sludge and also 100% of the returned solid content of the inlet flow result in relatively high sludge content containing low nitrification bacteria. The decrease in solid load for 50% is suggested for an increase in nitrification bacteria. If the returned solid content is 100%, then the surplus sludge should be stored for two weeks; here 52% of nitrification bacteria became available and if the returned solid load is 50% then the availability of the nitrification bacteria increases to 62%; where if the conditions are prepared for rapid growth of the bacteria, the removal of the nutrients is relatively complete.

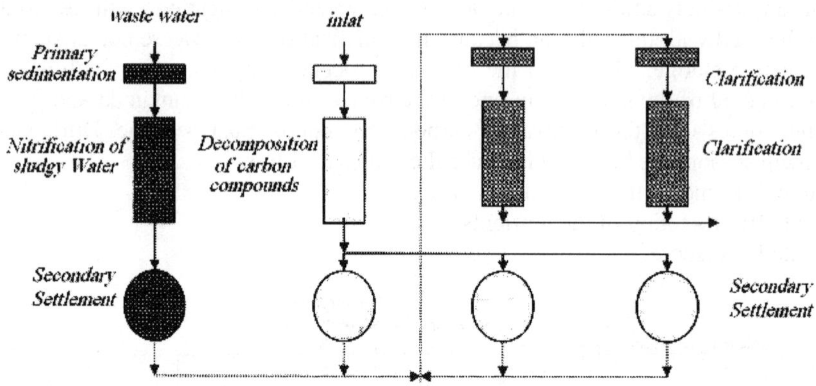

Figure 17: Schematic of exploitation from the facilities in growth season – second model.

It should be pointed out that during the growth phase of plants, a special route will be considered for nitrification of the dark sludgy water and, therefore, the nitrification bacteria are protected in the system.

4.2 Solution II: Wastewater treatment ponds

In this method, the solids are separated from fluid phase in two anaerobic ponds.

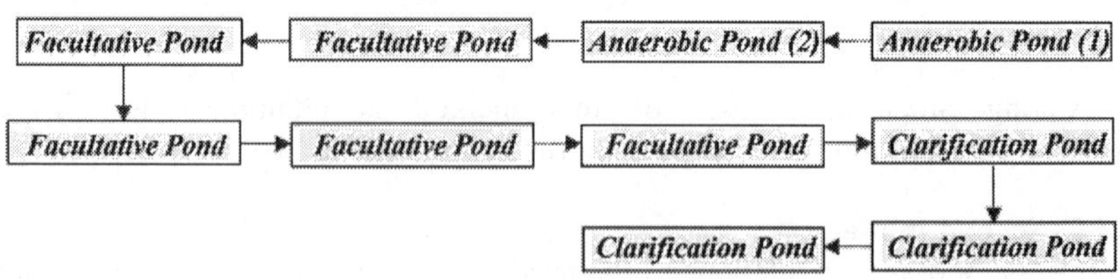

Figure 18: The schematic chart of wastewater treatment pond finally we suggest Accurate pilot studies for attainment to the best operation method of Wastewater treatment facilities.

For attainment to the target of the solution the special volume of the pond is considered 0.5 m^3/p and the pond depth is considered 4 m. The duration of the stay of the wastewater is 3 days, then the wastewater is discharged into optional aerobic ponds with 1.5 m depth. The duration of stay of the wastewater in these ponds will be 20 days. The clarification ponds are located after optional aerobic ponds which are used for eliminating microbial contaminators; the duration of the wastewater in these ponds is 5 days and their depth is 2 m. In the summer months unlimited use of wastewater of these ponds for irrigation is possible and the allowable range of coliform pollution is accessible.

5 Conclusions

Two pilot plants were designed and performed to investigate the impact of 14 kinds of selective plants irrigation using the effluent of wastewater treatment plants in Zahedan and Zabol, in arid region of Sistan and Baluchestan in southeast of Iran. According to FAO standards the quality variation in soil and plants growth is positive. Some solutions for the promotion of soil quality and also some related patterns to the adjustment and development of recent wastewater treatment plants are suggested. The results show the possibility of accessing water resources and waste management goals due to prevent contaminant diffusion in the area, environmental and social consequences and also decreasing the water shortage crises impact on the area.

Acknowledgements

Financial support was provided by the water and wastewater authority and also planning and budget organization of the province Sistan and Baluchestan. Special thanks to environmental protection and agricultural agency of the province for the support by laboratory works and information gathering.

References

[1] American Public Health Association (ed) (1976). Standard Methods for the Examination of Water and Wastewater. APHA, Washington D.C.

[2] Ayers R.S., Westcot D.W., (1985). Water quality for agriculture, FAO Irrigation and drainage paper 29, rev. 1, Rome

[3] Badalians Gholikandi G., Lashgari M., (2006). the final report on "Investigation of the Use of Treated Urban Wastewater for Irrigation of Green Space in Sistan and Balouchestan Province; Sample Pilots in Zahedan and Zabol Cities", Power and Water University of Technology (Shahid Abbaspour) University.

[4] Bandarian A., (1995). "Qualitative and Qualitative Assay of Water Resources in Sistan and Baluchestan Province", a research project sponsored by Sistan and Baluchestan University.

[5] Department-General of Groundwater (2004). Ministry of Water and Power Supply, Report on Study of Groundwater in Sistan and Baluchestan Province, Report No. 6.
[6] Department-General of Agriculture of Sistan and Baluchestan Province (1997). "Report on Agricultural Insight of the Province".
[7] Gried, A., (1999). Abwasserwiederverwendung in der landwirtschaft, WAR Schriftenreihe 116, Abwasserwiederverwendung in wasserarmen Regionen, Verein zur Foerderung des Institutes.
[8] Hosseinzadeh R., (1995). "Sistan 120-Day Winds", Geographic Researches Quarterly, issue No. 46.
[9] Khosravi M., (1989). "Adverse Bioclimatic Effects of the Natural Environment on Sistan Plain", Geographic Researches Quarterly, Issue No. 13.
[10] Office of Regional Planning, Management and Planning Organization (1988). "Study of Development of East of the Country", Volume II.
[11] Ministry of Agricultural Jihad, (2004), "Encyclopedia of Native Plants of Sistan and Baluchestan Province".
[12] Ministry of Agricultural Jihad (2004). "Climatic Studies of Sistan and Baluchestan Province".
[13] Office of Regional Planning, Management and Planning Organization (1988). "Study of Development of East of the Country", Volume II.
[14] Parsi B., (1997). "Study of Damaging the Agricultural Products by the Cold in Sistan and Baluchestan Province and the Suggested Controls", Department-General of Agriculture of Sistan and Baluchestan Province.
[15] Pettygrove A.S., Asano T., (1985). Irrigation with reclaimed municipal wastewater – a guidance manual. Lewis, Chelsea.
[16] Poepel (1999). Abpassung und Betrieb von Abwasserbehandlungsanlagen, War, TU Darmstadt.
[17] Sabeti H., (1995). "Bioclimatic Zoning of Iran (Study of Bioclimatic Zones of Iran), Tehran University Press, issue No. 1231.
[18] Shuval HI., Adin A., Fattal B., Rawitz E., Yekutiel P. (1986). Wastewater irrigation in developing countries – health effects and technical solutions. World Bank Tech Pap 51, Washington, D.C.

Integrating alternative water sources in urbanised environments

M. Schwecke, B. Simmons, B. Maheshwari & G. Ramsay
Co-operative Research Centre for Irrigation Futures,
School of Natural Sciences University of Western Sydney, Australia

Abstract

The need to use alternative sources of water for irrigation has arisen from the current shortages of potable water within Australia. As a result local governments are investigating alternative water sources (treated effluent, stormwater and groundwater) for irrigating urban playing fields and open spaces to cope with times of water shortages.

In this paper, using Manly Local Government as a case study, the quality of treated effluent, stormwater, groundwater and the receiving water was analysed, to determine the potential impacts of capturing and using these alternative sources on the local environment. Systems analysis was used to identify the connectivity between the various water sources.

It was observed that the injection of stormwater and effluent into the local water cycle could potentially pose a risk to the environmental, including ground and surface waters which are estuarine influenced systems. Stormwater was found to be of a variable quality and therefore requires constant monitoring to ensure no contamination events (such as road spill) enter the system. Groundwater salinity levels also required constant monitoring to prevent ingress of saline estuarine water. It also appears that as long as appropriate safeguards are put in place, the use of alternative water sources could help improve the local water cycle processes of urbanised catchments.

Keywords: environmental connectivity, environmental susceptibility, groundwater, stormwater, systems analysis, treated effluent, urbanised environments.

www.witpress.com, ISSN 1743-3541 (on-line)
doi:10.2495/SI080341

1 Introduction

Continuing drought conditions in much of Australia has led to an increase in the overall demand for water in many of the major cities and towns across the country. As a result of the increased water demand, water restrictions have been imposed by the state government in most parts of Australia. In the Sydney Metropolitan Area water restrictions have been in place since October of 2003 [1]. As these water restrictions limit the amount of potable water that can be used to irrigate local playing fields, local government agencies have started to examine alternative water sources such as treated effluent, stormwater and groundwater to irrigate their urban playing fields as well as the open spaces (such as parks, trees and garden beds).

Manly Council is one such local government agency which has started to explore the use of alternative water sources for irrigation on urban playing fields and open spaces. The Manly Local Government Area (LGA) is located approximately 11 kilometres (km) North-East of the Sydney Central Business District (CBD), within NSW. The LGA is approximately 15.14km² and includes 520 hectares of parks, reserves and other forms of open spaces representing 32 percent of the local area. Manly LGA is surrounded by 32.9 km of shoreline (Manly Lagoon to the North, Manly Beach to the East, and Middle Harbour to the South) [2] as seen in figure 1.

In response to increasing water shortages and bans on the use of potable supplies Manly Council and Manly Golf Club in 1998 [3] began investigation of alternative water sources to maintain their playing surfaces. This included the use of ground water, stormwater and treated effluent.

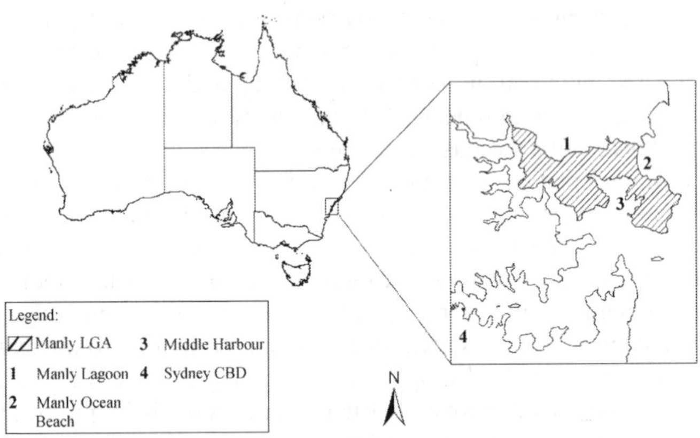

Figure 1: Location map of Manly [4].

2 Research design

In 2005 Manly Council commenced a joint research project with the University of Western Sydney and the Cooperative Research Centre for Irrigation Futures. The project was established to determine whether the use of alternative water sources such as stormwater, treated effluent and groundwater would affect the fragile environment of the receiving waterway (Manly Lagoon). The research study sites were Manly Golf Course and two surrounding playing fields - LM Grahams and Keirle Park, where the current use of groundwater and potable water will be supplemented with stormwater and treated effluent for irrigation use.

The current water sources for irrigation at the three sites vary, as do their size and water use (as seen in table 1).

Table 1: Water used to irrigate open spaces.

Open Space	LM Grahams	Keirle Park	Manly Golf Course
Ownership	Manly Council	Manly Council	Manly Golf Club
Area (Ha)	5.1ha	1.6ha	42.9ha
Source of Irrigation Water	Potable & Groundwater	Potable	Groundwater
Average Water Use (ML/Year)	7.21 ML/Year	3.74 ML/Year	218.4ML/Year
Average Water Use (per ha)	1.41 ML/ha/year	2.34 ML/ha/year	5.09 ML/ha/year

From table 1, it is clear that the largest consumer of water is Manly Golf Course which uses 5.09ML of water per hectare per year to irrigate its course. It can also be seen that Manly Golf Course currently only uses one source of water for irrigation – groundwater. This source however has started to become unusable as over extraction appears to be drawing saline water from the nearby lagoon into the groundwater aquifer.

Manly Council has commenced a project in co-operation with Sydney Water and Manly Golf Club to utilise treated effluent at the nearby treatment plant and pipe this recycled water to the playing fields for irrigation use. As the treated effluent supply to the three study sites is still under the design phase, the quality data of the treated effluent was unavailable. On the other hand, the groundwater at the golf course and LM Grahams has been monitored for quality. The golf club has plans to use the stormwater entering its course, therefore past stormwater data gathered by Manly Council was sourced and analysed in the present study.

Between 2006 and 2007, the groundwater which Manly Golf Club uses to irrigate its golf course and Manly Council to irrigate its playing field (LM Grahams) was monitored for the quality of the groundwater. The main purpose of the monitoring was to see if the electrical conductivity (EC) (a measure of dissolved salt concentrations) levels were increasing with the use of the groundwater for irrigation.

In addition to the groundwater the receiving waterway Manly Lagoon was monitored to assess its water quality over a one year period (2006). Manly lagoon is one of the most polluted coastal lagoons on the east coast of Australia [5]; it is for that reason a very sensitive and fragile environment. The quality data from the lagoon was then compared to the Australian and New Zealand Environment Conservation Council (ANZECC) guidelines for the protection of aquatic ecosystems [6] to determine the health of the water way.

Stormwater data around Manly Golf Course and LM Grahams Reserve was collected through Manly Council's water cycle management team which have been monitoring the stormwater in Cemetery Creek (stormwater system which runs through Manly Golf Course) since 2005. The stormwater data was analysed for quality and also assessed against the irrigation guidelines [7].

Treated effluent to be used to irrigate the three sites, is regularly monitored by Sydney Water. The treated effluent is produced through a tertiary membrane treatment facility located within North Head Sewerage Treatment Plant (STP) one of the largest STP within Sydney, and which is owned and operated by Sydney Water [1]. Since the treated effluent is monitored by Sydney Water, water quality data for the effluent were not available for the present analysis. However, depending on the treatment, treated effluent can still retain high levels of nutrients and salt and needs to be accounted in the total water cycle management.

Systems analysis techniques allowed the formulation of detailed pictures of the current situation, through understanding the processes, structures and relationships with the environment [8]. Flood [8] explains that systems enable us to appreciate as well as 'sense out' the connections to the wider environment. Through utilising systems tools it allows the issues surrounding the use of these alternative water sources (treated effluent, stormwater and groundwater) to be investigated further. A picture of the current system was created to assess the response to different inputs and uses and to identify if the alternative water sources and the receiving environment could potentially contaminate one another.

3 Results

3.1 Systems analysis

In analysis of the Manly Golf Course system (incorporating LM Grahams Reserve and Keirle Park) the interconnections between the various alternative water sources were identified. Figure 2 illustrates these connections between the various alternative water sources within the system and identifies external influences that could cause potential problems to the receiving environment (Manly Lagoon). An example of this might involve the irrigated treated effluent overflowing into the lagoon. The effects that might occur could include eutrophication, loss of biodiversity and/or contamination. The systems diagram also illustrates the ease that pollutants can flow into the various components of the system if appropriate safeguards are not put into place.

Systems analysis was used to identify the potential problems that could affect the lagoon system including: increased nutrients in the soil (nutrient imbalance);

increase in pollutants entering the lagoon water body; and salinity impacts on the soil and groundwater system. Although these problems/impacts from the use of alternative water sources in this fragile system were deemed as a minor risk to the system. However, if the risk is not monitored properly, the consequences to the receiving environment would be amplified and may impact of sustainability of the water resources in the area. However, of particular note is the connectivity between the stormwater, lagoon and groundwater which indicates that contamination can be transferred to groundwater either by overland flow or by groundwater ingress.

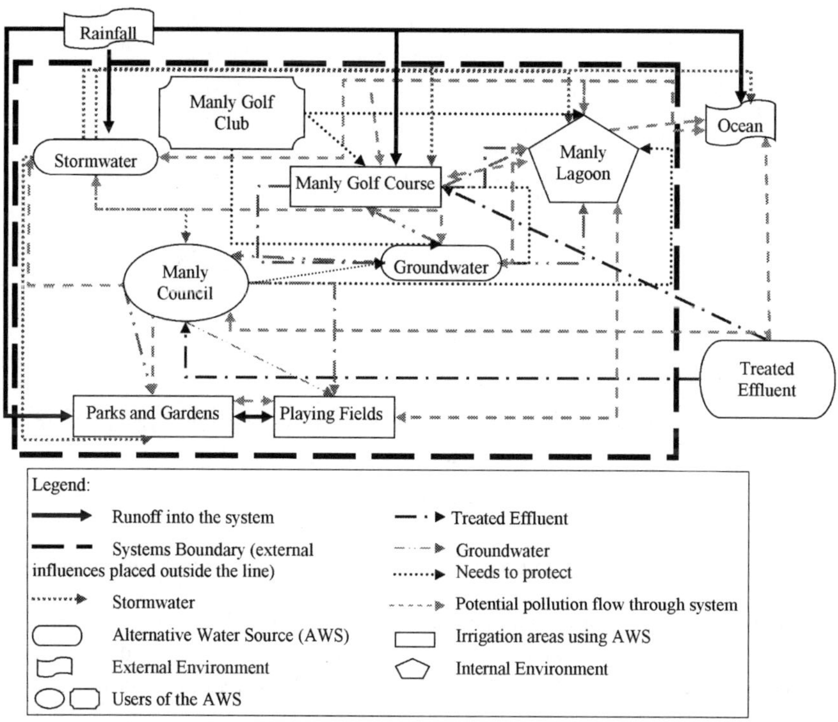

Figure 2: Manly Golf Course/Manly council water system.

3.2 Groundwater Monitoring

Monitoring of the groundwater that Manly Golf Club uses to irrigate the golf course in 2004 and 2006 showed that the EC levels were increasing the closer the location of the monitoring site to the Lagoon (as seen in figure 3). This increase suggests that the saline waters of Manly Lagoon are slowly being drawn into the groundwater system.

Monitoring in 2006 of the salinity profile bores in Manly Golf Course, revealed that the saline water which in June of 2006 could be seen at a depth of only 4m (EC levels increased from 1006μS/cm at 4m to 3850μS/cm at 4.5m), had encroached extensively in the six months between testing. In November of

2006 at the same salinity profile bore at a depth of only 0.5m the EC levels were already as high as 2032μS/cm and steadily increased with bore depth with EC values of 4060μS/cm at 10m in bore depth. If Manly Golf Club continues to overdraw on the groundwater the saline water from the lagoon will continue to increase the EC levels in the aquifer system hence making it unsuitable for irrigation use.

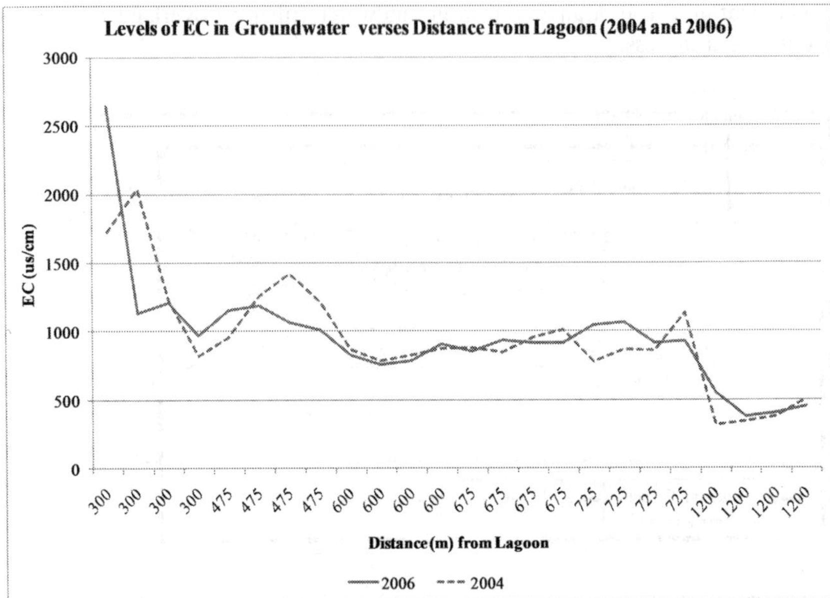

Figure 3: EC Levels in groundwater verses distance from Manly Lagoon.

Manly Council's LM Grahams Reserve bore, which although only established in early 2006, has seen an increase in EC levels (as seen in bold in table 2) between April 2006 and April 2007. An ongoing investigation and monitoring

Table 2: EC Levels in LM Grahams bore 2004–2008.

LM Graham Extraction Bore (EC – μS/cm)	
Date	Bore
7/04/2006	**1092**
6/09/2006	**947**
26/04/2007	**1523**
3/01/2008	*1431*
10/01/2008	*1450*
18/01/2008	*1226*

regime by Manly Council commenced in late 2007 in an attempt to determine why the EC levels were increasing. The investigation by Council determined that the LM Graham bore had been over extracted by the councils own parks and garden team for use elsewhere in the LGA [9]. This has led Council to reduce the amount of groundwater being extracted from the bore from 132,330L/week to 27,500L/week. Since that time the EC levels in the groundwater appear to be deceasing (as seen in italics in table 2) [9].

3.3 Stormwater data collection

Stormwater has been monitored by Manly Council on an irregular basis since 2005 from the Cemetery Creek Catchment which runs through Manly Golf Course into Manly Lagoon. Monitoring data obtained from Manly Council indicates that Stormwater running through the Cemetery Creek Catchment has low EC levels (all below 376 µS/cm). However, the Total Nitrogen (TN) and Total Phosphorus (TP) levels were elevated, regularly falling outside of the irrigation guidelines for Long Term Trigger Values (LTV) (as seen in table 3). The TN levels were compliant 95% of the time with the irrigation guidelines [7] except for a few events where the TN levels increased to as high as 182mg/L.

Table 3: TP and TN levels verses ANZECC [7] guidelines.

TP and TN levels within Cemetery Creek Stormwater				
	Min (mg/L)	Max (mg/L)	Mean (mg/L) N =33	ANZECC [7] Guidelines (LTV) (mg/L)
TP	0.1	0.85	1.1	<0.05
TN	0.36	182	4.9	5

3.4 Manly Lagoon monitoring

Monitoring of Manly Lagoon over a one year period (2006) indicated that this lagoon was an exceedingly degraded environment. The monitoring identified that the lagoon was severely stressed with elevated levels of heavy metals, and during rain events there is evidence of faecal coliforms entering the waterway. In addition to the heavy metals and faecal coliforms there were also elevated levels of TN and TP.

The lagoon's overall water quality health was also considerably degraded. The basic parameters which can give a straightforward indication of the current situation including pH, dissolved oxygen (DO), turbidity, electrical conductivity and temperature. The data collected during the study showed that the lagoons DO levels were outside the ANZECC [6] water quality guidelines for the protection of aquatic ecosystems the majority of the time (see table 4).

Comparing the water quality of Manly Lagoon with the ANZECC [6] water quality guidelines for the protection of aquatic ecosystems illustrated the fragile state of the receiving environment that could be further influenced/ degraded through the introduction of alternative water sources into the current system [10].

Table 4: DO levels recorded in Manly Lagoon 2006–2007.

Dissolved Oxygen (DO) levels recorded in Manly Lagoon 2006-2007			
Lowest level recorded	Highest level recorded	Mean DO level N=156	ANZECC [6] Guidelines for DO
1.1mg/l	8.8mg/l	4.7mg/l	>6mg/l

4 Discussion

An increase in the EC levels within the bores in Manly Golf Course may be due in part to the fact that the golf course has unsustainably managed their extraction, often extracting more water out of the bores than the aquifer can recharge. Consequently the groundwater supply cannot meet all the irrigation water needs of the golf course and for 107 days per year, the irrigation demand for the course exceeds the sustainable flow rate for the groundwater aquifer to recharge [11,12].

The increase in the EC levels at both Manly Council Recreational Parks and Manly Golf Club has forced them to think about supplementing their groundwater supplies for irrigation with alternative water sources such as stormwater and treated effluent.

The elevated nutrient readings in the stormwater illustrates the uncontrollable nature of stormwater quality that will have to be regularly monitored for quality control and have appropriate safeguards put in place to prevent TN and TP nutrients moving through the system into the fragile Manly Lagoon.

It is clear through comparing the quality of Manly Lagoon with the ANZECC [6] guidelines for the protection of aquatic ecosystems, that the lagoon was a very degraded receiving environment. Therefore the proposed alternative water sources for the irrigation of Manly Golf Course, LM Grahams and Keirle Park must not impact on the lagoon through runoff nor seepage into the lagoon. Through utilising system analysis techniques the connectivity of the Manly Lagoon system in relation to the hydro-ecological system were explored as a means of identifying if any problems may arise from the introduction of the different sources of water into this already fragile system.

5 Conclusion

Research over the past two years to obtain data relating to the quality of three types of alternative water sources: stormwater, groundwater and treated effluent for irrigation on three urban spaces was undertaken. Results from monitoring and data collection on both the alternative water sources and the receiving environment (Manly Lagoon) indicate that the quality of the alternative water would pose a risk to the local environment including the estuarine influenced groundwater and surface water systems surrounding the three study sites. The research illustrated that the hydro-ecological system surrounding the Manly area is quite complex and fragile, and as such needs to be constantly monitored to assess whether the system has adapted positively to the introduction of these new sources of water.

The research also indicated that there needs to be appropriate safeguards in place in order to minimise the variability of the quality of these water sources such as the stormwater to ensure no contamination events enter Manly lagoon. If these safeguards are installed and regular monitoring of the systems adaption to the new water sources are in place these alternative water sources could improve the availability of irrigation water and help improve the local water cycle process of urbanised catchments.

References

[1] New South Wales Government, *2006 Metropolitan Water Plan*, New South Wales Government: Sydney, pp. 17–22, 29–50, 2006.

[2] Manly Council. *2006-2009 Management Plan. Adopted 19 June 2006*, Manly Council: Manly, pp. 90–95, 2006.

[3] Morris, C., *Water Supply Needs. Formal Discussion Document*. GHD: Sydney, 1998.

[4] Galloway, M., *Urban Stormwater Pollutant Concentrations, Annual Loads and Current Treatment Train Success in Three Sydney Sub-catchments*. The University of New South Wales: Sydney, p. 10, 2005.

[5] Manly Environment Centre., *The Restoration of Manly (Queenscliff) Lagoon website*, Manly, http://www.mec.org.au/water/index.html .

[6] Australian and New Zealand Environment and Conservation Council (ANZECC). *Australian and New Zealand Environment and Conservation Council Guidelines for the Protection of Aquatic Ecosystems*, Australian and New Zealand Environment and Conservation Council: Canberra, 2000.

[7] Australian and New Zealand Environment and Conservation Council (ANZECC). *Australian and New Zealand Environment and Conservation Council Guidelines for Irrigation*, Australian and New Zealand Environment and Conservation Council: Canberra, 2000.

[8] Flood, R. L., *Rethinking the Fifth Discipline: Learning within the Unknowable*, Routledge: London, 1999.

[9] Galloway, M. Personal Communication, 10 January 2008, Catchment Project Officer, Manly Council, Manly, Australia.

[10] Schwecke. M., Understanding Issues of Irrigating Urban Playing Fields with Alternative Water Sources through Systems Approach. *Proc. of the 11th Int. Conf. on Environmental Research Event*, eds. P. Pulley, J. Carmody & K. Venkatraman, Environmental Research Event 2007 Organising Committee: Cairns, Australia, pp. 235–243, 2007.

[11] Patterson Britton and Partners. *Manly Golf Course Water Storage, Soil and Groundwater Investigation Including Implications and Way Forwards. Issue no.2 April 2005*, Patterson Britton and Partners Pty Ltd: North Sydney, 2005.

[12] Bradbury, M., *Minutes from Meeting with Manly Golf Club and Manly Council to discuss water source options for Manly Golf Course*. Manly Golf Club Ltd: Manly, 2003.

Filter and emitter performance of micro-irrigation systems using secondary and tertiary effluents

M. Duran-Ros[1], J. Puig-Bargués[1], J. Barragán[2], G. Arbat[1] & F. Ramírez de Cartagena[1]
[1]*Department of Chemical and Agricultural Engineering and Technology, University of Girona, Spain*
[2]*Department of Agricultural and Forestry Engineering, University of Lleida, Spain*

Abstract

The performance of four filtration systems (sand, screen, disc and a combination of screen and disc) and six emitter types (four pressure compensated and two non-pressure compensated), using secondary and tertiary effluents from a wastewater treatment plant, was studied for 1000 h. Only sand filtration significantly reduced turbidity and suspended solids. The best emission uniformity was obtained by the emitters placed after the sand filter and the screen filter with the secondary and tertiary effluent, respectively. On the other hand, emitters that operated with disc filters showed the worst emission uniformity for both effluents. Emitter type P2 was the only one achieving values of emission uniformity higher than 90% with all filtration systems and effluents except the screen filter and the tertiary effluent.
Keywords: wastewater, drip irrigation, filtration, clogging.

1 Introduction

The use of effluents in agriculture is a viable alternative in areas where water is scarce or there is strong competition for its use. The best way to apply effluents, from public health and environmental points of view, is micro-irrigation [1]. The main problem when using effluents in drip irrigation systems is emitter clogging [2] because the reduction of emitted flow affects water distribution and,

WIT Transactions on Ecology and the Environment, Vol 112, © 2008 WIT Press
www.witpress.com, ISSN 1743-3541 (on-line)
doi:10.2495/SI080351

consequently, yields [3]. Besides, clogging of filters and emitters makes micro-irrigation system management difficult. For this reason, several researchers have studied micro-irrigation system performance using different effluents [2, 4, 5]. However, the high variability observed in the results has made it necessary to increase the number of experiments with different effluents and irrigation equipment.

The objectives of this study were to analyse the performance of four filtration systems and six emitters when using two effluent qualities.

2 Material and methods

Secondary and tertiary effluents from the wastewater treatment plant (WWTP) of Celrà (Girona), which treated wastewater from the urban and industrial areas of this village, were used in the experiments. Secondary effluent was collected at a settling tank placed after a biological reactor-type oxidation ditch and operated by a sludge process. Tertiary effluent was obtained by filtration of the secondary effluent through a disc filter with a 130 μm filtration level and disinfection by ultraviolet radiation.

The performance of four filtration systems was studied. The first system was formed by two sand filters in parallel, both filled with 175 kg of sand as a single filtration layer. The effective diameter of the sand (screen opening that retains 90% of the sand) was 0.40 and 0.27 mm for the experiments with secondary and tertiary effluents, respectively. The uniformity coefficient (relationship between screen openings that retain 40% and 90% of the sand) was of 2.41 in the experiments with secondary effluent and 2.89 in those with tertiary effluent. The second filtration system had two disc filters in parallel, both with a filtration level of 130 μm. The third filtration system consisted of one 120 μm screen filter. The fourth filtration system consisted of one screen filter followed by two disc filters in parallel, with the same characteristics as the filters used in systems 2 and 3. The performance of six different emitters, whose main characteristics are shown in table 1, was also assessed.

Each filtration system supplied water to 24 drip-lines with a length of 87 m. Six types of drip-lines were used, each one with a different emitter and with four replications of every drip-line. Two experiments were carried out, both lasting 1000 h. The first experiment took place in the summer of 2005 with secondary effluent, while the second, during the summer of 2006, used tertiary effluent. In order to control and monitor the micro-irrigation system, a supervisory control and data acquisition (SCADA) system was used, which allowed both continuous collection of filter performance data and irrigation scheduling [6].

Water samples at filter inlets and outlets were taken periodically to characterize the effluents and to determine the effect of filtration on pH, electrical conductivity (EC), turbidity, total suspended solids (TSS), dissolved oxygen (DO) and particle number. With the values of every parameter at the filter inlet (N_o) and outlet (N), the removal efficiency (E) achieved in the filters was calculated with eqn. (1):

$$E = \frac{N_o - N}{N} \cdot 100 \tag{1}$$

Table 1: Main tested emitter and drip line characteristics, according to manufacturer specifications.

Characteristic	Emitter					
	UN	RM	P2	P8	TO	TI
Nominal flow (l/h)	2.30	2.30	2.00	8.50	1.75	2.00
Nominal pressure (kPa)	50-400	50-400	50-400	50-400	100	100
External diameter (mm)	17.0	17.0	16.0	16.0	16.6	16.1
Distance between emitters (m)	0.40	0.75	1.00	1.00	1.50	0.75
Drip line flow (l/h)	499	267	174	740	102	232
Number of emitters	217	116	87	87	58	116
Flow exponent x	0	0	0	0	0.48	0.46
Discharge coefficient K	2.30	2.30	2.00	8.50	0.58	0.69
Pressure compensation	Yes	Yes	Yes	Yes	No	No
Manufacturer variation coefficient	< 3%	< 3%	< 3%	< 3%	< 3%	< 3%

Emission flow uniformity (*UE*) was evaluated seven times in every experiment using the Merriam and Keller [7] method, modified by Vermeiren and Jobling [8]. Using this method, two contiguous drippers were selected in four drip lines (with the same emitter type and filtration system) at four locations in each emitter line (at the beginning, at 1/3 of the length, at 2/3 of the length and at the end of the emitter line). The working pressure at each location was measured by means of a digital manometer (± 0.07% accuracy). The water delivered for each selected emitter was collected for five minutes to measure the flow of the emitters. The data obtained from field measurements were used to calculate the percentage of totally clogged emitters of the sample and also the emission uniformity by means of eqn. (2):

$$UE = \frac{q_{25}}{\overline{q}} \cdot 100 \tag{2}$$

q_{25} being the average emitted flow of 25% of the emitters with the lowest flow rate (l/h) and $\overline{q}$ the average emitted flow of all the measured emitters (l/h). On the other hand, pressure uniformity (U_p) was computed as:

$$U_p = \left(\frac{p_{25}}{\overline{p}}\right)^x \cdot 100 \tag{3}$$

where p_{25} is the average pressure of 25% of the emitters with the lowest pressure (kPa), $\overline{p}$ the average pressure of all the tested emitters (kPa) and x the emitter flow exponent.

3 Results and discussion

3.1 Filtration system performance

Table 2 shows the values of the physical and chemical parameters at the filter inlet during the experiments. Both effluents had, according to the Bucks *et al.*

Table 2: Mean and standard deviation of physical and chemical parameters of secondary and tertiary effluents. For every parameter, different letters mean significant differences (P<0.05).

Effluent	Parameter					
	pH	CE (dS/m)	DO (mg/l)	Turbidity (FTU)	TSS (mg/l)	Particles/ml
Secondary	7.50±0.08[a]	5.43±0.80	2.79±0.57[a]	6.23±2.25	10.19±3.07	27559±12533
Tertiary	7.34±0.08[b]	5.11±1.10	1.78±0.17[b]	4.12±2.79	6.53±2.87	37111±21385

classification [1], a minor physical clogging hazard by total suspended solids and a moderate chemical clogging hazard with regard to pH. Only pH and dissolved oxygen were significantly greater (P<0.05) in the secondary effluent than in the tertiary one. The fact that dissolved oxygen was 1 mg/l smaller with tertiary effluent revealed that this effluent had more organic contamination than the secondary effluent. The explanation could be in the variability of effluents, because the experiments were not simultaneous. EC, turbidity and TSS were greater with secondary effluent, but showed no statistical differences with tertiary effluent. The tertiary effluent had a particle count 36% higher than the secondary one.

No statistical differences (P>0.05) were found between the effluent used by the different filtration units. However, statistical differences (P<0.05) were found between sampling days, which indicates the variability that the same effluent could have throughout the experiment.

Removal efficiencies achieved with both effluents by the different filtration systems are shown in table 3. It should be pointed out that sand filter reduced turbidity from secondary and tertiary effluents 57% and 66%, respectively, and total suspended solids, 47% and 66%, respectively. The performance of sand filters in turbidity and TSS removal was significantly different (P<0.05) from the other filtration systems, which only had slight, or even negative removal efficiencies, probably due to detachment of solids from the filter cake [9]. The low reductions in TSS achieved by screen and disc filters agree with those observed by other authors [2, 10, 11]. On the other hand, sand filter was the only one that reduced the number of particles, but without significant differences with the increments produced in other filters, as Adin and Alon [9] observed in screen filters.

The 5% removal efficiency of dissolved oxygen using a combination of screen and disc filters with tertiary effluent was significantly higher than that achieved with the other filtration systems.

3.2 Emitter performance

As the pressure uniformity (U_p) was above 90% in the different irrigation units for both studied effluents, the pressure distribution along drip lines was correct. Taking into account that the manufacturer's coefficients of variation were smaller than 3% (table 1), emission flow uniformity variations can only be produced by emitter clogging. Fig. 1 shows *UE* at 1000 h of irrigation in function of effluent, filtration system and emitter.

Table 3: Mean and standard deviation of removal efficiency (%) of the different effluent parameters by filtration system. Negative values show a parameter increment. For each parameter and effluent, a different letter means significant differences ($P<0.05$) among filtration systems.

Parameter	Effluent	Filtration system			
		Sand	Screen and disc	Disc	Screen
pH	Secondary	0.25 ± 0.57	0.32 ± 0.53	0.18 ± 0.31	0.57 ± 1.95
	Tertiary	-0.08 ± 0.94	0.43 ± 0.78	-0.26 ± 0.59	-0.22 ± 0.72
CE (dS/m)	Secondary	-0.29 ± 0.30	-0.28 ± 0.23	-0.37 ± 0.47	-0.16 ± 1.03
	Tertiary	0.17 ± 1.13	-0.07 ± 0.96	-0.33 ± 0.52	-0.41 ±0.74
DO (mg/l)	Secondary	0.49 ± 4.59	-0.12 ± 4.45	-1.17 ± 7.24	-2.69 ± 8.61
	Tertiary	-2.24 ± 13.59 [b]	5.07 ± 10.00 [a]	-2.32 ± 8.65 [b]	0.23 ± 7.83 [b]
Turbidity (FTU)	Secondary	57.57 ± 21.97 [a]	1.69 ± 11.16 [b]	-10.46 ± 13.95 [b]	-1.64 ± 15.72 [b]
	Tertiary	66.38 ± 20.23 [a]	12.42 ± 23.53 [b]	3.87 ± 24.58 [b]	7.14 ± 26.01 [b]
TSS (mg/l)	Secondary	47.30 ± 39.59 [a]	-0.46 ± 27.89 [b]	-0.40 ± 17.38 [b]	-0.19 ±22.51 [b]
	Tertiary	66.63 ± 14.22 [a]	8.48 ± 18.36 [b]	3.32 ± 31.29 [b]	-2.73 ±23.43 [b]
Particles/ml	Secondary	17.13 ± 52.58	-68.98 ± 158.45	-81.68 ± 204.4	-39.17 ± 65.76
	Tertiary	6.12 ± 51.63	-23.98 ± 100.91	-38.81 ± 72.59	-17.79 ± 95.37

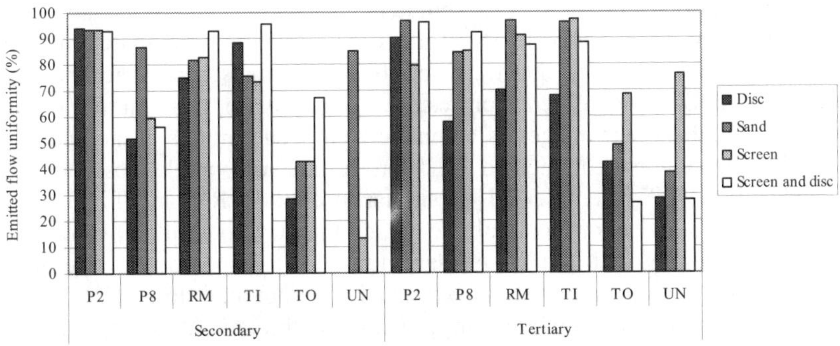

Figure 1: Emission flow uniformity regarding effluent, emitter and filtration system after 1000 h of operation.

The variation of *UE* due to the emitter type was greater than the variation caused by the filtration system by both secondary and tertiary effluents. While the best average *UE* with all six types of emitter was achieved by the sand filter with secondary effluent (77.5% ± 18.1) and by the screen filter with tertiary effluent (82.9% ± 10.5), the worst was found by the disc filter with secondary effluent (56.3% ± 36.8) and tertiary effluent (59.4% ± 21.9). Greater clogging of emitters using effluents filtered through disc filters has been observed previously in other secondary effluents [5].

Table 4 presents the *UE* after 1000 h of operation for every emitter and effluent, considering the average value with the four filtration systems.

Table 4: Mean and standard deviation of *UE* of every emitter after 1000 h of irrigation with secondary and tertiary effluents.

Effluent	Emitter					
	P2	P8	RM	TI	TO	UN
Secondary	93.30±0.44	63.51±15.84	83.10±7.30	83.22±10.69	45.25±15.94	31.57±37.52
Tertiary	90.47±8.01	79.96±15.11	86.23±11.54	87.45±13.54	46.50±17.27	42.69±22.94

Emitters P2, RM and TI achieved the best *UE* with both effluents. Emitter P2 stood out with an *UE* greater than 90%. In contrast, emitters TO and UN had the smallest *UE*, being lower than 45% with emitter UN. Nevertheless, emitter UN showed an acceptable performance with the sand filter and the secondary effluent (*UE* of 85%) and with the screen filter and the tertiary effluent (*UE* of 76%), which reveals the importance of the combination of filter, emitter and effluent. The low *UE* observed with emitter UN could be explained by the fact that this emitter had a higher number of emitters per drip line, as shown in table 1. Thus, the drip lines with emitter UN had higher amounts of water and higher particle quantities, increasing the emitter clogging probability. Besides, the formation of a thick film at the end of the drip lines, which was observed with all filters and effluents, penalized the *UE* values of emitter UN, because the Merriam and Keller [7] method, modified by Vermeiren and Jobling [8], requires the determination of the flow of two emitters placed at the end of the lateral. Due to a distance of 0.4 m between UN emitters, it was not possible to determine *UE* in drip line distal locations that were not affected by the sediment accumulation, which had an obvious influence on results. To be precise, this emitter presented one of the highest percentages of completely clogged emitters at the end of the drip line, as will be commented on later. If the last sample position is not considered, the *UE* of emitter UN is similar to emitters RM and TI. Drip lines with emitter P8 had an effluent volume greater than UN, but did not show such a low *UE*, especially with tertiary effluent. Emitter P8 stands out for the highest resistance to clogging among emitters with a higher nominal flow [12].

The distribution of completely clogged emitters throughout the experiments at the different test locations with secondary and tertiary effluents is shown in figs. 2 and 3, respectively. It was observed that the percentage of totally clogged emitters increased with operation time, but at a different rhythm. Thus, up to 825 h, progressive increases between 0.15 and 1.18% were produced, but at 1000 h the increases were of 3%. Emitter location along the drip line was also a significant factor ($P<0.05$) for the presence of completely clogged emitters at the end of the experiments, because, if at the end of the drip line around 22% of the emitters were totally clogged, in the other positions this percentage was lower than 3%. The highest incidence of clogging at the end of drip lines [2] is attributed to lower flow velocities at these points, which facilitate particle

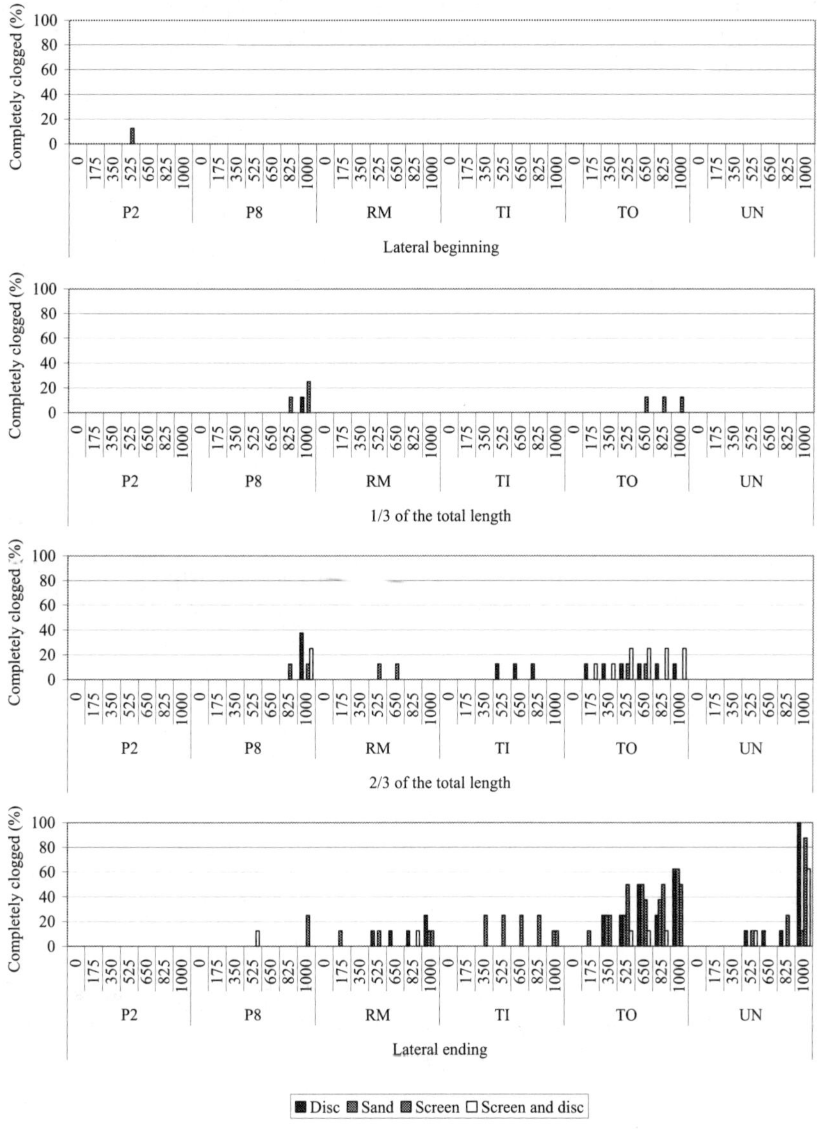

Figure 2: Percentage of emitters completely clogged with secondary effluent in sample locations for measuring *UE* throughout the experiment in function of location, emitter and filtration system.

settling [13]. Whether or not the different locations are taken into consideration, emitters TO and UN showed the highest average amounts of clogged emitters at the end of the experiment: 14% with emitter TO for both effluents and 16% and 12% with emitter UN and secondary and tertiary effluents, respectively. With secondary effluent, 9% of P8 emitters were completely clogged, but with tertiary effluent no totally clogged emitters were found. Emitters RM, P2 and TI, with less than 4% of completely clogged emitters, had the lowest incidence of totally clogged emitters with both the effluents used.

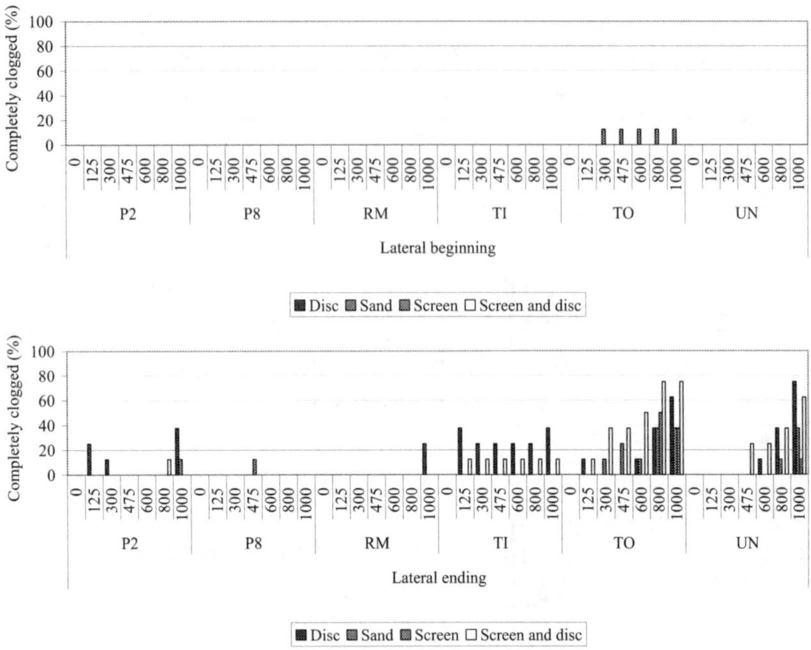

Figure 3: Percentage of emitters completely clogged with secondary effluent in sample locations for measuring *UE* throughout the experiment in function of location, emitter and filtration system. At locations 1/3 and 2/3 of lateral length, no clogged emitters were found.

It is important to point out that, during the experiment, some emitters experienced a reversion in their clogging status. In this sense, Ravina *et al.* [2] pointed out that emitter clogging does not necessarily have to be permanent, because emitters could be self-cleaning.

4 Conclusions

Only sand filtration significantly reduced turbidity and suspended solids from both secondary and tertiary effluents. With emitter P2 the best emission uniformities were obtained at the end of experiments, being higher than 90% for all combinations of effluent and filtration systems, except with screen filters and tertiary effluent. On the other hand, emission uniformity was lower than 68% with emitter TO for all filtration systems and effluents, and 40% with emitter UN, but with this last emitter, better values were observed with some other combinations of effluent and filter.

The best emission uniformity was obtained by the emitters placed after the sand filter (76%) and the screen filter (83%) with the secondary and tertiary effluents, respectively. Emitters that operated with disc filters showed the worst emission uniformity for both effluents. Clogged emitters were located mainly at the end of drip lines, where around 22% of the emitters were completely clogged after 1000 h of irrigation.

Acknowledgements

The authors would like to express their gratitude to the Spanish Ministry of Science and Technology for their financial support for this experiment through grant REN2002-00690/HID. The authors would also like to thank Regaber and the Municipality of Celrà for their help in carrying out this experiment.

References

[1] Bucks D.A., Nakayama F.S. & Gilbert R.G., Trickle irrigation water quality and preventive maintenance. *Agricultural Water Management*, **2(2)**, pp. 149–162, 1979.

[2] Ravina I., Paz E., Sofer Z., Marcu A., Schischa A., Sagi G., Yechialy Z., Lev Y., Control of clogging in drip irrigation with stored treated municipal sewage effluent. *Agricultural Water Management*, **33(2-3)**, pp. 127–137, 1997.

[3] Tajrishy M.A., Hills D.J., Tchobanoglous G., Pretreatment of secondary effluent for drip irrigation. *Journal of Irrigation and Drainage Engineering*, **120(4)**, pp. 716–731, 1994.

[4] Capra A., Scicolone B., Emitter and filter test for wastewater reuse by drip irrigation. *Agricultural Water Management*, **68(2)**, pp. 135–149, 2004.

[5] Puig-Bargués J., Arbat G., Barragán J., Ramírez de Cartagena F., Hydraulic performance of drip irrigation subunits using WWTP effluents. *Agricultural Water Management*, 77(1-3), pp. 249–262, 2005.

[6] Duran-Ros M., Puig-Bargués J., Barragán J., Ramírez de Cartagena F., Monitorización y control de una estación de riego por goteo mediante un sistema SCADA. *Actas del III Congreso de Agroingeniería.* eds. P. Aguado, J. Andrés, J.M. Morán, Universidad de León: León, pp. 59–60, 2005.

[7] Merriam J.L., Keller J., *Farm irrigation evaluation: A guide for management*, Department of Agricultural and Irrigation Engineering, Utah State University: Logan, Utah, 1978.

[8] Vermeirien L., Jobling G.A. *Localized irrigation,* Irrigation and Drainage Paper 36, FAO: Rome, 1986.

[9] Adin A., Alon G., Mechanisms and process parameters of filter screens. *Journal of Irrigation and Drainage Engineering*, **112(4)**, pp. 293–304, 1986.

[10] Taylor H.D., Bastos R.K.X., Pearson H.W., Mara, D.D., Drip irrigation with waste stabilisation pond effluents: solving the problem of emitter fouling. *Water Science and Technology*, **31(12)**, pp. 417–424, 1995.
[11] Puig-Bargués J., Barragán J., Ramírez de Cartagena F., Filtration of effluents for microirrigation systems. *Transactions of the ASAE*, **48(3)**, pp. 969–978, 2005.
[12] Ravina I., Paz E., Sofer Z., Marcu A., Schischa A., Sagi G., Control of emitter clogging in drip irrigation with reclaimed wastewater. *Irrigation Science*, **13(3)**, pp. 129–139, 1992.
[13] Shannon W.M., James L., Basset D.L., Mih W.C., Sediment transport and deposition in trickle irrigation laterals. *Transactions of the ASAE*, **25(1)**, pp. 160–164, 1982.

The use of treated effluent for agricultural irrigation: current status in the Bottelary catchment (South Africa)

N. Z. Jovanovic
CSIR, Natural Resources and Environment, Stellenbosch, South Africa

Abstract

The Bottelary river is located in a Mediterranean climate region (Western Cape, South Africa), where the agricultural sector plays an important role in the economy. During the dry summer season, there is not enough precipitation to meet the agricultural irrigation requirements. The objectives of this study were to investigate effluent quality, farmers' perception and the potential extent of irrigation of crops with treated effluent from the Scottsdene Wastewater Treatment Works (WWTW). The research made use of historic effluent quality data (from 2001 to 2004) to determine the suitability of this water for irrigation, a questionnaire to determine the farmers' perception, and the SAPWAT model for crop water requirements to determine the potential area that could be irrigated with treated effluent. The effluent quality analysis indicated that this water generally complied with international guidelines and national standards, and it is suitable for irrigation of most crops. The results of the questionnaire indicated that farmers are generally willing to use treated wastewater for irrigation, provided that infrastructure for wastewater supply is accessible and acceptable effluent quality is ensured by the WWTW to minimize possible negative impacts on health, crops, soils and waters. Estimation of crop water requirements indicated that treated effluent could become an additional water resource in the region, in particular during the dry summer months. In order to implement the use of treated effluent on a large scale, it is necessary to establish the required infrastructure, monitoring systems to control negative impacts on human health, crops, soils, surface and groundwaters, as well as extended public participation.

Keywords: Bottelary river, crop water requirements, effluent reuse, farmers' perception, irrigation, SAPWAT, Scottsdene, wastewater treatment works, water quality.

www.witpress.com, ISSN 1743-3541 (on-line)
doi:10.2495/SI080361

1 Introduction

Reuse of wastewater for irrigation is a practice that is already applied around the world for many years. Pescod [1] gave an overview of wastewater characteristics and quality parameters for reuse in agriculture and aquaculture, aquifer recharge, treatment technologies, irrigation management as well as policy issues. The advantages of reusing wastewater for irrigation were listed in several publications [2–4] (e.g. reduced demand on potable sources of fresh water; wastewater is a large and reliable water source rich in nutrients; soils and vegetation act as bio-filters with consequent lower treatment requirements and limited impact on the aquatic environment and downstream users; the system requires relatively low capital, operation and maintenance costs). The main disadvantage of this practice is the unknown impact of wastewater on crops and vegetation, soils, surface and groundwaters, as well as human and animal health, in particular in the long term. The main water quality issues were singled out to be pathogens, inorganic salts, heavy metals, pharmaceuticals, carcinogenic substances and endocrine disruptors [4]. The reuse of sewage water for irrigation is also constrained by management and socio-economic factors, like for example the availability of irrigable land in the vicinity of wastewater treatment plants, psychological repugnance and religion, distance to potable water supply wells, community size and relative location to other communities for trade-off opportunities, clarity on ownership of wastewater, and difficulty in quantifying opportunity costs of irrigation with wastewater. The seasonal character of wastewater reuse and peak period of demand may also result in overloading wastewater treatment plants during the rainy season and shortage of water supply during the dry periods of high water requirements [2].

Due to the population growth and extensive development in the Western Cape (South Africa), in particular in the Cape Flats in the vicinity of Cape Town, new and alternative strategies are required in order to manage treated sewage water. The total water supply within the Cape Metropolitan Area (CMA) is about 800 ML d^{-1} and the water is used by different sectors. There are twenty wastewater treatment works (WWTW) and three sea outfalls in the CMA. The total volume of treated wastewater is 539 ML d^{-1}, but only 9% is being reused for summer irrigation (6%), industrial processing (0.6%–1.5%) and aquifer recharge (1.7%–2.5%). The Western Cape was subject to serious water shortages due to below-average rainfall in the past years (in particular 2003 and 2004) that lead to necessary water restrictions being imposed on the population. If the treated effluent reuse can be extended to a wider area, portion of the fresh water supply will be freed for high value uses in domestic supply, commerce and industry.

In this study, an assessment of the current status of irrigation with treated effluent was carried out in the Bottelary catchment. The Bottelary river is a tributary of the Kuils river, it is 14 km long and its catchment covers an area of about 80 km^2. The catchment is located in the peri-urban area of the Cape Flats and it is, in many ways, representative of the whole CMA (fig. 1). Agriculture, predominantly winery and vegetable farming, is the major activity. The Scottsdene WWTW are located in this area. In summer, the effluent forms the entire water flow of the Bottelary river. The treated effluent is therefore a potential water resource for agricultural irrigation. The reuse of treated effluent provides an opportunity for the local farmers to sustain their livelihood especially during the dry summer season, and irrigation is already taking place on some vegetable farms adjacent to the discharge canal.

The specific objectives of the study were: i) to assess the quality of the effluent from Scottsdene WWTW and its suitability for agricultural irrigation in the Bottelary river catchment; ii) to investigate the farmers' perception of using treated effluent for agricultural irrigation; and iii) to assess the potential areas that could be irrigated with the effluent from Scottsdene WWTW.

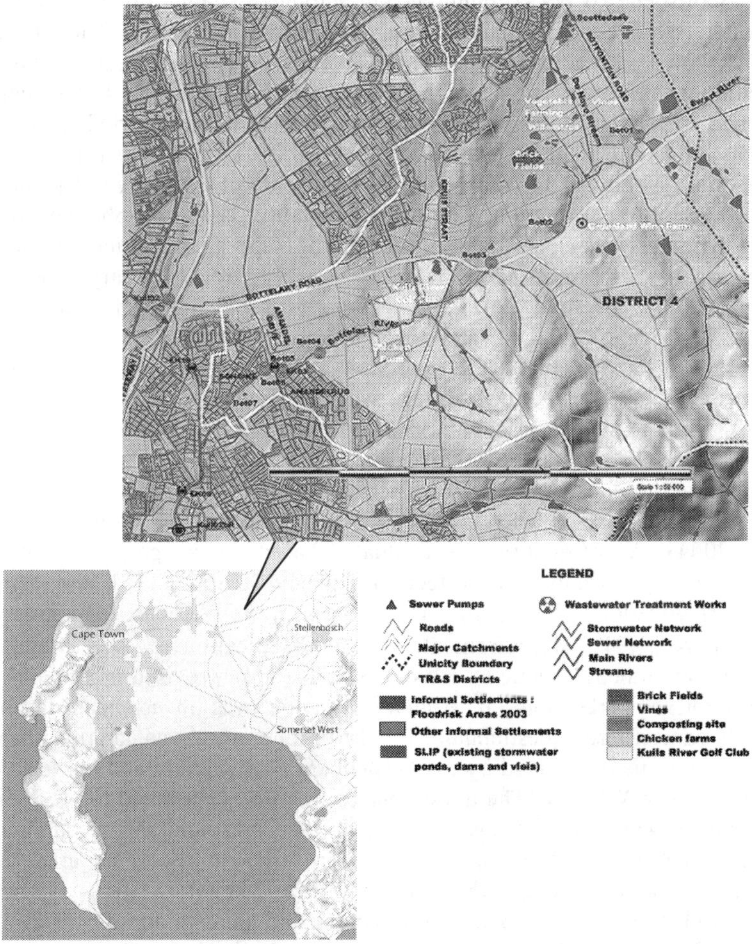

Figure 1: Location of the study area downstream of the Scottsdene wastewater treatment works and in relation to the Cape Peninsula.

2 Material and methods

The study area is located in a Mediterranean climate region, with hot dry summers and cool wet winters. The mean annual precipitation is approximately 600 mm, with 80% of the annual rain falling in the winter season between March and September. The upper streams of the Kuils and Bottelary rivers are located on the Malmesbury group, which includes quartzites, phyllite, greywacke and shales of Pre-cambrian age. The rocks are thinly covered by recently deposited turf and loam. These geological features result in high surface runoff and small subsurface flow. The sandy marine deposits downstream of the confluence of the two rivers result in little surface runoff.

The Scottsdene WWTW employs an activated sludge process to treat wastewater, which comes mainly from domestic water users. The process makes use of a separator for grit removal and a series of aerobic, anoxic, sedimentation and maturation ponds. Wastewater is chlorinated before being discharged to the receiving water body, whilst the sludge is dried in drying beds. No industrial effluent is fed into the sewage system. Historic data of effluent quality were used to determine the suitability of the treated effluent for reuse in agricultural irrigation. The effluent quality data were obtained from the Scientific Services Department of the City of Cape Town, which monitored effluent quality of Scottsdene WWTW weekly from January 2001 to November 2004. In total, 201 observations were available for statistical analysis. The water quality variables were total suspended solids (TSS), chemical oxygen demand (COD), ammonia (NH_3), nitrate (NO_3), ortho-phosphorus (PO_4), fecal coliform as well as flow rate. Seasonal trends, averages, medians, standard deviations, minimum and maximum values were determined for each variable. Regression analyses were done for each pair of water quality variables. Averages of electrical conductivity (EC) and pH were calculated for the periods when these variables were measured (from July 2002 to June 2004). Water quality was evaluated based on the guidelines of the United States Environmental Protection Agency (USEPA) [5], the World Health Organization (WHO) [6] as well as South African standards of Regulation No. 991 – Requirements for the Purification of Wastewater or Effluent (Government Gazette issued on 18 May 1984) – revised in 2005.

The farmers' perception on effluent irrigation was investigated through personally administered questionnaires. A list of nine farms was compiled based on the cadastral map provided by the Oostenberg Municipality and the location of the Scottsdene WWTW. The questionnaire was first distributed to these nine farms, most of which were vineyard farms. In a second round, the questionnaire was distributed to seven additional vegetable farms in the same area. The questionnaire included 26 questions split into four categories (socio-economic factors, production factors, behavioral factors and perception of wastewater irrigation). Most questions were organized into multiple-choice format in order to facilitate data processing and interpretation. Eleven questionnaires were returned out of 16 that were distributed. This small population sample allowed to gain useful information, although it wasn't suitable for statistical inferences.

The potential area that could be irrigated with effluent from the Scottsdene WWTW was calculated using data of cropping systems obtained from the questionnaires, effluent flow and irrigation requirements estimated with the SAPWAT computer model v. 2.6.0 [7]. SAPWAT is computer software for estimation of crop water requirements and irrigation planning. It includes databases of South African climates and crops. In this study, climate data were selected from the SAPWAT database (Klein Bottelary weather station). The software estimates crop factors from crop type, geographical region (winter rainfall), planting dates and crop canopy cover at full growth. Other input data are type of irrigation system and efficiency, frequency of wetting and wetting area. The SAPWAT model outputs monthly crop evapotranspiration, rainfall, effective rainfall and irrigation water requirements. Irrigation water requirements are calculated as the difference between crop evapotranspiration and effective rainfall. It was assumed that storage capacity is available to store the effluent during periods of low crop water requirements in winter.

3 Results

Table 1 summarizes the average, standard deviation, maximum and minimum values of effluent flow and quality variables during the study period. The high standard deviations indicated that values fluctuated widely. The average treating capacity was 7.8 ML d^{-1}, which indicated that the Scottsdene WWTW worked beyond its capacity of 7.5 ML d^{-1}. The peak flow was due to stormwater inflow

Table 1: Statistical data of effluent flow and quality (from January 2001 to November 2004) for the Scottsdene wastewater treatment works, and water quality guidelines.

Variable	Flow (ML/d)	TSS (mg/L)	COD (mg/L)	NH_3 (mgN/L)	NO_3 (mgN/L)	PO_4 (mgP/L)	Fecal coliform (counts/ 100 mL)
Average	7.8	13.4	51.2	7.8	2.0	7.2	33,372
Median	7.5	10.0	45.0	7.1	1.0	7.0	450
Standard Deviation	2.3	14.5	24.4	4.5	2.5	2.0	74,492
Maximum	17.1	190.0	279.0	24.0	14.0	15.0	510,000
Minimum	2.6	1.0	22.0	0.9	0.0	1.9	5
USEPA [5]	-	≤2 NTU[a,b] ≤30[c]	≤10 BOD[b,d] ≤30 BOD[c,d]	-	-	-	None[b] ≤200[c]
WHO [6]	-	<50[e]	-	TN[f] < 5 mg/L[e]		-	≤10^6 DALY[g]
S.A. standards	-	18	65	3.0	15	1.0	1,000[h]

[a] NTU - Nephelometric Turbidity Units
[b] For food crops not commercially processed
[c] For food crops commercially processed and non-food crops
[d] BOD – Biochemical Oxygen Demand
[e] No restriction on use [8]
[f] TN – Total Nitrogen
[g] DALY - Disability Adjusted Life Years per person per year as health-based target
[h] Proposed standard to be implemented in 2010 is 100 counts/100 mL

in the winter season. Generally, the flow in summer was lower because of lack of precipitation, but the seasonal trends were weak.

Values of TSS generally complied with international guidelines and local standards. In terms of the peak value, the possible reason could have been either poor water sampling or the breaking up of the sludge in the maturation pond. Concerning COD, the peak value (279 mg/L) appeared on 15 October 2003, which was exactly on the same day as the peak value of TSS. Values of COD were generally higher in summer due to the dilution effect of urban stormwater in winter, but the average and median were lower than the local standard of 65 mg/L. There is no generalized correlation between COD and BOD for comparison with international guidelines. Organic speciation can also pose a risk to human health. Maximum permissible health-related pollutant concentrations of specific organic compounds in receiving soils were given by WHO [6]. Weak seasonal trends were observed for NH_3, NO_3 and ortho-PO_4 concentrations. Concentrations of NH_3 were generally higher than those given by local standards, whilst NO_3 concentration complied with local guidelines. Concentration of ortho-PO_4 in treated effluent did not comply with the local standard. However, both elevated N and P in irrigation water could be beneficial to crops at certain growth stages. Most of the fecal coliform data complied with local standards as the average exceeded the standard value, but the median did not (table 1). This was due to a period from October 2003 to October 2004, when high fecal coliform levels were recorded because the chlorine tank at the Scottsdene WWTW was not operating.

Values of pH during the period of measurement ranged from 6.9 to 8.7 (average 7.4), whilst EC varied between 40.5 and 60.5 mS/m (average 50 mS/m). Local effluent standards indicate an optimal pH range of 5.5-9.5 and no requirements for EC. South African Water Quality Guidelines [9] provide a target EC value of 40 mS/m for agricultural use. If the range is between 40 and 90 mS/m, 5% yield losses can be expected for salt sensitive crops [9].

The squared coefficient R^2 between TSS and COD was found to be relatively high (fig. 2). This may indicate that suspended solids in water adsorb organic compounds and transport them. The R^2 values for all other pairs of variables were found to be < 0.2.

Given the quality of the effluent was found to be generally suitable for agricultural irrigation, the second step of the study involved the investigation of farmers' perception of this practice. The majority of the farmers (7 out of 11) were between 41 and 60 years, and they were all males. Five respondents matriculated or had lower education. The remaining six respondents attained a tertiary educational qualification and two of them have an agriculture-related diploma or degree. The smallest farm size was 8 ha, and the largest 192 ha. Six farms were < 50 ha, three farms were between 50 and 100 ha, whilst two farms were 172 ha and 192 ha. The finding was that bigger farms irrigate larger areas, so they need more irrigation water to meet the demand. Big farms are therefore more likely to adopt an unconventional water source such as treated effluent compared to small farms.

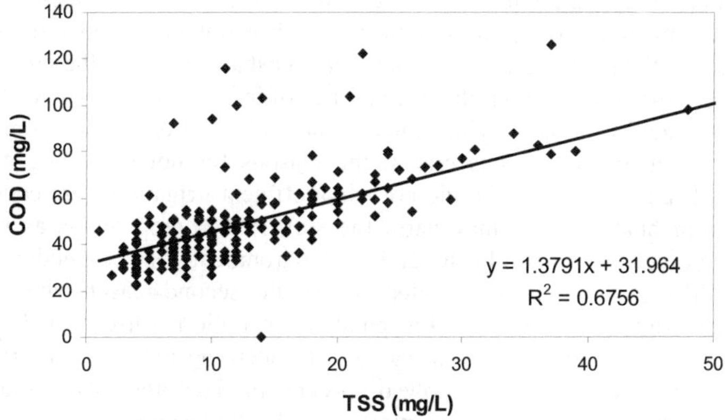

Figure 2: Regression analysis between chemical oxygen demand (COD) and total suspended solids (TSS) in effluent from the Scottsdene wastewater treatment works.

The main crops grown by the respondents in the Bottelary catchment were beans (6 farmers), carrot, cabbage and cauliflower (5), grapes (4), potato and lettuce (3), chilli, pumpkin, pepper, radish and beetroot (2), and grasses (1). Some farmers grow more than one crop. According to the data of crop salt tolerance published by Maas [10], the most popular crops amongst the respondents (beans and carrot) are sensitive to salinity. Cabbage and cauliflower were also widely grown in the study area and they are moderately sensitive to salinity. The average EC of the Scottsdene WWTW final effluent was 50 mS m^{-1}. This saline water gets more concentrated in the soil solution due to crop water uptake and it may affect crop yield, although no dramatic yield reduction can be expected, provided appropriate irrigation scheduling is practiced. Some of the salts are expected to be washed out of the root zone through rainfall, in particular during the rainy winter season. It should be noted that, due to high investment and management costs in vineyards, high quality effluent needs to be guaranteed and specific management practices need to be ensured on vine farms in order to avoid losses of profit that could be extremely detrimental to the business. Based on the questionnaire, all farmers used sprinkler irrigation systems (11). Other irrigation methods adopted were drippers (4 farmers), microjets (2) and subsurface irrigation (1). Some farmers used more than one irrigation method. If the effluent quality is poor, sprinkler irrigation can cause potential hazards to farm workers, crop leaves and soil. The main negative impact of effluent irrigation with drip systems is the clogging of emitters.

Eight respondents agreed that there is a water shortage problem in the area. All vineyard farmers strongly supported this point. Three respondents, who have small size farms, did not agree that there is a water shortage problem. Six respondents were using treated effluent for irrigation. These were mainly vegetable farmers, whilst no vineyard farmer used treated effluent. Six respondents were willing to increase the area irrigated with treated effluent, whilst the remaining five preferred fresh or other waters. Compared to other water sources, the treated effluent has a relative cost advantage to the farmers.

All respondents pointed out that the absence of infrastructure, such as the lack of pumping facility for conveying water to the farm land or no pipe connecting treated effluent to the users is one of the reasons for not practicing effluent irrigation. The respondents who did not adopt effluent irrigation were concerned that the poor quality water may harm the soil, reduce crop yields as well as impact negatively on human health and the environment. One respondent stated that "considering the consumers' interest" was the second reason why he used fresh water irrigation. However, all respondents who did not use treated effluent were willing to accept it, but with caution. Concerning the effluent irrigators, four out of the six respondents showed no concern. The other two respondents were concerned that effluent irrigation may bear health hazards.

Based on the promising feedback from the first two stages of the study (effluent quality and farmers' perception), it was considered that the area irrigated with treated effluent could potentially be increased to a large scale. The first step of the procedure for estimation of the potential area that can be irrigated involved the selection of a representative cropping system. Based on the questionnaire survey, a generic cropping system including winter and summer vegetables was selected in SAPWAT. Planting dates of crops were 1 October for summer-planted vegetables and 1 March for winter-planted vegetables. The crop canopy cover at full growth was selected to be 90%, and the wetting frequency was seven days throughout the season, based on common practices in the area. The wetted area was 100%, representing sprinkler irrigation with an application efficiency of 85%. Based on these inputs, the SAPWAT model was used to calculate monthly irrigation requirements (table 2).

For summer-planted vegetables, the default growth season was from October to February. The total seasonal water requirement was the sum of the five months' irrigation requirements, namely 684 mm. The total effluent flow was 1,120 ML (the sum of the five months' average effluent flow). By dividing the total available effluent flow by the total irrigation requirement for this period, the total potential area that could be irrigated with treated effluent for summer planted vegetables was calculated to be 164 ha. The potential extent of effluent irrigation of summer-planted vegetables was smaller than the total farming area of effluent users in the surveyed group (213 ha). This indicates that the volume of effluent available may not be enough to satisfy crop water requirements, should the current effluent users want to irrigate the whole area of their farms. As the treated effluent in summer makes up the entire river flow, this means that there is potential to dry out the Bottelary river in summer, should an extended area be irrigated.

For winter-planted vegetables, the growth season is from March to August. The total irrigation requirement was the sum of the irrigation requirements from March to May, namely 154 mm. During June, July and August, no irrigation requirements were calculated as rainfall is expected to fully satisfy crop water requirements. The total effluent flow from March to May was 760 ML. Effluent could be stored during periods of no irrigation requirements, or when the effluent flow is higher than the irrigation requirements. This stored effluent could then be used during the months of peak irrigation requirements. Storage reservoirs could also act as maturation ponds and buffer the effluent quality. In table 2, the balance between the surplus effluent in winter and the surplus demand in summer was determined for generic vegetables irrigated with a sprinkler irrigation system. Assuming full storage of effluent from March to May, it was calculated that the potential area that could be irrigated is 224.5 ha. A storage capacity of 315.13 ML would be required under these conditions. The storage facility would be filled with water in October (table 2), before the start of the summer season. The period of no irrigation requirements for this cropping system (June, July and August) was not included in the calculation, as there is currently no capacity or area to store such volumes of effluent.

It should be noted that water losses accounted for in the SAPWAT model are related only to the efficiency of the irrigation system, and do not include losses due to evaporation and seepage from storage facilities, conveyance losses etc. These additional losses would further decrease the efficiency of the system resulting in smaller areas that could potentially be irrigated.

Table 2: The balance of excess winter effluent flow and excess summer effluent demand for generic vegetables irrigated with a sprinkler irrigation system.

Month	Effluent flow (EF) (ML/month)	Irrigation requirements (mm/month)	Effluent requirement (ER) (ML/month)	Cumulative difference between ER and EF (ML)	Area (ha)
January	279	211	473.70	194.70	224.5
February	190	109	244.71	249.40	
March	217	69	154.91	187.31	
April	252	62	139.19	74.50	
May	291	23	51.64	-164.87	
October	267	52	116.74	-315.13	
November	183	92	206.54	-291.59	
December	202	220	493.90	0.31	

4 Conclusions

The reuse of treated effluent for agricultural irrigation can release pressure on fresh water resources and holds a promising future. However, the indiscriminate use of treated wastewater for irrigation could cause the problems associated with sewage reuse to merely shift from surface waters to soils, vegetation and groundwater, or from one water user to another. Also, the impact of this practice on the environment could take a long time to manifest and regular monitoring is essential. Socio-economic factors also play an important role in the implementation of this technology. Education, information and training of farmers are essential for operations with effluent irrigation. The infrastructure for effluent irrigation of a broader area could be planned. Farmers' involvement is important in this process. The building of storage dams is also needed in order to balance the effluent demand and supply. The findings of this study could be useful to other regions that have similar problems of water shortage and treated effluent production.

Acknowledgements

The author wishes to acknowledge the contributions of Ms Li Rui (MSc student at the Department of Earth Sciences, University of the Western Cape), Mr Alfred Mbewe (City of Cape Town) who provided water quality data, Prof Lincoln Raitt (Department of Biodiversity and Conservation Biology, University of the Western Cape) for his valuable inputs during questionnaire distribution and collection, Mr Charles Crosby for his assistance in the use of the SAPWAT software, as well as the farmers in the study area for their collaboration.

References

[1] Pescod, M.B., *Wastewater Treatment and Use in Agriculture*, FAO Irrigation and Drainage Bulletin No. 47, FAO: Rome, Italy, 1992.

[2] United Nations Environment Programme (UNEP), *Source Book of Alternative Technologies for Freshwater Augmentation in Latin America and the Caribbean*. International Environmental Technology Centre, United Nations Environment Programme: Osaka/Shiga, Japan, 1997.

[3] Angelakis, A.N. & Bontoux, L., Wastewater reclamation and reuse in Eureau countries. *Water Policy,* **3(1)**, pp. 47–59, 2001.

[4] Toze, S., Reuse of effluent water - benefits and risks. *Agricultural Water Management,* **80(1-3)**, pp.147–159, 2006.

[5] United States Environmental Protection Agency (USEPA), *Manual Guidelines for Water Reuse*, EPA/625/R-92/004, Environmental Protection Agency: Washington, D.C., U.S., 245 pp., 1992.

[6] World Health Organization (WHO), *WHO Guidelines for the Safe Use of Wastewater, Excreta and Greywater*, ISBN 92 4 154683 2, World Health Organization: Geneva, Switzerland, 2006.

[7] Crosby, C.T., *SAPWAT 1.0 - A Computer Program for Estimating Irrigation Requirements in Southern Africa*, Report No. 379/1/96, Water Research Commission: Pretoria, South Africa, 1996.

[8] Ayers, R.S. & Westcot, D.W., *Water Quality for Agriculture*, FAO Irrigation and Drainage Paper No. 29, Food and Agriculture Organization of the United Nations: Rome, Italy, 1985.

[9] Department of Water Affairs and Forestry, *South African Water Quality Guidelines. Volume 4: Agricultural Water Use*, Department of Water Affairs and Forestry, 2nd ed.: Pretoria, South Africa, 1996.

[10] Maas, E.V., Salt tolerance of plants. *Applied Agricultural Research*, **1(1)**, pp. 12–26, 1986.

Author Index